# Permeable Reactive Barrier

## Sustainable Groundwater Remediation

T0315218

# ADVANCES IN TRACE ELEMENTS IN THE ENVIRONMENT

## Series Editor: H. Magdi Selim

Louisiana State University, Baton Rouge, USA

**Permeable Reactive Barrier: Sustainable Groundwater Remediation**
edited by Ravi Naidu and Volker Birke

**Phosphate in Soils: Interaction with Micronutrients, Radionuclides and Heavy Metals**
edited by H. Magdi Selim

# Permeable Reactive Barrier

## Sustainable Groundwater Remediation

Edited by
### Ravi Naidu
### Volker Birke

CRC Press
Taylor & Francis Group
Boca Raton  London  New York

CRC Press is an imprint of the
Taylor & Francis Group, an **informa** business

CRC Press
Taylor & Francis Group
6000 Broken Sound Parkway NW, Suite 300
Boca Raton, FL 33487-2742

First issued in hardback 2020

© 2015 by Taylor & Francis Group, LLC
CRC Press is an imprint of Taylor & Francis Group, an Informa business

No claim to original U.S. Government works

ISBN-13: 978-1-4822-2447-4 (hbk)

**Visit the Taylor & Francis Web site at
http://www.taylorandfrancis.com**

**and the CRC Press Web site at
http://www.crcpress.com**

# Contents

# *Preface*

The past century has witnessed dramatic rates of industrialization around the world, with average annual economic growth rates exceeding 7% in some countries. One of the prices to be paid for such rapid growth, however, is environmental deterioration. Air, water, and soil pollutions have been of serious concern for decades in the United States, Canada, the United Kingdom, France, Germany, and other developed nations in Europe.

Over the last three decades, the spectacular economic growth of Latin America, China, India, Korea, and other Asian countries has generated an increasing number of contaminated sites and waste disposal problems. These represent a global challenge. The world's estimated five million potentially contaminated sites are both a major lost economic opportunity and also a threat to the health and well-being of the community and the wider environment. Common contaminants include petroleum hydrocarbons, chlorinated hydrocarbons, pesticides, inorganics such as toxic metal(loids), and radioactive wastes. These are frequently found at a variety of sites, such as oil, gas, and petrochemical operations, mines, industrial sites, waterways and harbors, fuel storage farms, workshops, munitions factories, and so on. Although site contamination has been recognized as an issue for more than 70 years, fewer than a tenth of all contaminated sites have been remediated due to the complex and challenging nature of contamination, the highly complex and heterogeneous subsurface that may vary from site to site, and the high costs of clean-up. Most of these contaminated sites have associated groundwater contamination problems that prevent their effective and reliable remediation and pose risks to the health of communities sometimes quite distant from the original site. Remediation of groundwater is often challenging due to the heterogeneity of the subsurface environment, difficulties with delineating contaminant plume, and the slow release and diffusion of contaminants from fractured rock and from sorbed phases. For these reasons, many groundwater remediation techniques currently in use have delivered only transient success.

A number of different techniques have been used for the remediation of groundwater with the most cost-effective strategy being a risk-based approach that is commonly practiced in Australia, where the state of Victoria's legislation requires clean-up of groundwater to the extent practicable (CUTEP). Similar laws have been adopted in other states where site remediators clean groundwater using technologies that may not fully remediate groundwater, given the technological limitations or other environmental constraints. Under these circumstances, the site is cleaned as far as practicable, and the groundwater is then monitored over a sustained period to demonstrate natural attenuation of the contaminants. In this approach, natural attenuation of groundwater is the primary strategy for remediation. Other less passive approaches include

- Pump and treat
- Bioventing
- Chemical oxidation (*in situ*)
- Permeable reactive barrier (more recent)

Pump and treat was one of the first techniques to be trialed in the United States. This involves pumping groundwater through an *ex situ* reactor and the cleansed water then reinjected back into the aquifer. Although pump and treat has often been unsuccessful and is prohibitively expensive, it is still widely used by large companies seeking to comply with the demands of regulators. Both bioventing and chemical oxidation techniques are also widely used and have proven similarly expensive and unattractive to owners of contaminated sites. For an overview of remediation techniques, readers are referred to a recent paper by Naidu (2013). Given the high cost of pump and treat technology, a host of other technologies for hydrocarbon remediation is also being tested in the field. Among these techniques is permeable reactive barrier (PRB) technology, which allows groundwater to pass through a buried porous barrier that either captures the contaminants or breaks them down. This approach is gaining popularity in the United States, Europe, and Australia. This book covers

- Two decades of experience in PRB applications
- Design criteria
- Predictive modeling to assist the design of PRBs
- Application to contaminants beyond petroleum hydrocarbons, including inorganics and radionuclides
- New areas of research

The book is intended for individuals responsible for the management of site contamination programs, regulators, remediators, and postgraduate students.

**Ravi Naidu**
**Volker Birke**

## Reference

Naidu, R. 2013. Recent advances in contaminated site remediation. *Water, Air, & Soil Pollution*, 224, 1–11.

# Editors

**Ravi Naidu** is the chief executive officer, managing director, and chief scientist of the Cooperative Research Centre for Contamination Assessment and Remediation of the Environment (CRC CARE). Professor Naidu is also the founding director of the Centre for Environmental Risk Assessment and Remediation (CERAR). He has researched environmental contaminants, bioavailability, and remediation for over 25 years. Professor Naidu has coauthored 414 refereed journal articles and 7 patents and coedited 11 books and 66 book chapters in the field of soil and environmental sciences. He has also supervised over 30 PhD completions. Professor Naidu was instrumental in developing a network of scientists working on contamination—Soil Contamination Research Australasia Pacific (SCRAP). As part of this network he has independently raised funds for research and training in the Asia region. Over the last 10 years he has conducted more than 20 workshops, 5 international conferences, and raised more than $500M (includes CRC CARE funding) for research in this region. The network has led to the establishment of similar groupings of people within the regional countries and now has over 4000 members across the region.

Professor Naidu's current research focuses on contaminated soil and water and potential impacts of contaminants on human health. Professor Naidu's vision is to expand his current research on the environment to China and the Asia region through collaboration and the development of a Centre of Excellence, an environmental risk assessment and remediation in China. The focus of such a center will be the development and worldwide marketing of environment technology.

In recognition of his contribution to environmental research he was awarded a Gold Medal in environmental science in 1998 by Tamil Nadu Agricultural University (TNAU). He is an elected Fellow of the Soil Science Societies of America (2000), New Zealand (2004), and the Agronomy Society of America (2006). In 2012 he was chosen as a winner of the Soil Science Society of America's International Soil Science Award, and in 2013 he was elected a Fellow of the American Association for the Advancement of Science. He is chair of the International Committee on Bioavailability and Risk Assessment and was chair of the Standards Australia Technical Committee on Sampling and Analyses of Contaminated Soils (1999–2000), chair of the International Union of Soil Sciences Commission for Soil Degradation Control, Remediation and Reclamation (2002–2010), president of the International Society on Trace Element Biogeochemistry (2005–2007), and sitting member of the Victorian EPA Contaminated Sites Auditor panel. In recognition of his contributions to agricultural and allied sciences, he was awarded an honorary DSc by TNAU in December 2013 and the inaugural Banksia CEO award during the same year for his outstanding contribution to environmental sustainability research.

**Volker Birke** has around 25 years of experience in organic and environmental chemistry, particularly in green chemistry, the destruction of hazardous wastes, and toxic compounds in the environment. He has experience in remediation of contaminated sites, hazardous waste management, and especially groundwater remediation, with 12 years as coordinator of the large-scale German R&D program "RUBIN" for the application of permeable reactive barriers (PRBs) to groundwater pollution, 25 years in the development and application of innovative *ex situ* and *in situ* remediation technologies to persistent organic pollutants (POPs), particularly mechanochemical dehalogenation (MCD) regarding *ex situ* treatment, as well as innovative groundwater remediation technologies such as PRBs or nano and microscale reactive particles for *in situ* treatment of polluted groundwater. Another focus covers environmental analytics, especially regarding POPs such as polychlorinated biphenyls (PCBs) or dibenzodioxins (PCDD).

He has been senior scientist, project leader, and senior lecturer in environmental chemistry and engineering at the Faculty of Civil and Environmental Engineering at the Ostfalia University of Applied Sciences, Campus Suderburg, Germany, since 2000. Since 2009, he has been commissioned to administer the professorship of chemistry at the faculty. He teaches the management of contaminated sites, remediation technologies, chemistry, physics, and mathematics for engineers and environmental scientists as well as occupational health and safety, with a focus on special applications and requirements for working in areas contaminated by hazardous compounds and at polluted sites.

Dr. Birke has been involved in the management of about 50 remedial investigations, remedial designs, and clean up of polluted sites in Germany as well as abroad (Europe, the United States, India, Australia). He earned a PhD in organic chemistry at the University of Hanover. Dr. Birke has published about 50 papers focusing on innovative *ex situ* and *in situ* remediation technologies such as mechanochemical dehalogenation and PRBs and is the main author of one book on remediation of contaminated sites using PRBs in Germany.

# Contributors

**Steven D. Acree**
U.S. Environmental Protection
  Agency
Office of Research and Development
National Risk Management
  Research Laboratory
Ada, Oklahoma

**Cherri Adair**
U.S. Environmental Protection
  Agency
National Risk Management
  Research Laboratory
Ada, Oklahoma

**Dawit N. Bekele**
Centre for Environmental Risk
  Assessment and Remediation
  (CERAR)
and
CRC for Contamination Assessment
  and Remediation of the
  Environment (CRC CARE)
University of South Australia
Adelaide, Australia

**Volker Birke**
Ostfalia University of Applied
  Sciences Braunschweig/
  Wolfenbuettel
Wolfenbuettel, Germany

**Harald Burmeier**
Ostfalia University of Applied
  Sciences
Suderburg, Germany

**Sreenivasulu Chadalavada**
Centre for Environmental Risk
  Assessment and Remediation
  (CERAR)
and
CRC for Contamination Assessment
  and Remediation of the
  Environment (CRC CARE)
University of South Australia
Adelaide, Australia

**Frank Ingolf Engelmann**
Sensatec
Fehrbellin, Germany

**Hans-Jürgen Friedrich**
Verein für Kernverfahrenstechnik
  und Analytik Rossendorf
  (VKTA)
Dresden, Germany

**Ronald Giese**
GFI Groundwater Research
  Institute
Dresden, Germany

**Tony R. Lee**
U.S. Environmental Protection
  Agency
Office of Research and Development
National Risk Management
  Research Laboratory
Ada, Oklahoma

**Ludwig Luckner**
GFI Groundwater Research Institute
Dresden, Germany

**Ralph D. Ludwig**
U.S. Environmental Protection
    Agency
Office of Research and
    Development
National Risk Management
    Research Laboratory
Ada, Oklahoma

**Janet Macmillan**
Western Australian Department
    of Environment and
    Conservation
Perth, Western Australia, Australia

**Jan Sebastian Mänz**
Institute of Sustainable Chemistry
    and Environmental Chemistry
Leuphana University Lüneburg
Lüneburg, Germany

**Ben McCarthy**
LandCorp
Perth, Western Australia, Australia

**Mary Sue McNeil**
U.S. Environmental Protection
    Agency
National Risk Management
    Research Laboratory
Ada, Oklahoma

**Tamás Meggyes**
UK and ICP Hungária
University of Wolverhampton
Budapest, Hungary

and

BAM Federal Institute for
    Materials Research and
    Testing (Formerly)
Berlin, Germany

**Keely Mundle**
Golder Associates Pty Ltd.
Western Australia, Australia

**Ravi Naidu**
Centre for Environmental Risk
    Assessment and Remediation
    (CERAR)
and
CRC for Contamination Assessment
    and Remediation of the
    Environment (CRC CARE)
University of South Australia
Adelaide, Australia

**Wolf-Ulrich Palm**
Institute of Sustainable Chemistry
    and Environmental Chemistry
Leuphana University Lüneburg
Lüneburg, Germany

**Randall R. Ross**
U.S. Environmental Protection
    Agency
Office of Research and
    Development
National Risk Management
    Research Laboratory
Ada, Oklahoma

**Wolfgang Ruck**
Institute of Sustainable Chemistry
    and Environmental Chemistry
Leuphana University
    Lüneburg
Lüneburg, Germany

**Christine Schuett**
Ostfalia University of Applied
    Sciences Braunschweig/
    Wolfenbuettel
Wolfenbuettel, Germany

**Franz-Georg Simon**
Department of Material and
    Environment
BAM Federal Institute for Materials
    Research and Testing
Berlin, Germany

**James Stening**
Orica Limited
New South Wales, Australia

**Chunming Su**
U.S. Environmental Protection
    Agency
National Risk Management
    Research Laboratory
Ada, Oklahoma

**Dietrich Swaboda**
Chunming Su
GFI Groundwater Research Institute
Dresden, Germany

**Uli Uhlig**
GFI Groundwater Research
    Institute
Dresden, Germany

**Scott D. Warner**
ENVIRON International
    Corporation
Emeryville, California

**Martin Wegner**
M&P GEONOVA GmbH
Hannover, Germany

**Richard T. Wilkin**
U.S. Environmental Protection
    Agency
National Risk Management
    Research Laboratory
Ada, Oklahoma

# 1

## Permeable Reactive Barriers: Cost-Effective and Sustainable Remediation of Groundwater

Ravi Naidu, Dawit N. Bekele, and Volker Birke

### CONTENTS

## 1.1 Introduction

Contaminated sites represent a major challenge for the long-term sustainability of the environment. In addition to their potential adverse impacts on human health, surface and groundwater quality, and ecological processes, they also represent a lost economic opportunity. Sources of contaminants include those arising from anthropogenic activities such as industrial and agricultural practices, mining activities, accidental spillages, and so on (Barzi et al., 1996; Naidu, 1996), and natural geogenic processes (Naidu et al., 2006), with the latter largely associated with metals and metalloids such as arsenic, lead, cadmium, and mercury. Irrespective of the source of contaminants, they will interact with soil colloidal particles and moisture in the near-surface leachable zone. They can, therefore, be dissolved/solubilized into water infiltrating through any unsaturated zone present in the soil profile. They can penetrate below the water table and subsequently migrate laterally in flowing groundwater and transported off-site, thus posing a serious risk to groundwater quality.

This chapter provides a brief overview of permeable reactive barrier (PRB) technologies for groundwater remediation. We hope that it will encourage further reading by providing a selection of references covering the now extensive literature in the field of PRB groundwater remediation technologies.

## 1.2 Groundwater Contamination

The risks that are posed to human health and to the environment by exposure to groundwater contamination are well recognized by regulatory bodies, owners of potentially contaminated sites, the local community, and the public at large. As a consequence, regulatory guidelines have been developed to both protect the environment and, where necessary, to clean a contaminated environment to the required level based upon these guideline values. The remediation endpoints required by regulators, the scope of monitoring programs, and the assignment of legal/financial liability for remediation efforts all vary greatly from country to country (Rao et al., 1996). Environmental literacy and public perception of the relative risks of soil and groundwater contamination (in comparison to other hazards) can influence regulatory policy, and the acceptable levels of contamination, as well as the expectations of any required cleanup of contaminated sites. Such issues play a dominant role in identifying soil and groundwater contaminated sites (Rao et al., 1996).

Millions of potentially contaminated sites have been identified globally—and these require cleanup (Singh and Naidu, 2012). According to the NRC (1994) and Rao et al. (1996), there are between 300,000 and 400,000 contaminated sites in the United States with a wide variety of toxic chemicals identified. Total cleanup costs were estimated to be in the range of $500 billion to $1 trillion. More recent estimates, however, show that the number of contaminated sites could be as many as 500,000 (Table 1.1) with many of these experiencing groundwater contamination with complex mixtures of chlorinated solvents, fuels, metals, and/or radioactive materials. Inclusion of sites contaminated in other industrialized countries with those in United States, Europe, and Australia suggests that there are in excess of 1 million potentially contaminated sites (Table 1.1). While there are no data available for potentially contaminated sites in developing countries from Asia, one estimate suggests the existence of millions of such sites (Naidu, 2013) in both rural regions as well as in urban areas. Whereas contaminated sites in the urban environment constrain urban renewal, those present in the rural environments pose risks to the "clean and green" image of a country in addition to posing risks to human health. In 2005, the Canadian Environment Industry (CEI) identified more than 30,000 contaminated sites in Canada (CEI, 2005) concluding that such sites

**TABLE 1.1**

Global Estimate of Potentially Contaminated Sites

| Country | Number of Potentially Contaminated Sites | Value of Current Market | Future Potential | Major Market Drivers |
|---|---|---|---|---|
| USA | 450,000 to 500,000 | US $10+ billion per year (1/3 of global demand | Estimated at US $650 billion over 30–35 years | US superfund law; small business liability relief and Brownfields revitalization act; new underground storage tank regulations; real estate development activity; federal cleanup programs |
| Western Europe | 600,000+ | An estimated €50 billion, timeframe unspecified | 0.5%–1.5% of GDP is likely to be spent per annum | Strict regulatory approach; permitting process for industrial sites, liabilities in mining, civil, building, regional, and urban planning codes, soils conservation acts |
| Japan | 500,000+ | $1.2 billion+, timeframe unspecified | Estimated to grow to $3 billion by 2010 | Soil contamination counter-measures law; real estate appraisal standards and the law of housing site and house transactions; some prefectural and municipal governments have incentive programs for foreign businesses |
| Australia | 160,000 | >$3 billion per annum | Unassessed | Guidelines for the assessment and management of contaminated sites; provincial acts such as the contaminated sites act 2003 (Western Australia); restrictions on landfills; increasing environmental liabilities in business and property transactions |
| Asia region | >3,000,000 | Unassessed | Unassessed | Unassessed |

*Source:* Modified from CEI. 2005. Soil remediation technologies: Assessment, clean-up, decommissioning, rehabilitation. Canadian Environmental Industries (Energy and Environmental Industries Branch), available at: http://www.ic.gc.ca/eic/site/ea-ae.nsf/eng/ea02201.html.

represent a lost economic opportunity and threaten the economic well being of Canadians and the environment. As a consequence, the industry felt that there was a growing need for soil remediation, which was poised to become a large driver of technology, products, and services for years to come. The CEI estimate suggests that the remediation industry is worth billions of dollars (Table 1.1) and that it is an industry which is rapidly growing. A similar report in 2002 by Aus Industry suggests that the remediation industry in Australia will grow by 27% per annum. This is well reflected in the current market for contaminated site assessment and remediation with the industry growing from $300 million per annum in 1999 to >$3 billion per annum in 2012.

Contamination frequently affects more than surface soils. Although the focus of the CEI was contaminated soils, it is well recognized that once contaminants are in soils, they can leach into the groundwater with recharge waters (i.e., rainfall or surface water) when they come into contact with contaminated soil. Subsequently, contaminants travel in a more horizontal direction creating a dispersion plume. Shallow aquifers are usually important sources of groundwater. These upper aquifers are also the most susceptible to contamination. Contaminants may enter an upper aquifer in one of the following ways: (1) artificial recharge or leakage through wells; (2) infiltration from precipitation or irrigation return flow through the vadose zone above the water table; (3) induced recharge from influent streams and lakes or other surface water bodies; (4) inflow through aquifer boundaries and leakage from overlying or underlying formations; and (5) leakage or seepage from impoundments, landfills, or miscellaneous spills.

Groundwater contamination is now well recognized as an integral component of contaminated sites and its assessment and remediation can pose significant technical and financial challenges. Sources of groundwater contamination include

a. Leaching of contaminants from contaminated soils especially due to inadvertent releases, spills, or leaks of liquid wastes
b. Leaking underground storage tanks
c. Landfills that were not engineered or designed to hold leachates
d. Poorly constructed injection wells
e. Anthropogenic activities that enhance release of geogenic contaminants such as arsenic

## 1.2.1 Groundwater Contamination Fate and Transport

Contaminant interactions with phases in the subsurface may reduce the rate of their transport; and because most geologic materials have surfaces that possess a net negative charge, contaminants in cationic form are frequently observed to interact with solid surfaces, at least to some degree. The fate and transport of contaminants through soil to groundwater is influenced by many variables such as properties of the contaminant itself, soil conditions, and climatic factors. Some organic contaminants can undergo chemical changes or degrade into products that may be more or less toxic than the original compound. Metallic and metalloid contaminants cannot break down, but their characteristics and chemical states may change.

There are two basic processes by which contaminants move from the earth's surface through soils and groundwater. These processes are diffusion and mass flow (advection and dispersion).

Diffusion and mass flow are affected by properties of the contaminant, the soil, the intermediate vadose zone (the area below a crop root zone and above the permanent water table) and the aquifer; climatological factors and vegetation patterns:

- Properties of contaminants that determine their movement and potential threat to water quality include water solubility, any tendency to adhere to soil materials, persistency, and toxicity.
- Properties of soil, the intermediate vadose zone, and the aquifer that affect rate of contaminant movement include infiltration characteristics, pore size distribution, microbial population density and diversity, organic matter content, total porosity, ion exchange capacity, hydraulic properties, pH, and redox status.
- Climatic factors include temperature, wind speed, solar radiation and frequency, intensity, and duration of rainfall.
- Vegetation may act as a sink for contaminants by uptake or assimilation, thus reducing the amount of contaminant available for transport to groundwater.

All these properties interact to determine the rate and amount of movement of contaminants in soils and groundwater. Groundwater contamination proves to be most challenging from assessment and remediation perspectives as it depends on both the nature of contaminants and regional hydrogeology.

Once in the subsurface environment as part of the aquifer, contaminants are transported either in the dissolved phase or bound to nanocolloid particles, thus resulting in a contaminant plume away from the source zone. The plume composition varies with time and distance as its size increases. Based on the plume composition at a particular contaminated site, it is convenient to separate the plume into three regions (a) a near-field or source region, (b) a transition zone, and (c) a far-field or dissolved plume region. Rather than distance from the contaminant source, the criterion employed to designate these regions is the chemistry of the contaminant mixture (Rao et al., 1996).

The rate at which contaminants move in groundwater may vary between fractions of a cm to a few cm per year, forming under certain idealized conditions, an elliptical plume of contamination with well-defined boundaries. In a recent study conducted by CRC CARE in Adelaide, Australia, a trichloroethylene plume was found to extend nearly 300 m away from the source zone despite the groundwater flow being slow, only 5 cm per year (Chadalavada et al., 2011). Where geological formations include fractured rocks, some of

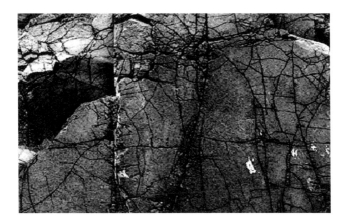

**FIGURE 1.1**
Dynamics of fluids in fractured rock. (Photograph by Dr. Jessica Winder.)

the contaminant species may migrate in rock fractures (Figure 1.1). Once distributed in rock factures, contaminant assessment, delineation, and remediation become an extremely challenging and expensive process. This is further constrained by a lack of regulatory policies dealing with endpoints for remediation of such contaminated sites. Often where such sites have been remediated, rebound from rock fractures has also been a major challenge.

The schematic diagram in Figure 1.2 shows a dense NAPL completely filling pores in the subsurface soil/groundwater environment and also coating soil particles, which makes delineation of the contaminant plume challenging and often very difficult.

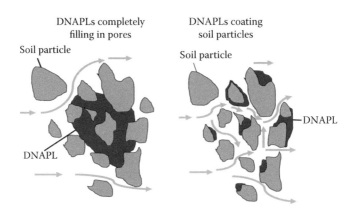

**FIGURE 1.2**
Dense nonaqueous phase liquid (DNAPL) interactions with soils in vadose zone.

## 1.3 Groundwater Remediation

Whilst soil remediation itself is recognized as a significant challenge (Naidu, 2013), the presence of contaminants in groundwater together with the dynamic nature of subsurface environment makes groundwater remediation among the most challenging and expensive environmental problems, and often the primary factor limiting closure of contaminated sites. Many reports suggest that despite years of operation focusing on remediation, it has proven difficult and costly to meet applicable cleanup standards (Scherer et al., 2000). The most common technology used for the remediation of groundwater has been *ex situ* pump and treat systems (Rao et al., 1996). While such systems have been found to be suitable for certain site-specific scenarios, limitations of this technology have also been recognized. One such limitation being the cost associated with the technology and the long-term nature of remediation that also makes it an energy-intensive technology.

Although a variety of remediation technologies are available for the remediation of contaminated groundwater (Table 1.2), no single technology has been found to be appropriate for all contaminant types and the variety of site-specific conditions that exist at different contaminated sites. In an excellent review, Khan et al. (2004) conclude that site conditions, contaminant types, contaminant source, source control measures, and the potential impact of the possible remedial measures determine the choice of a remediation strategy and technology employed. Often more than one remediation technology is needed to effectively address most contaminated site problems. Multiple technologies applied concurrently or sequentially are referred to as "treatment trains," and are often formed to address an overall site remediation strategy. It is recognized that no single specific remediation technology may be considered as a solution for all contaminated site problems (FRTR, 2007b; Khan et al., 2004). Remediation technology implemented at most contaminated sites is not a stand-alone, or one-size-fits-all remedy, but is generally part of a "treatment train." Hence, tailored approaches and remediation techniques are needed on a site-by-site basis.

Over the last couple of decades, depending on the nature of contaminants, many different technologies have been tested for groundwater remediation. The list of remediation technologies and their applications has grown, driven by improved technical knowledge, innovative ideas, technical and economic resources, and more complex site issues (Hyman and Dupont, 2001; Leeson et al., 2013; Naidu, 2013). The past 20 years of research have focused heavily on the development of *in situ* treatment technologies. Among recent technologies, groundwater circulation wells and PRBs are increasingly being used in the United States and Europe (USEPA, 1999a,b). The process of groundwater circulation continuously removes volatile organic compounds

**TABLE 1.2**

Groundwater Remediation Technologies and Approximate Costs Based on Case Studies Available

| Groundwater Remediation Technology | Contaminant | Site If Applicable | Cost, US $ Per Cubic Meter | References |
|---|---|---|---|---|
| *In Situ* | | | | |
| Air sparging | VOCs | | $15 per foot for injection wells $5000–$25,000 for air injection pump | Hyman and Dupont (2001) |
| Air stripping | VOCs | | Capital cost: $250–$400 per gpm through up to 100 gpm Operating cost: $20–$50 per pound contaminant removed | Hyman and Dupont (2001) |
| Bioaugmentation | Trichloroethene (TCE), *cis*-dichloroethene (cDCE) | Fort Dix, NJ, USA | $875 per cubic yard | FRTR (2011) |
| Bioaugmentation | Tetrachloroethene (PCE) | Dover National Test Site (DNTS), Dover, DE, USA | $11,000 per kg of PCE addressed | FRTR (2008) |
| ZVI PRB system | Arsenic (As) (20 mg/L) groundwater plume size 450 feet wide and extends 2100 feet | East Helena, MT, USA | $325,000 to construct (no operating and maintenance cost) | FRTR (2007c) |
| ZVI: sand (1:1 ratio) PRB system $W \times L \times D$ (1 × 375 × 29 ft) | TCE, *cis*-1,2-DCE, Vinylchloride (VC) | Industrial Site, SC, USA | $360,000 to construct (no operating and maintenance cost) | USEPA (2001) |

| Peat, funnel-and-gate PRB system Funnel trench W × L × D (0.6 × 27 × 5 m), gate W × L (2 × 4 m) variable depth (slope surface) | Petroleum hydrocarbon | South eastern Australia | $71,500 to construct (no operating and maintenance cost) | Guerin et al. (2002) |
|---|---|---|---|---|
| Nitrate-enhanced bioremediation | Jet fuel (toluene, xylene, ethylbenzene) | Air force demonstration, EglinAir Force Base, FL, USA | $160–$230/gal fuel removed | DuTeaux et al. (1996) |
| Electronically induced redox barriers | TCE | F.E. Warren Air Force Base, WY, USA | Normalized to the cross-sectional area of the e-barrier capital costs: $409/ft$^3$/year and operating and maintenance cost:$10/ft$^3$/year | FRTR (2007a) |
| Intrinsic remediation or natural attenuation | Natural attenuation is confirmed (unfavorable for heavy metals) | | No capital or operating and maintenance costs other than long-term monitoring | |
| Source remediation via nano ZVI injection | Dechlorination of DNAPL contaminants | | Cost varies depending on depth, nature and quantity of contaminant, and local geology: $250,000–$300,000 per site (based on pilot trials) | Cundy et al. (2008) |
| Pump-and-treat technology (P&T) system augmented with *in situ* bioremediation. The treated groundwater was oxygenated and amended with nitrogen and phosphorus before reinjection | 1,2-DCA, VC, and benzene | French Ltd., TX, USA | Capital cost per volume of groundwater treated Per Year $200/1000 gallons; Average annual operating cost per volume of groundwater treated per year $43/1000 gallons | USEPA (1999a,b) |

*continued*

**TABLE 1.2 (continued)**

Groundwater Remediation Technologies and Approximate Costs Based on Case Studies Available

| Groundwater Remediation Technology | Contaminant | Site If Applicable | Cost, US $ Per Cubic Meter | References |
|---|---|---|---|---|
| Granular zero-valent iron approximately 450 tons PRB system W × L × D (2 × 152 × 22 ft) | VOCs and metals | USCG Center, NC, USA | Capital cost per volume of groundwater treated per Year $190/1000 gallons; Average annual operating cost per volume of groundwater treated per year $33/1000 gallons | USEPA (1999a,b) |
| *Ex Situ* | | | | |
| P&T: Type of *ex situ* treatment: Filtration with oil/water separation and Granular activated carbon adsorption. | Metals/VOCs, As, PCP, Cr, total PAHs | MSWP, AR, USA | Capital cost per volume of groundwater treated per year $38/1000 gallons; Average annual operating cost per volume of groundwater treated per year $7.4/1000 gallons | USEPA (1999a,b) |
| P&T: Type of *ex situ* treatment: Filtration with oil/water separation and Biological treatment | NAPL | Libby, MT, USA | Capital cost per volume of groundwater treated per year $1000/1000 gallons; Average annual operating cost per volume of groundwater treated per Year $36/1000 gallons | USEPA (1999a,b) |

| | | | | |
|---|---|---|---|---|
| P&T: Type of *ex situ* treatment: GAC and Strip | PCBs, TCE, 1,2-DCE, 1,1,1-TCA, VC, 1,1-DCA, PCE | LaSalle, IL, USA | Capital cost: $1400 per 1000 gL of groundwater treated per year Operating cost: $40 per 1000 gL of treated groundwater | USEPA (2001) |
| P&T: Type of *ex situ* treatment: GAC oxidation and strip | Benzene, benzidine, 2-chloroaniline, 1,2-DCE, TCE, 3,3-dichlorobenzidene, aniline, VC | Bofors Nobel, OU 1, MI, USA | Capital cost: $70 per 1000 gL of groundwater treated per year Operating cost: $4.3 per 1000 gL of treated groundwater | USEPA (2001) |
| P&T: Type of *ex situ* treatment: Physical or chemical removal of metal | Chromium | United Chrome, OR, USA | Capital cost per volume of groundwater treated per year $460/1000 gallons; Average annual operating cost per volume of groundwater treated per year $13/1000 gallons | USEPA (1999a,b) |

(VOCs) without bringing it to the surface. The target contaminant groups for this technology are halogenated VOCs, semivolatile organic compounds (SVOCs), and fuels. A circulation pattern is created in an aquifer by drawing water into and pumping it through a well and then reintroducing it without reaching the surface (Khan et al., 2004). PRB is a passive remediation technology installed *in situ*, allowing groundwater to flow under the natural gradient through a reactive cell where a reactive medium degrades or captures the contaminants (Thiruvenkatachari et al., 2008). Pump-and-treat systems (PTS) have been used for more than 20 years in the remediation of groundwater contamination. In these systems, the contaminated groundwater is extracted from the ground, treated above ground, and finally discharged or reinjected (Naidu, 2013; USEPA, 1999a,b).

Other groundwater remediation technologies are summarized in Table 1.2. These vary from passive to active remediation technologies and have significant differences in their nature of operation and cost (FRTR, 1998).

For a variety of reasons summarized below, it is currently difficult to develop accurate comparisons of remediation technology costs in many situations (National Research Council, 1997).

- Costs reported under a set of local conditions (technology costs are sensitive to site-specific geological, geochemical, and contaminant conditions, especially for *in situ* technologies).
- Technology vendors report costs using a variety of different metrics that cannot be compared directly (costs may be reported as $ per volume treated, reduction in contaminant concentrations achieved, contaminant mobility reduction achieved, mass of contaminant removed or surface area treated).
- Technology providers do not report the variable costs just "up and running" costs are given. This may be acceptable if the user only wants to compare the cost of installed operations, but the user is usually interested in the overall project cost. If certain remediation technologies have large and variable initial costs, they may not be competitive, even if the "up and running" costs appear competitive.
- Inconsistencies in the way costs are derived. Comparisons of unit costs have little meaning unless there is uniformity in the underlying methodologies and assumptions used in calculating the costs. For example, if different interest rates are used to estimate the costs of a cleanup system over its entire life cycle, conclusions about the cost competitiveness of a technology can vary widely.

This loss of compiled cost information greatly hinders dissemination of consistent cost data and makes it difficult for a new technology provider to develop comparative cost information. Furthermore, even where information regarding cost is made available to private users, it is extremely rare to see detailed cost breakdowns that would allow the reviewer to judge the realism of the cost elements.

Thus, the costs of remediation summarized in Table 1.1 are approximate only and are likely to vary with the contaminated site, including the local hydrogeology of the region and the depth to groundwater.

The potential for aquifer restoration decreased with increasing complexity of the aquifer. In 1993, the US Environmental Protection Agency issued technical impracticability (TI) waiver guidance (USEPA, 1993). This guidance specified that the waivers were appropriate for sites where the agency deems that restoration of groundwater to drinking water standards is technically impracticable. As of August 2012, the USEPA issued 91 TI waivers, 85 of which applied to groundwater (USEPA, 2012). The majority (67%) of the TI waivers granted were related to VOC contamination. In Australia, regulatory authorities have approached groundwater remediation using a risk-based strategy where Remediation to the Extent Necessary (SA EPA) or CleanUp to the Extent Practicable (EPA Vic) is recognized as the most reliable and reasonable approach to *in situ* manage contaminated groundwater. Where groundwater is remediated to the extent necessary, natural attenuation of contaminants becomes the strategy driving remediation. There has nevertheless been significant concern with regard to residual groundwater contamination postremediation and strategies that need to be put in place to minimize the potential impact of contaminants remaining in the environment, and in particular, the endpoints for remediation of sites contaminated with NAPL. This has led to recognition of the need for new innovative solutions for the treatment of contaminated groundwater.

## 1.4 PRBs: Emerging Technology for Groundwater Remediation

During the past decade and a half, PRBs have been emerging as an alternative passive *in situ* but effective remediation technology in the United States and Europe. As a consequence, this technology has also found its way to Australia. It is based on a relatively simple concept which includes underground construction of a vertical treatment wall using a reactive material in the subsurface vadose zone at a location that intercepts the groundwater contaminant plume (Figure 1.1) (Baciocchi et al., 2003; Birke et al., 2003; Meza, 2009; Thiruvenkatachari et al., 2008). Groundwater remediation technologies listed in Table 1.1 range from "least green" (pump-and-treat) to "most green" (monitoring natural attenuation [MNA]). ITRC (2011) reported PRBs lie next to MNA in terms of their green characteristics and are considered particularly sustainable when used for 10 years or more. PRBs are

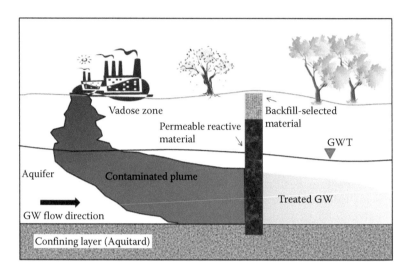

**FIGURE 1.3**
PRB intercepting a contaminant plume.

designed to be more permeable than the surrounding aquifer materials so that contaminants are treated as groundwater readily flows through without significantly altering its hydrogeology. As groundwater passes through the wall under a natural gradient, contaminants are either trapped by the reactive barrier or removed via chemical degradation and transformed into harmless substances that flow out of the wall (Figure 1.3).

Although the applications of PRBs are limited to certain site conditions, where applicable, they are most favored particularly in the urban environment and in built-up areas due to no obstruction of aboveground activities. A major benefit of PRB techniques is that as an *in situ* technology, an extensive range of contaminants can be treated, often reducing their concentration to below their detection limits. Research on PRBs increased significantly during the 1990s leading to a number of new approaches in terms of PRB design, suitable reactive materials, and target contaminants (ITRC, 2011). There are two main types of configuration for the installation of a PRB in the field and these include (a) continuous reactive barriers enabling a flow through its full cross section, and (b) "funnel-and-gate" systems where only part of PRB wall gates are permeable to contaminated groundwater (Roehl et al., 2005a). Cut-off walls (the funnel) modify flow patterns so that groundwater primarily flows through high permeability wall (the gate). The material used for constructing the funnel includes slurry walls, sheet piles, a geotextile membrane, or a soil admixture applied by soil mixing or jet grouting (Figure 1.4).

Understanding the groundwater flow regime on a more localized scale (i.e., tidal and/or seasonal variation, aquifer heterogeneity, the evapotranspiration

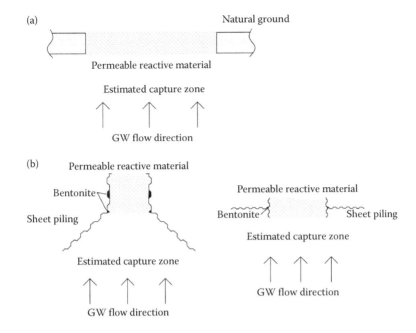

**FIGURE 1.4**
PRB configuration: (a) continuous barrier and (b) "funnel-and-gate" system.

by large trees, recharge from nearby water body) is a vital step in the design of a PRB system to ensure that the PRB is oriented perpendicular to the flow so that it captures the maximum volume of groundwater (Puls, 2006). Generally, the recommended approach is to conduct high-resolution site characterization along with groundwater and solute transport modeling to simulate possible case scenarios and design the orientation and dimensions of the PRB.

The reactive material used for the construction of the permeable wall varies with the type and concentrations of contaminants, the total mass of contaminants, and the groundwater composition (Table 1.2) (Birke et al., 2003; Thangavadivel et al., 2013). Feasibility studies are crucial for the design of PRB systems including choice of the reactive material, laboratory column experiments, estimation of required residence time, and calculation of reactive zone thickness (Roehl et al., 2005a). Since the first PRB trial in Canada (Gillham and O'Hannesin, 1992, 1994), a range of different contaminant types have been remediated using reactive materials that vary considerably in their chemical composition (Table 1.2) and their interactions and mechanism of contaminant removal. Zero-valent iron (ZVI) is the most common reactive material that generates low redox potential in groundwater, resulting in the precipitation and removal of both inorganic (metallic) and organic contaminants (Meza, 2009; Xenidis et al., 2002). Permeable reactive subsurface barriers constructed using granular elemental iron (ZVI (Fe$^o$) or Fe$^{2+}$) have been used successfully in North America and Europe for the elimination of a variety of contaminants including chlorinated solvents, heavy metals, and radionuclides (Roehl et al., 2005a). The most commonly used mechanisms are redox and sorption reactions. PRBs allow the passage of groundwater through the reactive zone of the barrier, but either immobilize (i.e., precipitation, sorption, ion-exchange, surface complexation, solid-solution formation) or chemically transform (i.e., oxidation, reduction, and degradation) contaminants to a more desirable (i.e., less toxic, more readily biodegradable) state (USEPA, 1999a,b).

Apart from iron-based materials, other types of reactive materials suitable for use in PRBs for the removal of inorganic and organic compounds from groundwater are available in a number of publications. These are briefly summarized in Table 1.3.

Readers interested in case examples of the use of PRBs are directed to Chapters 2, 4, 6, and 8 in this book. As will become evident, variants of this technology are being trialed globally and while this approach to remediation has been found to be attractive, the technology has failed in a number of instances. Failures have been attributed to the clogging of walls either due to chemical reactions resulting in precipitation of insoluble compounds or enhanced microbial activity leading to the growth of algae, or other products that clog the permeable pores.

### 1.4.1 Potential Problems Associated with the Long-Term Performance of PRBs

Designing a PRB system for a given contaminant, should include a feasibility study (e.g., column experiment) to determine the flow velocity in the barrier and effect geochemistry of groundwater on reactive material, and the retention time required to treat the groundwater. The design and selection of the reactive material has to be thick enough to allow a decrease in the contaminant concentrations to an acceptable level (the remediation target), and a longer lifetime before breakthrough. While PRBs have proven to be quite effective with the remediation of a range of contaminants, long-term sustainability and efficacy of the barriers have been major uncertainties. Despite these being addressed by a number of researchers (Gu et al., 2002; Henderson and Demond, 2007; Liang et al., 2003), microgeochemical processes within the barrier could play an influential role in the long-term performance, as they have been found to affect contaminant removal both within and downgradient of the barrier matrix. For example, PRB demonstration studies in Germany and Australia show that depending on aquifer chemistry, microorganisms can reduce the porosity of the barrier by forming a biofilm (Taylor and Jaffe, 1990) which reduces pore space by clogging (Vandevivere and Baveye, 1992) or by contributing to mineral precipitation, or by producing gas bubbles that restrict water flow.

**TABLE 1.3**

Review of Reactive Material Suitable for the Construction of the PRB Wall

| PRB Reactive Material | Target Chemical | References |
|---|---|---|
| Organic-based materials: activated carbon, leaf, peat, sewage sludge, sawdust, compost, wood chips, chitin, lignin, and so on | Metals and metalloids: arsenic (As), hexavalent chromium, Cr(VI), cadmium (Cd), mercury (II), uranium (U) and molybdenum (Mo) Aromatic compounds | Blowes et al. (2000), Guerin et al. (2002), Han et al. (2000), Meggyes and Simon (2000), Meza (2009), Scherer et al. (2000) |
| Alkaline-complexing agents: hydrated lime, ferrous sulfate, etc. | Petroleum hydrocarbons and metals | Roehl et al. (2005b) |
| Phosphate minerals: hydroxyapatite and biogenic apatite (e.g., fish bone) | Divalent heavy metal ions | Admassu and Breese (1999), Arey et al. (1999), Leyva et al. (2001) |
| Surfactant-modified zeolites (SMZs): Natural zeolites coated with hexadecyltrimethylammonium, Clinoptiloite-rich zeolite | BTEX and other gasoline by-products, perchloroethylene (PCE), radionuclides, inorganic oxyanions (sulfate, chromate, and selenate) | Haggerty and Bowman (1994), Li et al. (1998), Xenidis et al. (2002) |
| Colloidal: Iron size (1–3 μm) or nano size (1–00 nm = 0.001–0.1 μm) | Carbothiolate herbicide, molinate, dechlorination of TCE and PCBs | Joo et al. (2004), Wang and Zhang (1997) |
| Metal oxides: (Iron/calcium oxides, and fine-grained activated aluminum oxide, elemental copper (CuO) | Phosphorus attenuation, removing mercury ($Hg^{2+}$), chlorinated hydrocarbon, and some aromatics | Baker et al. (1998), Huttenloch et al. (2003), Tratnyek et al. (2003) |
| Alkaline materials: recycled concrete, limestone, calcite-bearing zeolitic breccia, blast furnace slag, lime, organo-clay, and fly ash, and so on | Leachate from acid sulfate soils | Ake et al. (2001), Golab et al. (2006), Waybrant et al. (1998) |
| Microorganisms | Chlorinated solvents | USEPA (2000) |
| Polymers: polyacryloamidoxime resin derived from polyacrylonitrile, which is deposited from solution onto the surface of quartz sand to form a thin film coating | Uranium contaminated groundwater, carbon tetrachloride, copper ($Cu^{2+}$), nitrobenzene, 4-nitroacetophenone, and chromate ($CrO_4^{2-}$) | Shimotori et al. (2004), Stewart et al. (2006) |

The longevity of PRBs denote their ability to sustain their function (hydraulic capture, residence time, and reactivity) in the years and decades following installation. It is strongly dependent on the groundwater chemistry and flow rates, and contaminant concentrations at the remediation site. The observed mineral phases at different PRB sites are controlled by the groundwater constituents that naturally vary depending on the biogeochemical setting of the site (Roehl et al., 2005a). PRBs require a much better understanding of site characteristics than conventional pump-and-treat-type remedies. The efficacy of the PRB material is greatly influenced by hydrogeology, microbiology (i.e., iron-reducing, sulfate-reducing, and/or methanogenic bacteria) and geo-chemistry properties (i.e., the concentration, solubility and speciation of the contaminants and cosolvents, and the prevalent, pH and Eh condition). The lifetime of a PRB is a problem encountered as a consequence of precipitation of minerals and/or the growth of microbial populations. These can lead to cementation and a reduction in the porosity of reactive media resulting in a decreasing permeability and the formation of a coating on the reactive surface area of the PRB material. Puls (2006) reported that more than 100 PRBs have been installed worldwide, but little data have been collected on the long-term performance and consequences of the rate of formation of surface precipi-tates, bio-fouling, and remobilization of adsorbed contaminants. Since iron-based reactive materials are most commonly used in PRBs, their long-term performance is well recorded compared to others listed in Table 1.3.

Key issues associated with the design of a PRB wall during the design phase include residence time in the reaction zone, the reaction zone size for optimal life span, the impact of the reaction medium on the groundwater quality, and the ultimate fate of PRB walls. Once the capacity of the medium is exhausted, contaminant breakthrough will occur. For example, Stehmeier (1989) reported breakthrough of the PRB with peat reactive material used to remove dissolved and free phase petroleum hydrocarbons. Furthermore, understanding the chemical reaction mechanism is critical in order to evalu-ate the potential for release/remobilization of sequestered contaminants and for improved design of the reactive material.

A failure in a PRB system has been reported on many occasions meet-ing performance standards because of an inadequate understanding of the groundwater flow system that will exist after the PRB has been installed. Numerical modeling of various PRB design scenarios and evaluations of the resulting groundwater flow systems can aid in determining the appropri-ateness of the PRB for specific site conditions and finalization of the pre-construction design (Scott and Folkes, 2000). A PRB design should include development of an adequate network of monitoring wells and an appropri-ate frequency of sampling to document the performance objective and to assess long-term operation and maintenance requirements. The well screen interval should be adequate to monitor the saturated zone treated by the PRB, particularly at the high flow zone or the highest contaminant concen-tration area to monitor treatment along the preferential pathway (ITRC, 2011). Where a lower confining layer (Aquitard) is not present, monitoring wells screened at deeper depths may be required to ensure that there is no contaminant bypass beneath the PRB walls. ITRC (2011) recommends that, depending on the width of the reactive zone and the reactive media used, it is useful to have monitoring locations within the PRB itself.

## 1.5 Conclusion

A PRB fits the concept of a green and sustainable groundwater remediation technology; it is the most green remediation technique after monitored natural attenuations. PRB, where applicable, is less expensive as compared to ex situ technology as it involves no operational and maintenance costs. Reactive materials are frequently waste products (e.g., mulch, saw dust, some iron ore slags) or are recycled materials (e.g., iron scrap). The performance and longevity of a PRB is reliant on the design stage as there will be little or no possibility of correcting the system after construction. Consequently, detailed high-resolution site characterization such as the nature and extent of the contaminant plume(s), selection of the reactive material, the hydraulic design, and the vertical extent of contamination are particularly important criteria. PRB remediation technology is still evolving with new and innovative reactive materials introduced to treat different contaminants utilizing innovative new construction methods.

## References

Admassu W., Breese T. 1999. Feasibility of using natural fish bone apatite as a substitute for hydroxyapatite in remediating aqueous heavy metals. *Journal of Hazardous Materials* 69:187–196.

Ake C.L., Mayura K., Huebner H., Bratton G.R., Phillips T.D. 2001. Development of porous clay-based composites for the sorption of lead from water. *Journal of Toxicology and Environmental Health Part A* 63:459–475.

Arey J.S., Seaman J.C., Bertsch P.M. 1999. Immobilization of uranium in contaminated sediments by hydroxyapatite addition. *Environmental Science & Technology* 33:337–342.

Baciocchi R., Boni M.R., D'Aprile L. 2003. Characterization and performance of granular iron as reactive media for TCE degradation by permeable reactive barriers. *Water, Air, and Soil Pollution* 149:211–226.

Baker M.J., Blowes D.W., Ptacek C.J. 1998. Laboratory development of permeable reactive mixtures for the removal of phosphorus from onsite waste water disposal systems. *Environmental Science & Technology* 32:2308–2316. DOI:10.1021/es970934w.

Barzi F., Naidu R., McLaughlin M.J. 1996. Contaminants and the Australian soil environment, In: R. Naidu et al. (Eds.), *Contaminants and the Soil Environment in the Australasia-Pacific Region*, Springer Netherlands. pp. 451–484.

Birke V., Burmeier H., Rosenau D. 2003. Design, construction, and operation of tailored permeable reactive barriers. *Practice Periodical of Hazardous, Toxic, and Radioactive Waste Management* 7:264–280.

Blowes D.W., Ptacek C.J., Benner S.G., McRae C.W.T., Bennett T.A., Puls R.W. 2000. Treatment of inorganic contaminants using permeable reactive barriers. *Journal of Contaminant Hydrology* 45:123–137.

CEI. 2005. Soil remediation technologies: Assessment, clean-up, decommissioning, rehabilitation. Canadian Environmental Industries (Energy and Environmental Industries Branch), available at: http://www.ic.gc.ca/eic/site/ea-ae.nsf/eng/ea02201.html.

Chadalavada S., Datta B., Naidu R. 2011. Uncertainty based optimal monitoring network design for a chlorinated hydrocarbon contaminated site. *Environmental Monitoring and Assessment*, 173:929–940.

Cundy A.B., Hopkinson L., Whitby R.L.D. 2008. Use of iron-based technologies in contaminated land and ground water remediation: A review. *Science of the Total Environment* 400:42–51. DOI:http://dx.doi.org/10.1016/j.scitotenv.2008.07.002.

Davis G.B., Merrick N., Mc Laughlan R. 2006. Protocols and techniques for characterising sites with subsurface petroleum hydrocarbons—A review, CRCCARE Technical Report no. 2, CRC for Contamination Assessment and Remediation of the Environment, Adelaide, Australia.

DuTeaux S.B. 1996. *A Compendium of Cost Data for Environmental Remediation Technologies.* United States. Department of Energy. Office of Science Technology. Los Alamos National Laboratory.

FRTR. 1998. Guide to documenting and managing cost and performance information for remediation projects. Prepared by members agencies of the federal remediation technology roundtable (FRTR), www.frtr.gov, EPA 542-B-98-007.

FRTR. 2007a. Electronically induced Redox Barriers. Federal Remediation Technologies Roundtable. USEPA, Washington, DC http://costperformance.org/profile.cfm?ID=403&CaseID=403.

FRTR. 2007b. *In Situ Chemical Oxidation, Soil Vapor Extraction, and in situ Bio-Stimulation at Hanner's Dry Cleaners, Pompano Beach, Florida.* Federal Remediation Technologies Roundtable. USEPA, Washington, DC. http://costperformance.org/profile.cfm?ID=408&CaseID=406.

FRTR. 2007c. Permeable Reactive Barrier. Federal Remediation Technologies Roundtable. USEPA, Washington, DC. http://costperformance.org/profile.cfm?ID=395&CaseID=395.

FRTR. 2008. Bioaugmentation. Federal Remediation Technologies Roundtable. USEPA, Washington, DC. http://costperformance.org/profile.cfm?ID=424&CaseID=422.

FRTR. 2011. Bioaugmentation. Federal Remediation Technologies Roundtable. USEPA, Washington, DC. http://costperformance.org/profile.cfm?ID=408&CaseID=406.

Gillham R.W., O'Hannesin S.F. 1992. Metal-catalysed abiotic degradation of halogenated organic compounds. Univ., Waterloo Centre for Groundwater Research.

Gillham R.W., O'Hannesin S.F. 1994. Enhanced degradation of halogenated aliphatics by zero-valent iron. *Groundwater* 32:958–967.

Golab A.N., Peterson M., Indraratna B. 2006. Selection of potential reactive materials for a permeable reactive barrier for remediating acidic groundwater in acid sulphate soil terrains. *Quarterly Journal of Engineering Geology and Hydrogeology* 39:209–223.

Gu B., Watson D.B., Wu L., Phillips D.H., White D.C., Zhou J. 2002. Microbiological characteristics in a zero-valent iron reactive barrier. *Environmental Monitoring and Assessment* 77:293–309.

Guerin T.F., Horner S., Mc Govern T., Davey B. 2002. An application of permeable reactive barrier technology to petroleum hydrocarbon contaminated groundwater. *Water Research* 36:15–24. DOI: http://dx.doi.org/10.1016/S0043-1354(01)00233-0.

Haggerty G.M., Bowman R.S. 1994. Sorption of chromate and other inorganic anions by organo-zeolite. *Environmental Science & Technology* 28:452–458.

Han I., Schlautman M.A., Batchelor B. 2000. Removal of hexavalent chromium from groundwater by granular activated carbon. *Water Environment Research* 72(1): 29–39, Water Environment Federation.

Henderson A.D., Demond A.H. 2007. Long-term performance of zero-valent iron permeable reactive barriers: A critical review. *Environmental Engineering Science* 24:401–423.

Huttenloch P., Roehl K.E., Czurda K. 2003. Use of copper shaving store move mercury from contaminated groundwater or waste water by amalgamation. *Environmental Science & Technology* 37:4269–4273. DOI:10.1021/es020237q.

Hyman M., Dupont R.R. 2001. *Groundwater and Soil Remediation: Process Design and Cost Estimating of Proven Technologies*. ASCE Publications, Reston, VA. Doi: 10.1061/9780784404270.

ITRC. 2011. Permeable Reactive Barrier: Technology Update, The Interstate Technology & Regulatory Council PRB: Technology Update Team. Washington, DC. http://www.itrcweb.org/Guidance/GetDocument?documentID=69.

Joo S.H., Feitz A.J., Waite T.D. 2004. Oxidative degradation of the carbothioate herbicide, molinate, using nanoscale zero-valent iron. *Environmental Science & Technology* 38:2242–2247. DOI:10.1021/es035157g.

Khan F.I., Husain T., Hejazi R. 2004. An overview and analysis of site remediation technologies. *Journal of Environmental Management* 71:95–122. DOI:http://dx.doi.org/10.1016/j.jenvman.2004.02.003.

Leeson A., Stroo H.F., Johnson P.C. 2013. Groundwater remediation today and challenges and opportunities for the future. *Groundwater* 51:175–179. DOI:10.1111/gwat.12039.

Leyva A.G., Marrero J., Smichowski P., Cicerone D. 2001. Sorption of antimony on to hydroxyapatite. *Environmental Science and Technology* 35:3669–3675.

Li Z., Roy S.J., Zou Y., Bowman R.S. 1998. Long-term chemical and biological stability of surfactant-modified zeolite. *Environmental Science & Technology* 32:2628–2632.

Liang L., Sullivan A.B., West O.R., Moline G.R., Kamolpornwijit W. 2003. Predicting the precipitation of mineral phases in permeable reactive barriers. *Environmental Engineering Science* 20:635–653.

McLaughlan R., Merrick N., Davis G.B. 2006. Natural attenuation: A scoping review, CRCCARE Technical Report no. 2, CRC for Contamination Assessment and Remediation of the Environment, Adelaide, Australia.

Meggyes T., Simon F.-G. 2000. Removal of organic and inorganic pollutants from groundwater using permeable reactive barriers. *Land Contamination & Reclamation* 8:3.

Meza I. 2009. The use of PRBs (permeable reactive barriers) for attenuation of cadmium and hexavalent chromium from industrial contaminated soil. Thesis submitted to the Graduate School, Ball State University.

Naidu R. 1996. Contaminants and the soil environment in the Australasia-Pacific region. Proceedings of the First Australasia-Pacific Conference on Contaminants and Soil Environment in the Australasia-Pacific Region, held in Adelaide, Australia, February 18–23, 1996, 1st Australasia-Pacific Conference on Contaminants and

Soil Environment in the Australasia-Pacific Region, Adelaide, S. Aust. (USA), 1996, Kluwer Academic.

Naidu R. 2013. Recent advances in contaminated site remediation. *Water, Air, & Soil Pollution* 224:1–11. DOI:10.1007/s11270–013–1705-z.

Naidu R., Smith E., Owens G., Bhattacharya P. 2006. *Managing Arsenic in the Environment: From Soil to Human Health*. CSIRO Publishing, Victoria, Australia.

National Research Council. 1997. *Innovations in Groundwater and Soil Cleanup: From Concept to Commercialization*. The National Academies Press, Washington, DC.

NRC. 1994. Alternatives for Groundwater Cleanup. Alternatives, C.G.W.C. Commission on Geosciences, E.R. Studies, D.E.L. Council, N.R. National Academies Press.

Puls R. 2006. Long-term performance of permeable reactive barriers: Lessons learned on design, contaminant treatment, longevity, performance monitoring and cost—An overview, In: I. Twardowska, H. Allen, M. Häggblom, and S. Stefaniak (Eds.), *Soil and Water Pollution Monitoring, Protection and Remediation*, NATO Science Series. Springer, Dordrecht, The Netherlands. pp. 221–229.

Rao P.S.C., Davis G.B., Johnston C.D. 1996. Technologies for enhanced remediation of contaminated soils and aquifers: Overview, analysis, and case studies, In: R. Naidu, R.S. Kookana, D.P. Oliver, S. Rogers, and M.J. McLaughlin (Eds.), *Contaminants and the Soil Environment in the Australasia-Pacific Region*, Kluwer Academic Publishers, London, pp. 361–410.

Roehl K.E., Czurda K., Meggyes T., Simon F.-G., Stewart D. 2005a. Permeable reactive barriers. *Trace Metals and Other Contaminants in the Environment* 7:1–25.

Roehl K.E., Meggyes T., Simon F., Stewart D. 2005b. Long-term performance of permeable reactive barriers. Access online via Elsevier.

Scherer M.M., Richter S., Valentine R.L., Alvarez P.J. 2000. Chemistry and microbiology of permeable reactive barriers for in situ groundwater cleanup. *Critical Reviews in Microbiology* 26:221–264.

Scott K.C., Folkes D.J. 2000. Groundwater modeling of a permeable reactive barrier to enhance system performance. *Proceedings of the 2000 Conference on Hazardous Waste Research*, HSRC.

Shimotori T., Nuxoll E.E., Cussler E.L., Arnold W.A. 2004. A polymer membrane containing FeO as a contaminant barrier. *Environmental Science & Technology* 38:2264–2270.

Singh B.K., Naidu R. 2012. Cleaning contaminated environment: A growing challenge. *Biodegradation* 23(6):785–786.

Stehmeier L. 1989. Development of anoclanosorb (TM) (peat) filter for separating hydrocarbons from bilge water. Calgary, CA: Nova Husky Research Corporation: 12.

Stewart D.I., Csõvári M., Barton C.S., Morris K., Bryant D.E. 2006. Performance of a functionalised polymer-coated silica at treating uranium contaminated groundwater from a Hungarian mine site. *Engineering Geology* 85:174–183.

Taylor S.W., Jaffe P.R. 1990. Biofilm growth and the related changes in the physical properties of a porous medium: 3. Dispersivity and model verification. *Water Resources Research* 26:2171–2180.

Thangavadivel K., Wang W., Birke V., Naidu R. 2013. A comparative study of trichloroethylene (TCE) degradation in contaminated groundwater (GW) and TCE-spiked deionised water using zero valent iron (ZVI) under various mass transport conditions. *Water, Air, & Soil Pollution* 224:1–9. DOI:10.1007/ s11270-013-1718-7.

Thiruvenkatachari R., Vigneswaran S., Naidu R. 2008. Permeable reactive barrier for groundwater remediation. *Journal of Industrial and Engineering Chemistry* 14:145– 156. DOI:http://dx.doi.org/10.1016/j.jiec.2007.10.001.

Tratnyek P.G., Scherer M.M., Johnson T.J., Matheson L.J. 2003. Permeable reactive barriers of iron and other zero-valent metals. In: M.A. Tarr (Ed.), *Chemical Degradation Methods for Wastes and Pollutants: Environmental and Industrial Applications*. Environmental Science & Pollution. CRC Press, Marcel Dekker: New York, pp. 371–421.

USEPA 1993. Guidance For Evaluating the Technical Impracticability of Groundwater Restoration, Interim Final. United States OSWER Directive 9234.2-25, Washington, DC. http://www.epa.gov/superfund/health/conmedia/gwdocs/pdfs/tech1 .pdf.

USEPA 1999a. Groundwater Cleanup: Overview of Operating Experience at 28 Sites. EPA542-R-99-006, Washington, DC 20460. http://epa.gov/tio/download/remed /ovopex.pdf.

USEPA 1999b. Field Applications of in situ Remediation Technologies: Permeable Reactive Barriers, U.S. Environmental Protection Agency, Office of Solid Waste and Emergency Response Technology Innovation Office, Washington, DC.

USEPA 2000. Engineered approaches to in situ bioremediation of chlorinated solvents: Fundamentals and field applications. U.S. Environmental Protection Agency Office of Solid Waste and Emergency Response Technology Innovation Office Washington, DC. EPA 542-R-00-008. July 2000 (revised).

USEPA 2001. Remediation Technology cost compendium year 2000. EPA-542-R-01 -009, Solid waste and emergency response (5102G). http://epa.gov/tio/down load/remed/542r01009.pdf.

USEPA 2012. Summary of Technical Impracticability Waivers at National Priorities List Sites. United States Environmental Protection Agency. OSWER Directive 9230.2-24. http://www.epa.gov/superfund/health/conmedia/gwdocs/pdfs /TI_waiver_report%2009Aug2012.pdf.

Vandevivere P., Baveye P. 1992. Effect of bacterial extracellular polymers on the saturated hydraulic conductivity of sand columns. *Applied and Environmental Microbiology* 58:1690–1698.

Wang C.-B., Zhang W.-x. 1997. Synthesizing nanoscale iron particles for rapid and completed echlorination of TCE and PCBs. *Environmental Science & Technology* 31:2154–2156. DOI:10.1021/es970039c.

Waybrant K.R., Blowes D.W., Ptacek C.J. 1998. Selection of reactive mixtures for use in permeable reactive walls for treatment of mine drainage. *Environmental Science & Technology* 32:1972–1979.

Xenidis A., Moirou A., Paspaliaris I. 2002. Reactive materials and attenuation processes for permeable reactive barriers. Mineral? *Wealth* 123:35–49.

# 2

## Two Decades of Application of Permeable Reactive Barriers to Groundwater Remediation

**Scott D. Warner**

### CONTENTS

## 2.1 Introduction

The year 2011 marked the 20th year anniversary of the first pilot testing of the permeable reactive barrier (PRB)* as an *in situ* groundwater remedy by University of Waterloo researchers at the Canadian Forces Base (CFB) Borden site in Ontario, Canada (Gillham and O'Hannesin, 1994). Over the ensuing 20 years, the PRB concept would evolve from its standing as an "innovative" remedy for chemically impacted groundwater first commercially applied at a former semiconductor manufacturing facility in northern California, USA, to a "developed" technology that has been installed at sites around the globe, as well as being identified as one of the most sustainable groundwater treatment remedies available. Furthermore, over the past two decades, this remediation concept has grown to be the subject

---

\* Composed of zero-valent iron (ZVI).

of research at many academic institutions; it has become a common-place addition to feasibility studies and alternative analysis documents for remediation projects. As a consequence, PRBs have been the specific focus of technical short courses and conferences, spanning the 1995 special session on PRBs at the American Chemical Society's annual meeting in California, USA, to Internet-based short-courses sponsored by the U.S. Environmental Protection Agency (USEPA), to international meetings in the United Kingdom, Germany, and Italy, and to the September 2011 Clean-Up conference in Adelaide, Australia. Considering that the first full-scale commercial PRB composed of zero-valent iron (ZVI) was installed in November 1994 (Yamane, 1995) and continues to function, is the proof that this remedial approach is truly one of the more sustainable and resource conservative treatment concepts for chemically affected groundwater.

Although exact wording has morphed over the past two decades, the generally accepted definition of a PRB, as modified from Interstate Technology and Regulatory Council (ITRC, 2005) is

> an engineered, continuous, in situ permeable treatment zone designed to intercept and remediate a contaminant plume. (Figure 2.1)

Historically, the PRB has taken additional descriptive names including "permeable treatment zone," "applied reactive treatment zone," "permeable reactive treatment zone," and so on. From these names and definition, it is clear that the treatment of contaminated groundwater occurs within the PRB, or immediately adjacent to it (e.g., the evolution of hydrogen from a ZVI-based PRB may enhance biodegradation processes a short distance down gradient from the PRB). This recognition is important because the PRB, by itself, would not completely remediate a contaminant plume until

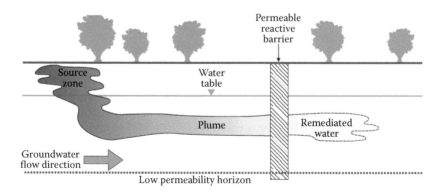

**FIGURE 2.1**
General concept of a PRB for treating a plume of contaminated groundwater.

all affected groundwater at a site flows through the PRB. More accurately, the primary use of a PRB is to eliminate or substantially reduce the mass flux of the target contaminant(s) to areas located immediately downgradient of the PRB alignment. The PRB is not, in its typical intent, used as a *source remediation* technology; but is more correctly a *source control* remediation technology. The PRB may also be used as a *receptor protection* technology if its primary purpose is to protect sensitive receptors located either near or far from a potential contaminant source.

As remediation technologies for contaminated groundwater have progressed since the 1980 genesis of the U.S. Superfund Program, the intended movement away from energy-intensive groundwater pump and treat, to hydraulically passive treat-in-place, or *in situ*, methods, has allowed technologies such as the PRB to develop from purely a research-and-development exercise to a proven remediation technology used worldwide. While pump-and-treat still is used for many groundwater remediation projects, the general progression from active to passive technologies has occurred for a number of reasons including the intent to find less expensive, more focused, more resource conservative methods to protect environmental receptors from being negatively impacted by groundwater contaminants.

As introduced above, the PRB technology, which has been the focus of hundreds if not thousands of technical articles and publications is considered, for example, a "remediation concept" whereby affected groundwater is allowed to flow through, or is routed through, an emplaced subsurface zone of treatment media that has geochemical or biochemical properties appropriate for either destroying, immobilizing, or altering the molecular form of the contaminant sufficiently enough to render it harmless.

Early versions of the PRB concept were used to neutralize acidic water using limestone filters off of mine tailings piles (Pearson and McDonnell 1975); the first commercial PRB using granular iron metal was constructed in 1994 to destroy chlorinated aliphatic compounds in groundwater (Warner et al., 1998) following the development of the technology by researchers at the University of Waterloo (Gillham and O'Hannesin 1994), and a number of pilot tests of PRBs for treating radioactive constituents in groundwater were attempted in the mid- to late 1990s (Naftz et al., 2002). Reasonably, it makes sense that since the early 1990s, several hundred pilot tests and fewer, though still a substantial number, of the full-scale remedies involving the PRB concept have been implemented worldwide. Today, with the focus on sustainable and green remediation, the PRB is arguably one of the most sustainable approaches considering that the oldest systems, now more than 15 years later, have functioned successfully since their installation to treat contaminated groundwater *in situ* without the need to apply energy or pump water—a claim that few other treatment methods can make.

## 2.2 General Considerations of the PRB

The PRB concept has been the focus of hundreds if not thousands of technical articles and publications (e.g., ITRC, 1999, 2005; Naftz et al., 2002; RTDF, 1998; Roehl et al., 2005, Thiruvenkatachari et al., 2008) as well as Internet web sites (e.g., the Oregon Graduate Institute [now part of the Oregon Environmental Health University] database for the use of ZVI as a treatment media within a PRB—http://cgr.ebs.ogi.edu/ironrefs/). A paper describing the 10-year performance history of this remedial concept was provided by Warner and Sorel (2003). This article provides a brief update to what was learned then and what is understood today. However, this article is not an exhaustive review of PRB technology. The reader is instead referred to other examples, including the newly published ITRC guidance document (ITRC, 2011), which provides several hundred technical references on PRB development and performance, and the CRC CARE Technical report on PRB Guidance (CRC CARE Technical report Number 25) (Perlmutter et al., 2013).

Since the concept of either destroying dissolved contaminants or rendering them immobile or less toxic within a subsurface treatment zone was described by McMurty and Elton (1985), several hundred formally recognized PRBs, and perhaps several hundred more similar type remedies have been installed at a variety of sites including industrial, mining, and retail petroleum facilities. Simply, the goal of the PRB is to introduce, or enhance, geochemical or biological reactions that afford the necessary treatment of the target contaminants within a subsurface engineered zone. The key and perhaps most important aspect of the PRB, and its greatest asset, is that the treatment is performed under hydraulically ambient conditions; that is, affected groundwater flows and contaminants migrate through the PRB under natural gradient conditions without the addition of pumping or other energy-induced methods.

An important acknowledgment proven for over 20 years is that each PRB application is unique in all characteristics—design, objectives, construction, economics, monitoring, and specific performance. One supporting reason for this is that each remediation site is unique with respect to site geology and hydrogeology, contaminant occurrence and distribution, land-use issues, and regulatory drivers. This uniqueness is both an advantage and disadvantage for PRB application. As an advantage, because the PRB is not an "off-the-shelf" remedy, it must be designed to suit the needs of the site (Shoemaker et al., 1995). This has led to the identification and development of multiple installation methods (e.g., excavate and fill, injection and jetting, single pass trenching and replacement, large diameter boring emplacements) and geometries (e.g., continuous wall, funnel and gates, multidiscrete depth application, dual walls), and the identification and use of different treatment media (beyond ZVI) for a variety of chemicals (for and beyond chlorinated hydrocarbons). For example, PRB treatment materials in addition to ZVI have ranged from bi-active mulch, to crushed limestone, to zeolite minerals, with contaminants treated ranging from chlorinated hydrocarbons, to dissolved reducible metals, to radioactive constituents such as strontium-90. Research continues in various institutions to develop a better technical understanding of many of the specific treatment mechanisms to not only develop treatments for more chemical types, but also to better understand longevity of a given treatment type so that appropriate contingencies can be invoked before PRB performance is reduced beyond the objectives of a project.

Two principal design components of the PRB include: the treatment matrix and the hydraulic control system. The treatment matrix may be composed of any one, or more, of a variety of materials shown to have the capability of directly destroying, enhancing natural treatment processes, or immobilizing target chemicals (see Table 2.1). The hydraulic system is

**TABLE 2.1**

Example Reactive Treatment Materials for Use in PRBs

| Constituents of Concern | Granular Iron Metal | Biological (C, H, O) | Phosphatic (Apatite) | Zeolite (Ion-Exchange) | Organophilic Clay |
|---|---|---|---|---|---|
| Chlorinated aliphatic compounds (e.g., TCE) | X | X | | | |
| Chlorinated methanes | | | | | |
| Chlorinated pesticides | | X | | | |
| Chlorofluorohydrocarbons | X | | | | |
| Nitrobenzene | X | | | | |
| Benzene, toluene, ethylbenzene, and xylenes (BTEX) | | X | | | |
| Polycyclic aromatic hydrocarbons (PAHs) | | | | | X |
| Energetics | X | X | | | |
| Perchlorate | | X | X | X | |
| Cationic metals (e.g., Cu, Ni, Zn) | X | X | X | | |
| Arsenic | X | X | | | |
| Chromium(VI) | X | | | X | |
| Uranium | X | X | X | | |
| Strontium-90 | | | X | X | |
| Selenium | X | X | | | |
| Nitrate | | X | | | |
| Ammonium | | X | | X | |
| Sulfate | | X | | | |
| Methyl tertiary butyl ether (MTBE) | | X | | | |

*Source:* International Technology Regulatory Council (ITRC). 2011. Permeable Reactive Barrier: Technology Update. *Prepared by the Interstate Technology Regulatory Council Permeable Reactive Barriers Team,* Washington, DC, http://www.itrcweb.org (accessed 2011).

*Note:* Biological PRB includes the following enhancements: C (carbon), H (hydrogen), and O (oxygen).

directly related to the emplaced treatment zone which may be emplaced by one of a number of means as described previously, including trench excavation (by conventional equipment of newly developed single-pass trenching), large diameter filled borings, injection of micron-sized material, or injection of chemical substrates that geochemically alter native subsurface materials to afford the required treatment conditions. Whichever way the treatment matrix is emplaced, the PRB system typically is intended to perform using hydraulically passive means; that is, the PRB is designed to allow groundwater and the target chemicals to flow through the PRB without mechanical assistance. Modern sustainable hydraulic enhancements may include the use of solar or wind-driven, low-rate groundwater pumps, or passive-siphon action to further control and route affected groundwater through a PRB. Aside from monitored natural attenuation or intrinsic remediation strategies, there may be no other primary groundwater remediation method that conveys the concept of a green and sustainable system as well as the PRB.

## 2.3 PRB Developments

A third relatively new design component focuses on the sustainability of the system. Considering that the firm formally recognized that a PRB was installed in late 1994 and continues to function more than 17 years after installation without any energy induced or water removed (except for monitoring activities), the PRB must naturally be considered one of the more sustainable groundwater remedies. Today, with the advent of calculations to assess the sustainability of treatment materials and construction methods, evaluating the sustainability of the PRB can take on a different meaning. However, compared to every active groundwater remedy, and most *in situ* remedies, perhaps with the exception of MNA and some bioremediation techniques, the PRB clearly is a leading sustainable remedy.

The following sections provide a discussion of developments in PRB technology. Again, an exhaustive review is not provided; however, several important considerations are provided herein.

### 2.3.1 Chemical Treatment

PRBs historically have been used to treat dissolved organic contaminants in groundwater. This includes the common halogenated compounds including tetrachloro- or perchloroethylene (PCE) and trichloroethylene (TCE), and metals including chromium and arsenic, which are treated using granular ZVI (chloro-reduction for the aliphatic compounds and geochemical reduction for the metals). New developments have shown that ZVI also is effective in treating energetic compounds including nitroamines and trinitrotoluene (TNT). Furthermore, ZVI is now recognized as an effective bioremediation-enhancing agent due to the ZVI corrosion reaction in water that provides a sustained (in years) flux of dissolved hydrogen to the aqueous system. PRBs have also been installed to include such new materials as zeolites to provide an ion-exchange treatment of radioactive strontium-90 (Warner et al., 2012), and biological materials to create biowalls (Bellis et al., 2010). Early PRBs sometimes utilized granular-activated carbon (GAC) within a PRB treatment cell to provide sorption of target organic compounds. More often used at installations in Europe than in North America, fewer systems using GAC appear to have been designed in recent years.

Biowalls using solid organic materials (mulch) have been greatly applied over the past 10 years to stimulate anaerobic degradation of chlorinated hydrocarbons as well as energetic and munitions compounds. These systems also have been employed to nitrate-impacted groundwater. The greater understanding of biogeochemical transformation processes in biowalls that are responsible for the resulting abiotic dechlorination of solvents. PRBs composed of "organic material" have evolved with respect to arsenic treatment, and new media including organic carbon-rich combinations, emulsified ZVI, and organophyllic clays have been applied as reactive treatment media.

## 2.3.2 Unintended Performance Issues

PRB performance, particularly when performance is seen as underachieving, is not often presented in technical literature. While most PRBs are assumed to have performed adequately, an inadequate performance can occur and typically results from inadequate hydraulic design. Experience has shown that an insufficient hydraulic design has its genesis in incomplete site characterization. Even if characterization is satisfactory, overly complex hydrogeologic conditions may preclude a cost-effective application of PRBs. Such characteristics may include high rates of groundwater flow, high permeability, extreme aquifer heterogeneity, undiscovered preferential flow paths, or excessive depth to groundwater. The nature and extent of the contaminant distribution must be well characterized to design an effective PRB and should consider the nature and anticipated persistence of the contaminant source. The vertical extent of contamination is particularly important. The contaminant discharge (mass flux) through the PRB should be sufficiently characterized so that the upgradient concentrations can be accommodated by the PRB design. It is also imperative to understand the plume shape and direction variability over time.

While the oldest PRB remains functional after 17 years, it is well understood that these systems age due to both the exhaustion of the treatment media, and the important effect of inorganic constituents that may lead to mineral precipitates in pore spaces. Calcium carbonate, iron carbonate, iron hydroxide, and iron sulfide precipitates may form in the media with pH

and reduction–oxidation shifts associated with the geochemical conditioning of the groundwater from the treatment reactions. Regarding ZVI, geochemical changes to ZVI from the presence of sulfate, nitrate, and oxygen are
widely observed and more research into prevention of performance loss is
an ongoing area of work. Generally, it is now recognized that excess nitrate
can passivate the ZVI corrosion reaction and render ZVI nearly useless as a
treatment material within a matter of months.

Effective remediation of groundwater contaminants using PRBs depends
on achieving appropriate conditions for the degradation reactions to occur
and having a reaction zone (size/thickness) that provides sufficient residence
time for contaminants to degrade to performance objectives. For biological
PRBs, insufficient residence time of the contaminants in the reaction zone
may result in accumulation of regulated intermediate degradation products.
The success of biological PRBs largely depends on the presence of microbes
that are capable of facilitating the requisite degradation reactions.

### 2.3.3 Monitoring Improvements and Longevity

The most important aspects of monitoring improvements with regard
to PRBs may be the development of alternative compliance monitoring
metrics—including mass discharge and toxicity reduction calculations—
and improved analytical monitoring tools, including compound-specific
isotope analysis (CSIA) and molecular biological tools (MBTs). Analysis of
iron and sulfide mineralogy to evaluate biogeochemical transformation processes has become important particularly for organic-media PRBs and the
assessment of precipitation reaction zones within ZVI treatment systems
provides an indication of aging progression with the ZVI PRB. Process monitoring and performance monitoring may require different analytical protocols, monitoring locations, and monitoring frequencies as the approaches
coincide, such as: (1) baseline characterization for performance comparison
to design considerations; (2) process monitoring to optimize system operation and performance and to evaluate the need for system modifications;
and (3) performance and compliance monitoring to evaluate and validate
the effectiveness of the system with regard to meeting remedial action
objectives (RAOs).

The improvements in monitoring strategies directly affects evaluation
of PRB longevity, where longevity refers to the ability of a PRB to sustain
hydraulic capture, residence time, and reactivity in the years and decades
following installation. Because PRBs are used to treat plumes that may persist for years or decades, regulators in particular are interested in determining how long PRBs will continue to retain a desirable minimum level of
hydraulic capture and reactivity without requiring major maintenance or
replacement of the reactive media. Of the several hundred PRBs that have
been installed since the first full-scale PRB application occurred in 1994,
many are reported to be performing acceptably, although the literature does

include examples of concerns related to PRB performance including, but not limited to: (1) permeability loss due to solids formation and gas buildup (e.g., Henderson and Demond, 2011); (2) insufficient hydraulic performance due to incomplete or inaccurate subsurface characterization (e.g., Henderson and Demond, 2007); (3) the negative impact of anion competition (e.g., nitrate and chloride) on the treatment mechanisms important to contaminant reduction by iron metal (e.g., Moore and Young, 2005); (4) flow reduction upgradient of a PRB potentially due to upgradient diffusion of hydrogen and/or guar-gum from the PRB installation (Johnson et al., 2008), and (5) biofouling.

Despite the vast collection of sites, longer-term performance aspects of PRBs are still a source of uncertainty in planning future applications. Sustained field data with sufficient detail to enable a relatively thorough evaluation of longer-term performance were available at very few sites. The primary factors limiting longevity of ZVI barriers are corrosion of iron and precipitation on iron surfaces of native inorganic constituents from groundwater. When excessive, these factors have led to reduction in reactivity of the ZVI, loss of porosity and permeability, hydraulic mounding, and plume bypass around the PRB. At some sites, this loss of performance has been fairly severe within 5 years of installation.

The effectiveness and longevity of biowall PRBs primarily depends on sustaining appropriate levels of bioavailable organic substrate in the biowall reactive zone and maintaining the permeability of the biowall trench. The primary factor limiting longevity in biowalls has been the depletion of the more easily biodegradable portion of the organic substrate in approximately 4–5 years after installation. Injection of a slow-release biodegradable substrate, such as vegetable oil, has been effective in extending the life of a biowall, although this periodic enhancement increases the life-cycle cost of the PRB and makes it more of a semipassive system. Longevity expectations for PRBs are closely tied with the economics of the application. In previous years, the economic comparison revolved around PRBs and pump-and-treat systems. While this comparison is still valid—especially where there is a potential to replace aging pump-and-treat systems with a PRB—fewer new sites are considering or installing pump-and-treat systems. At newer sites, technology selection and economic comparison usually revolves around PRBs and other *in situ* plume treatment or control options. In all these economic comparisons, how long the treatment media will last without the need for active replenishment or replacement is a key consideration.

### 2.3.4 Sustainability

PRB technology is widely considered a sustainable groundwater remediation method because: (1) the general intent of a PRB system is to perform under hydraulically passive means (i.e., no energy or mechanical input for routing chemically impacted groundwater through the PRB), (2) groundwater is not removed from the subsurface nor degraded through discharge with lower quality surface or waste water, and (3) the treatment media often consists of recycled or "waste" material such as mulch/compost. The intent of the PRB is to operate as a hydraulically passive *in situ* groundwater remedy. Once the system is constructed, advective groundwater flow is the primary driver moving groundwater contaminants through the PRB for chemical and/or biological treatment. The most energy- and carbon footprint-intensive activities with PRB use are related to construction, creation, and collection of the materials used in the PRB construction (e.g., reactive media, slurry wall material), and excavation and disposal of construction spoils.

### 2.3.5 Construction and Implementation

PRB construction and implementation methods have advanced over the past 15 years—from conventional trench-and-fill to single-pass trenching, and from pneumatic injection to large-diameter borehole-filled completions. Research to develop treatment media with greater reaction rates has been ongoing for the past 15 years, and the expectation is that reliable, fast-acting, sustainable materials will continue to be a goal in PRB development. For PRBs to become more reliable, usable, and sustainable, hydraulic design improvements must take center stage. Developing reactive media and PRB systems that can reliably treat mixed plumes and emerging contaminants will continue to be an important need. Key to the success of this approach will be sufficient testing to ensure complete treatment of chemicals that may have different geochemical stability signatures (oxidizing vs. reducing) and thus require different treatment mechanisms. Biological reactions in PRB systems that utilize facultative bacteria or cometabolic processes may be particularly promising in this context. Ensuring that PRBs provide long-term reliable treatment will be of even greater focus than we have seen over the past 15 years.

### 2.3.6 Cost

Representative cost information for PRBs, where cost includes both capital installation and long-term operation costs, is not widely available in the public record. Over the next few years, particularly with a greater emphasis on implementing green and sustainable remediation methods, the hope is that there will be less reluctance in providing cost data. For the future, given possible greater use of PRBs for metals including radioactive constituents, and as PRBs continue to age, it should be anticipated that closure plans will be requested more often by regulatory groups.

In some respects, providing cost data for previous installations or as rules of thumb, may inadvertently bias future design evaluations. This is because the cost of materials can change greatly due to market conditions (particularly for the use of granular ZVI) and construction methods are dependent on the unique conditions (location, depth, existing infrastructure). Most installations involving construction may involve substantial capital invest-ment. However, the benefit of low or negligible costs post construction allow a life cycle analysis to conclude that economic value comes from designing the PRB for successful operation over as long a time as possible.

An important lesson with respect to overall costs for a PRB is that it is much easier to test the functionality of PRB treatment media under labora-tory conditions than it is under field conditions. However, this does not pre-clude the occasional "investment" of a small-scale pilot test to assure a final PRB design that will meet long-term remedial objectives. We anticipate that cost performance data will be collected from a greater number of sites—in particular those involving novel reactive media—to better confirm the long-term efficacy of PRB technology for treating groundwater plumes. Finally, while a return on investment (ROI) is important for any application, past experience indicates that compared to active energy-intensive remedies, the ROI likely is only a few years compared to the overall several decade life of typical contaminated site.

## 2.4 Recent Advances

When considering what recent advances in PRB technology have been valu-able, what future improvements would be beneficial is also an important exercise. Because the PRB technology now is approaching 25 years in use, and many may consider the technology to be a "developed" remedial mea-sure, we may in fact find more utility in discussing potential future improve-ments. In this vein, a discussion of future improvements may follow the outline of advancement categories proposed herein and as considered by ITRC in its 2011 PRB Technology update (ITRC, 2011):

- Construction methods
- Reaction rates
- Hydraulic design improvements
- Monitoring improvements
- Treatment of mixed plumes and emerging contaminants
- Longevity enhancements
- Cost performance improvements
- Closure and decommissioning methods

Over the past 20 years, we have seen improvements in construction meth-ods by use of one-pass trenching, injection methods, and the use of recycled or "green" treatment materials. Reaction rates are generally not much different than when the PRB materials were first conceived in the 1990s and 2000s and research into newer and better PRB materials may have waned. However, there likely will be improvements primarily in concert into researching new PRB treatment materials for emerging contaminants. Hydraulic design improvements specifically will occur consistent with comprehensive site characterization programs. With the collection of thorough and complete subsurface information, higher performance hydraulic designs are more likely to be developed providing appropriate design techniques where both engineering and numerical analysis are used. Different geometric designs that help to capture contaminant plumes without alternating ambient flow and with less potential for creating hydraulic bypass conditions are not technological advances, but rather, are the result of developing a better three-dimensional understanding of the contaminant distribution and hydraulic characteristics of a site.

With regard to longevity, again, technological advances in materials are less likely to create longer-lasting PRBs compared to more fully understanding the issues (such as geochemical competition from inorganic constituents, i.e., nitrate as a cause of reducing performance by ZVI as a treatment media) that cause PRBs to age more quickly. Design considerations that reduce the potential plugging from carbonate-shift mineralization will result in a longer-lasting PRB. New advances in "demineralizing" PRB materials may be in the future, but most research to date has not provided evidence of effective and inexpensive rejuvenation techniques, with the exception of reloading liquid organic carbon into some types of bioremediation PRB systems.

What we still do not have a good handle on is closing and decommissioning PRBs, chiefly because few if any PRBs have been closed due to regulatory completeness of a remediation project. The expectation is that most PRBs will remain in place if the treatment material becomes exhausted or if the remediation program is completed. PRBs are composed primarily of earth materials or native components and should not negatively impact a groundwater resource even when exhausted. The exception may be PRBs used for remediation of radioactive plumes—in this case, it still may be more appropriate to keep the PRB in place until radioactivity is at a nonproblematic level, or maintain sufficient monitoring to avoid potential problems for as many years as necessary.

The most well received advances will likely continue to be in the areas of cost performance. The ability to construct PRBs from local, green, or abundant natural materials will allow the technology to be more universally used in economically disadvantaged areas to protect important and sensitive water resources and receptors. The use of green compost—native sorptive materials such as sawdust and woodchips—and inexpensive construction methods in areas downgradient of leaking landfills or septic systems will create a new market for inexpensive PRBs worldwide and thus, reduce the deleterious nature of contaminant migration for a large part of the world's population.

## 2.5 Conclusion

The ability to use PRBs to destroy or immobilize dissolved chemical constituents is dependent on many issues, including cost and constructability, but notably, effective and complete site characterization. Advances in PRB development, notably in the use of natural materials and the advanced use of traditional treatment materials (such as ZVI), will continue to develop and provide practitioners with a bevy of new options intended to be highly effective, and also highly cost efficient. Expected research and assessment into the effective longevity of PRB materials used within an application will likely be a primary focus over the next several years.

Because the PRB is a passive technology, which is not easily adjusted once installed, an accurate conceptual model must be developed and the most important data gaps must be filled before a PRB can take on a final design. If the PRB is designed based on incomplete or inaccurate site information, failure or unintended performance may result. The PRB may not be the correct remedy to select for many contaminant remediation schemes; however, the several decade experience that our profession has with the PRB technology opens up many more potential opportunities for reducing cost, impact, and secondary treatment issues that characterize the difficulty in mitigating radioactive plumes. Technologies for implementing PRBs will continue to improve and develop (such as injection and deep-well PRB installation); the benefits of not having to pump groundwater nor supply ongoing energy for treatment will be the primary drivers for the development of new technologies.

## References

Bellis, M.B., J.D. Chamberlain, L.M. Michaelczak, C. Dayton, and R.H. Frappa. 2010. *Mitigation of the Contaminated Groundwater Plume at the West Valley Demonstration Project*, New York, USA 10409. Waste Management Symposium Inc., Phoenix, Arizona. March 7, 2010.

Chamberlain, J., and L. Michalczak. 2011. Design and installation of a permeable treatment wall at the west valley demonstration project to mitigate expansion of strontium-90 contaminated groundwater 11138. *Waste Management Symposium Inc.*, Phoenix, Arizona. February 27–March 3, 2011.

Gillham, R.W., and S.F. O'Hannesin. 1994. Enhanced degradation of halogenated aliphatics by zero-valent iron. *Groundwater*, 32(6), 958–967.

Henderson, A., and A. Demond. 2007. Long-term performance of zero valent iron permeable reactive barriers: A critical review. *Environmental Engineering Science*, 24(4): 401–423.

Henderson, A., and A. Demond. 2011. Impact of solids formation and gas production on the permeability of ZVI PRBs. *Journal of Environmental Engineering*, 137 (8):689–696.

International Technology Regulatory Council (ITRC). 1999. Regulatory guidance for permeable reactive barriers designed to remediate chlorinated solvents. Prepared by the Interstate Technology Regulatory Council Permeable Reactive Barriers Team, Washington, DC, http://www.itrcweb.org (accessed 1999).

International Technology Regulatory Council (ITRC). 2005. Permeable reactive barriers: Lessons learned/new directions. Prepared by the Interstate Technology Regulatory Council Permeable Reactive Barriers Team, Washington, D.C., http://www.itrcweb.org (accessed 2005).

International Technology Regulatory Council (ITRC). 2011. Permeable reactive barrier: Technology update. Prepared by the Interstate Technology Regulatory Council Permeable Reactive Barriers Team, Washington, DC, http://www.itrcweb.org (accessed 2011).

Johnson, R.L., R.B. Thoms, R. O'Brien Johnson, J.T. Nurmi, and P.G. Tratnyek. 2008. Mineral precipitation upgradient from a zero-valent iron permeable reactive barrier. *Groundwater Monitoring and Remediation*, 28(3), 56–64.

McMurty, D.C. and M.O Elton. 1985. New approach to in-situ treatment of contaminated groundwaters, *Environmental Progress*, 4(3), 168–170.

Moore, A., and T. Young. 2005. Chloride interactions with iron surfaces: implications for perchlorate and nitrate remediation using permeable reactive barriers. *Journal of Environmental Engineering*, 131(6), 924–933.

Naftz, D.L., S.J. Morrison, J.A. Davis, and C.C. Fuller. 2002. *Handbook of Groundwater Remediation Using Permeable Reactive Barriers, Applications to Radionuclides, Trace Metals, and Nutrients*. Academic Press, San Diego, California.

Pearson, F.H., and A.J. McDonnell. 1975. Use of crushed limestone to neutralize acid wastes. *Journal of Environmental Engineering Division ASCE*, 101, 139–158.

Perlmutter, M., A. McKenna and D. Williamson. 2013. Guidance document: A framework for selecting, designing and implementing a permeable reactive barrier system, CRC CARE Technical Report no. 25, CRC for Contamination Assessment and Remediation of the Environment, Adelaide, Australia.

Rabideau, A.J., J. Van Nenschoten, A. Patel, and K. Bandilla. 2005. Performance assessment of a zeolite treatment wall for removing Sr-90 from groundwater. *Journal of Contaminant Hydrology*, 79(1–2), 1–24.

Remediation Technology Development Forum. 1998. *Permeable Reactive Barrier Technologies for Contaminant Remediation*. Office of Research and Development, EPA/600/R-98-125.

Roehl, K.E., T. Meggyes, F.G. Simon, and D.I. Stewart (Eds.) 2005. *Long-term Performance of Permeable Reactive Barriers*. Elsevier, Amsterdam, The Netherlands.

Shoemaker, S.S., J.F. Greiner, and R.W. Gillham. 1995. Permeable reactive barriers. In R.R. Rumer, and J.K. Mitchell (Eds.) *Assessment of Barrier Containment Technologies*, International Baltimore, Maryland, Containment Technology Workshop.

Thiruvenkatachari, R., S. Vigneswaran, and R. Naidu. 2008. Permeable reactive barrier for groundwater remediation. *Journal of Industrial and Engineering Chemistry*, 14, 145–156.

Warner, S., D. Bablitch, and R. Frappa. 2012. PRB for contaminated groundwater, *The Military Engineer*, Society of American Military Engineers, 104(675), 53–54, January-February 2012.

Warner, S.D. and D. Sorel. 2003. Ten years of permeable reactive barriers, lessons learned and future expectation. In *Chlorinated Solvent and DNAPL Remediation:*

*Innovative Strategies for Subsurface Cleanup*, ACS Symposium Series 837, American Chemical Society, pp. 36–50. January.

Warner, S.D., C.L. Yamane, N.T. Bice, F.S. Szerdy, J. Vogan, D.W. Major, and D.A. Hankins. 1998. Technical update: The first commercial subsurface permeable reactive treatment zone composed of granular zero-valent iron. In *Designing and Applying Treatment Technologies: Remediation of Chlorinated and Recalcitrant Compounds*, 145–150. Batelle Press, Columbus, Ohio.

Yamane, C.L., S.D. Warner, J.D. Gallinatti, F.S. Szerdy, T.A. Delfino, D.A. Hankins, and J.L. Vogan. 1995. Installation of a Subsurface Groundwater Treatment Wall Composed of Granular Zero-Valent Iron. Paper presented at American Chemical Society Annual Meeting, April 1995.

# 3

## Choosing the Best Design and Construction Technologies for Permeable Reactive Barriers

Dawit N. Bekele, Ravi Naidu, Volker Birke,
and Sreenivasulu Chadalavada

## CONTENTS

## 3.1 Introduction

At many sites, groundwater (GW) remediation is proving to be a much more difficult and persistent problem. Permeable reactive barriers (PRBs) employ an innovative technology that offers a passive system for addressing long-term GW contamination problems. Although PRBs were initially applied to treat chlorinated hydrocarbon (CHC) plumes, they have also been applied to treat or capture other inorganic contaminants such as trace metals (chromium, manganese, uranium, zinc, lead, etc.), and anionic contaminants (sulfate, nitrate, phosphate, arsenate) (Gavaskar et al. 2000). PRBs containing lime have also been used to raise the pH of GW with acid mine drainage near mining operations (Johnson and Hallberg 2005).

There are presently over 200 field-scale PRBs in operation throughout the world (see Chapter 1) and they have been in use since the early 1990s.

The salient features of a PRB system include (ITRC 2011; Gavaskar et al. 2002)

- Underground installation is limited by land use (i.e., large space requirement and not easy to use in an urban setting)
- The PRB serves as a barrier to contaminants but not to GW flow (should not alter the GW flow rate and direction)
- Once installed there is little possibility for rectification or maintenance
- Passive remediation can continue for many years even for decades
- Performance should not change under varying GW parameters, including those of the contaminant(s) of concern
- PRBs should not only remediate the parent contaminant but also its by-products
- Reactions in the PRBs should not introduce additional contaminants to the GW

Most field- and pilot-scale PRBs have affected their remediation objectives but a few are failing to meet their objectives (Richardson and Nicklow 2002). There are several causes for this including: failure to encompass potential shifts in hydraulic gradient, incorrect design of thickness of the reactive media, channeling effects, diminishing reactivity with time, smaller porosity of the media compared to the aquifer, and fast corrosion of the reactive media (RTDF 2001). Common causes of short-term failures of PRBs include: inadequate reactive material at some part of the barrier (e.g., heterogeneous contaminant source), an insufficient depth of barrier to an aquitard, leakage between the permeable and impermeable barrier joints (funnel-and-gate system), smaller porosity of the barrier, and hence channeling.

PRB is a capital intensive technology with limited possibilities for rectification after installation, so any form of failure needs to be anticipated and addressed at the design stage (Gavaskar et al. 2002). This chapter reviews the design and construction technologies for PRB, with particular emphasis on their application for remediation of CHC-contaminated GW. It aims to identify key issues that need to be addressed to increase the performance and longevity of PRB systems. While the chapter focuses on CHC and the use of zero-valent iron (ZVI), the parameters optimized for the design of PRBs are the same irrespective of the contaminant and material used as the reactive barrier.

## 3.2 PRB Engineering Design Methodology

Engineering design of PRBs comprises the selection of the most suitable barrier configuration, appropriate reactive medium, the size of the barrier, barrier construction technology, quality control of the barrier, and performance-monitoring GW well installation. The design needs to be tailored to provide a site-specific solution to meet the GW remediation objectives. A detailed knowledge of the site hydrogeology (e.g., an understanding of aquifer and aquitard), geochemistry (e.g., GW chemistry other than for the chemical of concern), and chemistry of contaminants (e.g., source, composition, and concentrations) is required for successful PRB design (Roehl et al. 2005). The site information required includes: activities near the proposed barrier; existing utilities lines and other obstructions; site soil quality; construction methodology and contractor's expertise in the area; and knowledge of the costs of reactive material.

### 3.2.1 Selection of Reactive Material

Following a detailed site characterization of the plume and GW information, a suitable type of design can be chosen to capture the plume and remediate it to meet the site remediation objectives—within a budget. The selection criteria for reactive media should incorporate (Gavaskar et al. 2000; ITRC 2011; Richardson and Nicklow 2002)

- *Reactivity.* Sufficient reactivity for an economical and practical barrier thickness.
- *Stability.* The ability to remain reactive for several years to decades under the site-specific geochemical and microbial conditions, with a minimal amount of maintenance.
- *Availability and Cost.* A cheaper medium is preferred especially if any differences in performance are reported to be slight.
- *Hydraulic Performance.* A practical size sufficiently greater than that of the aquifer material so that an effective capture zone can be created and maintained.

- *Environmental Compatibility.* Environmentally compatible by-products.
- *Construction Method.* Easy construction, that is, minimization of excavated material.

Common contaminants (e.g., trichloroethylene) and common reactive media (e.g., granular ZVI) are the most common PRBs applications used to date (Thiruvenkatachari et al. 2008; Gavaskar et al. 2000; RTDF 2001). Whereas the technology initially used ZVI as the reactive medium for the remediation of GW contaminated with CHC (with the first field trials in the early 1990s and the first commercial deployment in late 1994), recently a range of materials for the remediation of other organics have been deployed (Naidu 2013). For other materials used in PRB see Chapter 1. The use of granular iron PRB for treating dissolved CHC is rapidly gaining acceptance as a cost-effective technology due to it being a long-term and low-maintenance cost solution (O'Hannesin and Gillham 1998; Henderson and Demond 2007).

Following site characterization and identification of prospective reactive media, treatability testing is conducted to evaluate the performance of the reactive medium with GW from a specific site (see Figure 3.1). Batch tests can be conducted to initially screen prospective media, but column tests as illustrated in Figure 3.1 should also be performed (Gavaskar et al. 2000). Treatability testing serves the following purposes:

- Screening and selecting a suitable medium for the reactive cell.
- Estimating the half-life of the degradation reaction.

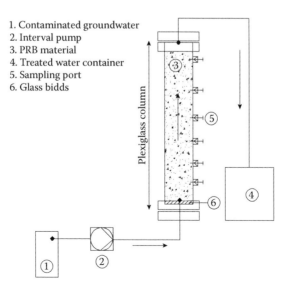

1. Contaminated groundwater
2. Interval pump
3. PRB material
4. Treated water container
5. Sampling port
6. Glass bidds

**FIGURE 3.1**
Laboratory column experiment setup for testing PRB reactive material.

- Determining the hydraulic properties of the reactive medium.
- Evaluating the longevity of the reactive medium.

### 3.2.1.1  Geochemistry of GW

Geochemistry here refers to the native constituents of the GW (constituents other than the target contaminants) that affect short- and long-term performance of a PRB in terms of mineral precipitation and microbial build-up. As with design criteria during site characterizations, monitoring of GW microbial communities and geochemistry that can cause fouling of the barrier via microbial growth and formation of a precipitate must be conducted, preferably over an extended period. Geochemical characterization of the GW, such as redox potential [Eh], pH, dissolved oxygen, and inorganic constituents (e.g., Ca, Fe, Mg, Mn, Ba, Cl, F, $SO_4^{2-}$, $NO_3^-$, silica, and carbonate species [alkalinity] etc.) should be conducted. Geochemical computer modeling codes (PHREEQC) can be used to determine the types of reactions and by-products that may be expected when GW contacts the reactive medium (Gavaskar et al. 1997; Richardson and Nicklow 2002). Unless the aquifer is relatively thin, GW microbial and geochemical parameters may vary with depth. For example, dissolved oxygen may vary by depth in the aquifer, leading to different degrees of iron corrosion in the ZVI reactive cell.

### 3.2.1.2  Reaction Kinetics

As mentioned in Section 3.2.1, CHC and ZVI are commonly used as a typical contaminant and reactive material for the barrier. The reaction kinetics and half-life of contaminants are two of several key parameters that require optimization for the design of an effective barrier. Iron concentrations in the interstitial water within the reactive barrier usually range from 0.5 to 14.8 mg/L. There is a high possibility for the blockage of the barrier due to Fe(III) mineral precipitation at higher pHs under aerobic conditions. Due to inherent very slow GW flow rates, lamina flow conditions are always present at the barrier and this needs consideration when designing the barrier. The commonly used ZVI has grain sizes that vary with construction, shown in Table 3.1.

The process involves the simultaneous oxidative corrosion of the reactive iron metal by both water and CHC in the presence of ZVI (Focht et al. 1996) illustrated in Equation 3.1:

$$Fe^0 + RCl + H^+ \rightarrow Fe^{2+} + RH + Cl^- \tag{3.1}$$

This is a first-order reaction of the ZVI barrier which acts as a plug flow reactor; it can be represented as in Equation 3.2:

$$C = C_0 e^{-kt} \tag{3.2}$$

**TABLE 3.1**

ZVI Gain Size Specification for Different Types of PRB Construction

| Construction Type | Iron Grain Size (mm) |
|---|---|
| Excavation | 2–0.25 |
| Vertical hydraulic fracture | 1–0.17 |
| High-pressure jetting | 0.59–0.21 |
| Pneumatic fracturing | 0.04–0.08 |

where
  C  = the outlet concentration
  $C_0$ = the initial concentration
  k  = the reaction kinetic
  t  = the residence time within the barrier

The reactive ZVI reaction kinetics (Equation 3.2) and formation of by-products requires investigation by conducting laboratory batch and column experiments (Figure 3.1), using contaminated GW from the remediation site (Thangavadivel et al. 2013). Such experiments also help to select an appropriate reactive material for the particular site application. The column experiments acts as a plug flow reactor and mimic the actual GW flow conditions within it. The reaction rate obtained from the column study can be used for sizing the PRB barrier wall thickness $L_B$ as shown in Figure 3.2 (see Section

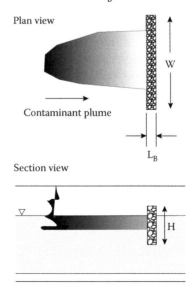

**FIGURE 3.2**
PRB dimensions and design parameters, where $L_B$ = PRB wall thickness, W = PRB wall width, and H = PRB wall height.

3.2.2.2). The reaction rate is generally computed after 10 pore volumes of water have passed through the column.

## 3.2.2 PRB Engineering Design

### 3.2.2.1 Hydrogeological Data for PRB Design

The primary physical function of the PRB is to capture the contaminant plume and allow sufficient residence time in the reactive media to achieve the desired cleanup goals. Understanding the GW flow regime is key to the physical design of a PRB system (ITRC 2005). It must be designed to encompass potential shifts in the direction and magnitude of the GW flow due to seasonal fluctuations (Richardson and Nicklow 2002). Seasonal variations and nearby above- or below-ground activities such as pumping GW may alter the GW flow and direction. Since almost all aquifers are heterogeneous, their permeability vary and hence, GW flow rates also vary in each aquifer. The average GW velocity and water table may not be adequate for optimal design of the barrier (ITRC 2011). The GW flow rate is the key to determining the barrier width to provide sufficient residence time. Consequently, accurate site characterizations of the GW flow rate and direction throughout the site and seasons variations are essential. To intercept the plume migration effectively the reactive wall is installed perpendicular to the GW flow direction. The depth to the GW table and underneath the lower permeable layer (Aquitard) determines the height of the PRB wall and therefore cost of the remediation project.

Although emphasis has been placed on losses of reactivity and permeability, inadequate hydraulic characterization has been the most common cause of the few PRB failures reported in the literature (Henderson and Demond 2007). Fate and transport hydrogeological modeling is implemented to assess PRB configurations, site parameters, and performance scenarios based on the information gathered from the site characterization and treatability study. For most applications, commonly used available computer codes such as MODFLOW, MODPATH, or FEFLOW are sufficient for developing a GW model as a design tool (Richardson and Nicklow 2002). Modeling the PRB system aids in optimizing the design parameters (ITRC 1999) to

1. Establish the width of the barrier and funnel walls (when funnel-and-gate is used) in relation to the plume size and estimate capture zone size
2. Determine the best location for the barrier and simulate various PRB configurations
3. Decide the best location for installing performance monitoring wells
4. Evaluate the effect of aquifer heterogeneity, buried utilities, buildings, land used, and seasonal fluctuations on the system
5. Assess the potential underflow, overflow, or flow around the barrier

### 3.2.2.2 Sizing the Reactive Material Wall

The design of the appropriate sizing for the reactive barrier wall is made subsequent to the results from laboratory trials (material selection), and hydraulic and geochemical modeling. Detailed site characterization and hydrogeological modeling are essential steps in PRB design and construction such as the capture zones (i.e., optimization of the length, best location and orientation, type and configuration of the barrier) and aquifer thickness and heterogeneity (determines depth/height of the barrier). However, sizing the barrier thickness involves understanding the removal mechanism and reaction kinetics with an appropriate safety factor for the uncertainties of the input design parameters, illustrated in Equation 3.3. The thickness of the barrier should ensure that the contaminant(s) of concern are treated to remediation targets at the downstream of the PRB wall, and thick enough to accommodate monitoring wells for evaluating the performance of the barrier.

The most important role of modeling is to evaluate and optimize different PRB types, configurations, and dimensions for a given set of design parameters (Gavaskar et al. 2000). For example, maximizing the hydraulic capture zone width increases the flow velocity and decreases the residence time. Consequently, the reactive barrier must be thicker and wider, therefore, increasing cost (Henderson and Demond 2007). The thickness of the reactive medium is governed by the residence time (half-life) of the target contaminant to the reactive medium and the velocity of GW flow through the barrier.

Empirical design equations determining flow through thickness of the barrier is expressed in Equation (3.3):

$$L_B = t_{res}^* V_b^* S_f \qquad (3.3)$$

where
  $L_B$ = PRB wall thickness (m)
  $t_{res}$ = time of residence in the wall (d)
  $V_b$ = velocity of the wall (m/d)
  $S_f$ = the safety factor

A safety factor may be incorporated to account for seasonal variations in the flow, potential loss of reactivity of ZVI over time, and any other field uncertainty. The expected GW velocity can be determined from hydrogeological modeling (see Section 3.2.2.1). The degradation rate from a laboratory kinetic study using a column experiment (see Section 3.2.1.2) requires some correction for field application and is dependent on temperature.

The GW temperature in the field is typically 10°C which is generally lower than the room temperature of the laboratory column tests (typically 20–25°C) which adversely affects the reaction kinetics. Consequently, the empirical residence time may need to be increased to account for the lower temperature. Field observations at a test site in New Jersey have shown that the

TCE degradation rate declines by a factor of 2–2.5 at temperatures of 8–10°C compared with rates measured in the laboratory. Similar results have been observed at other field sites (Gavaskar et al. 2000).

The bulk density of the reactive medium in the field is generally lower than that measured in the laboratory. Consequently, the surface area of reactive medium per unit volume in the field may be lower than the surface area measured during a column testing. Furthermore, the reaction rates (or half-lives) are proportional to the specific surface area of the reactive medium (Gillham 1996; Johnson et al. 1996). The field residence time must therefore be increased to account for the lower expected ratio of reactive surface area:volume of solution. Currently, there is no clear indication of how large the bulk density correction factor should be (Gavaskar et al. 2000). To a degree, the surface area of reactive medium per unit volume in the field will depend on the efficiency of the construction methodology and on how well the reactive medium consolidates after construction.

### 3.2.3 Type and Configuration of Barrier

The main requirement of the PRB design is to capture the contaminant plume throughout its life span and remediate the plume to meet the remediation target. Figure 3.3 shows various PRB types and configurations. The selection of an appropriate configuration should be made on a site-specific

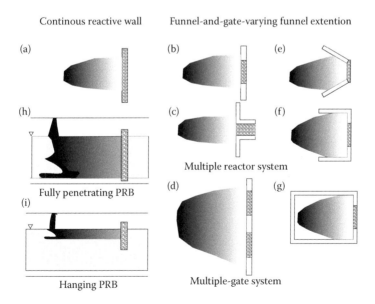

**FIGURE 3.3**
The continuous and various configurations of funnel-and-gate PRBs. (Adapted from Starr, R., and J.A. Cherry. In situ remediation of contaminated ground water: The funnel-and-gate system. *Ground Water*. 1994. 32(3):465–476. Copyright Wiley-VCH Verlag GmbH & Co. KGaA.)

basis depending on the depth, width, and saturated thickness of the plume which controls the overall dimensions of the system. Several other aspects of the subsurface construction procedure need to be considered such as the need for dewatering during excavation, the means and costs of GW and soil disposal, health and safety, and disruption to site activity. The continuous and funnel-and-gate PRB designs are the two PRB types which are currently being used for full-scale applications.

### 3.2.3.1 Continuous PRBs

For any site where GW flow and plume geometry are well understood and there is no construction constraint, the continuous barrier (Figure 3.3a) is the best design choice (Gavaskar et al. 2000). Continuous PRBs are the most common field installations operating today and they have relatively minimal impact on the natural GW flow conditions at a site. Since the barrier covers the full width of the plume, the continuous barrier design generally requires a large amount of reactive medium compared to the funnel-and-gate system. The prevailing ZVI cost can be a major factor influencing the capital cost and hence, the choice of design. Continuous barriers do not have to be buried in a below low permeability zone of the aquifer as long as the hydraulic conductivity of the saturated aquifer is less than that of the PRB (Figure 3.3h and i).

### 3.2.3.2 Funnel-and-Gate PRBs

In the funnel-and-gate PRB configuration with low permeability funnels, the GW is directed toward the reactive medium (the gate) (Figure 3.3b) (Birke et al. 2003). The funnel can be made of sheet piling, slurry walls, or some other material penetrating into an impermeable layer (aquitard) to prevent contaminant underflow (Gavaskar 1999) (see Section 3.3). Impermeable funnels are generally keyed up to 1.5 m into the aquitard. In an extremely large contaminant plume or highly heterogeneous aquifer the funnel-and-gate system can be modified to have multiple gates (Birke et al. 2003). Multiple reactive media in a separate vertical "treatment train" might be considered at a specific site with a combination of contaminants (Figure 3.3c). Care must be taken so that the reactions do not interfere or limit one another (Gavaskar 1999). GW velocity within the treatment zone is usually 2–5 times higher than that resulting from the natural gradient, depending on the funnel:treatment-zone ratio (Day et al. 1999). To ensure that GW flow does not occur beneath the system, funnel-and-gate systems must be keyed into an underlying low permeable zone (Lai et al. 2006).

It is very important to ensure that there is no gap between the permeable and impermeable wall funnel joints. Since the construction consists of both permeable and impermeable barriers, the construction cost is high compared to a continuous barrier design using less reactive materials (Gavaskar 1999). Typically, the ratio of funnel length:permeable treatment zone is <6. The discharge through the gate can be increased by increasing the width, length, and hydraulic conductivity of the gate, and the width of the funnel (Starr and Cherry 1994). Various funnel extensions (Figures 3.3e, f, and g) are occasionally used to capture plumes under site-specific conditions.

### 3.2.3.3 Reactive Vessels

For decontamination, GW can be routed through a natural or engineered preferential pathway to a reaction vessel. The reactive vessel PRB design is very similar to the funnel-and-gate barrier except that it replaces the gate with *in situ* reactive vessels. These can be lifted out from the ground for maintenance or to replace the reactive medium (Phillips et al. 2010). The reactive medium can readily be replaced when exhausted and its performance can be restored quickly. The design of the reactive vessel enables investigation of any problems and allows them to be fixed to restore performance. The design does not add much cost to the conventional funnel-and-gate system, but requires a permanent structure or provision for lifting the barrier at the site. In Europe, this design has become popular and several installations are now in place (Birke et al. 2007).

### 3.2.3.4 Caissons PRBs

Caissons (also known as *in situ* reactive vessels) are load-bearing enclosures that are used to protect an excavation. They are a relatively inexpensive way of installing reactive cells at depths inaccessible to a standard backhoe (Figure 3.4). Caissons can be of any shape in cross section and are made from common structural materials. Both ends of a caisson are open and are placed in the soil by a vibrating hammer. It is not economical to drive a caisson with a diameter larger than 2.5 m into the subsurface. The interior of the caisson is then excavated using a large auger to make room for the reactive medium (Gavaskar 1999). Once this has been loaded into it, the caisson is extracted which is a more difficult task than its installation. However, this is a relatively inexpensive way of installing reactive cells at depths inaccessible to a standard backhoe (Thiruvenkatachari et al. 2008). Ensuring a good seal

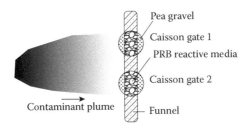

**FIGURE 3.4**
Caisson gate PRB configurations.

between the caisson gate and the funnel wall in a funnel-and-gate system is an important consideration.

### 3.2.4 Performance Monitoring

Performance monitoring GW well locations for the PRB are generally focused on the PRB system itself rather than on the entire site or the compliance boundaries (Figure 3.5). Traditional GW sampling approaches often result in withdrawal of a large volume of water that might compromise the PRB sampling objective. Consequently (Gavaskar et al. 2000), passive or semi-passive GW sampling approaches are recommended (Powell and Puls 1997). The performance monitoring program should be designed to detect changes in

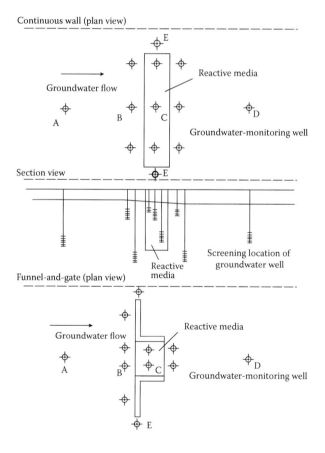

**FIGURE 3.5**
Locations of the performance monitoring groundwater well for continuous and funnel-and-gate PRBs configurations. (Reproduced with permission from Carey, M.A. et al. 2002. Guidance on the use of permeable reactive barriers for remediating contaminated groundwater. National Groundwater and Contaminated Land Centre report NC/01/51 P.69.)

reactivity, permeability, and contaminant under or over the PRB, or leakage through funnel walls if present as illustrated in Figure 3.5.

The location of monitoring wells is a critical element in determining whether the PRB is meeting compliance and performance criteria determined from hydrogeological modeling (see Section 3.2.2.1). Although site-specific conditions should always dictate the placement of monitoring wells, performance monitoring wells should be located up-gradient and down-gradient and, if possible, within the PRB. In addition, wells at each end of the PRB are necessary to verify hydraulic capture and evaluate potential plume bypass, particularly if impermeable funnels are used to intercept and control GW flow. Figure 3.5 illustrates idealized monitoring networks for continuous and funnel-and-gate PRB design, respectively (Carey et al. 2002). Regulatory requirements for location monitoring wells and sampling frequency vary. Therefore, regulatory-specific monitoring well requirements should be considered during well design and construction.

## 3.3 PRB Construction Techniques

The installation method for a PRB system is very critical for their successful performance and includes geotechnical/civil design considerations, the method of construction, service and infrastructure, waste management, and health and safety requirements. Important factors for suitable construction for a specific site include

- Soil quality at site and soil waste generation
- Designer and contractor's expertise
- Aquifer and aquitard characteristics
- Construction constraints at the site (space availability for construction of barriers)
- Construction budget

Excavation and injection methods are the two major types of construction techniques for PRB installations. The depth to the GW table plays a part in the selection of the construction techniques, that is, for a shallow depth, excavation techniques are suitable and for a deeper barrier, injection methods are suitable. Depending on the contaminants present and disposal methods, the soil generated also influences the choice of the construction technique. Contaminated site soil quality influences the type of excavation in terms of the side wall support. In general, the excavation technique generates more spoil than injection-based techniques, however, excavation techniques are relatively easy and straightforward. In most cases it is assumed the aquifer and aquitard are homogeneous, but they are always heterogeneous in nature. Detailed information on these techniques helps to choose the most appropriate construction methodology to install the barrier successfully.

### 3.3.1 Excavation Techniques

#### 3.3.1.1 Biopolymer Trenching

Biopolymer trenching is a cost-effective and versatile PRB construction method (Andromalos and Schindler 2005). During the excavation of a biopolymer trench (Figure 3.6), biopolymer slurry such as guar gum is pumped into the trench (Day et al. 1999) to give stability to the trench wall. The excavation work can then continue while the biopolymer slurry remains in the trench. As the trench wall is stabilized other forms of trench support such as trench boxes or hydraulic shores may not be required. Trenches up to 21 m in depth can be constructed. A standard backhoe can excavate down to 7.5 or 9 m and a modified backhoe down to 25 m (Gavaskar 1999). Backhoes have been used for many PRB installations and enable a rapid rate of excavation and generally require less skill to operate. After granular iron has been placed in the barrier, an enzyme is recirculated to degrade the biopolymer.

#### 3.3.1.2 Continuous Trenching

Continuous trenching excavation, as shown in Figure 3.7, involves filling with reactive medium and backfilling simultaneously, so that there is

**FIGURE 3.6**
Biopolymer trenching. (Reproduced with permission from Geo-Con®, Geo-Solutions Inc.)

**FIGURE 3.7**
Continuous trenching and filling of reactive media. (Reproduced from University of Waterloo.)

no open trench at any time. Continuous trenching is the most rapid PRB installation method. Continuous trenches can be dug to install a treatment zone from 0.3 to 0.9 m wide to a depth of 11 m. An arrangement of parallel trenches can be made to install thick walls. Although the cost of the continuous trenching equipment is relatively high, the fast trenching rate may make the overall installation cost economical for a larger barrier.

### 3.3.1.3 Unsupported and Supported Excavation

PRBs can be constructed without any sidewall supports (including biopolymer methods) if the trench can remain open for about 4 h without any significant caving in. This method is typically limited to depths of 6 m or less. A standard backhoe can be used for the excavation. If a trench wall collapses it is not possible to rectify it and a new trench needs to be constructed a few meters away, which may not be an ideal location for the PRB. Caution must therefore be exercised during unsupported excavation. Supported excavation for trench side walls can be constructed by trench boxes or hydraulic shores. The depth of the trench should be <6 m. Successful installation depends on an effective site layout, construction sequence and the selection of heavy equipment, flexibility in the construction method to accommodate unforeseen conditions, an understanding of backfill materials, the potential impact on the community and the environment, and the season for the construction work.

### 3.3.1.4 Sheet Piling

During trench construction sheet piles are driven around the perimeter of the PRB and the soil. Typically, internal bracing is required as depth increases. After backfilling is completed the sheet piling is removed and the GW is allowed to flow through the treatment zone, also known as "the cofferdam approach." A crane with a vibrating hammer is used to install the impermeable barrier sheet pile for the construction of the funnel of a funnel-and-gate

**FIGURE 3.8**
Sheet piles installation. (Reproduced with permission from Robertson GeoConsultants Inc.)

PRB (Figure 3.8) (Gavaskar 1999). Before driving the sheet into the subsurface by a drop hammer or vibrating hammer, the sheet piles are connected at their interlocking edges to form a cutoff wall. The installation of the sheet pile is relatively easy and quick. The University of Waterloo, Canada, has patented a technique for sealing adjoining sheet piles by pouring grout into the joints. The integrity of the sheet piles can be maintained to depths of about 15 m. Use of sheet piles generates much less soil than other methods and is a very useful construction technique when the barrier has to be installed where there are horizontal space limitations.

### 3.3.1.5 Slurry Wall

Slurry wall installation is generally used to construct the impermeable funnel of the funnel-and-gate PRB. Soil/bentonite, cement/bentonite, and composite slurry walls are most commonly used (Meggyes 2005), with soil/bentonite being the most common (Figure 3.9). A slurry trench is generally excavated with a backhoe, a modified backhoe, or a clamshell digger, depending on the required depth. Backhoes are used extensively in North America for PRB installation whereas in Europe a hydraulic grasp and crane is the most popular method. Appropriate slurry is placed in the trench to maintain its stability. As the slurry permeates into the sides of the excavation depending on the slurry used, a fully hydrated filter cake of bentonite or

**FIGURE 3.9**
Slurry wall construction mixing constituents. (Reproduced with permission from Geo-Con, Geo-Solutions Inc.)

composite is formed along the sides. Finally soil/bentonite backfill is placed in the trench (Geo-Solutions Inc.). The slurry wall construction process is time consuming and is more expensive than sheet piling.

### 3.3.2 Injection Techniques

Injection methods do not involve excavation and, therefore, have considerable potential to minimize health and safety issues (Geo-Solutions Inc.). However, the requirement of special equipment incurs higher operating and maintenance costs than conventional emplacement techniques. They also require especially skilled personnel to implement this technology on site.

#### 3.3.2.1 Hydraulic Fracturing

With this technology a series of wells are installed along the length of the proposed barrier. A controlled vertical fracture is created in the wells and an iron/gel mixture is injected into the well as a reactive barrier. In the case of an impermeable barrier installation, a soil/bentonite slurry or soil/cement or composite slurry is injected. The gel is used to suspend and transport the iron filings into the subsurface. This is a promising technique for the construction of deeper barriers; however, particular care needs to be taken in designing such gel mixtures.

Caldwell Trucking, northern New Jersey (USA), have used a gel to suspend and transport iron filings into the subsurface. They found that the gel must be of sufficiently high quality and purity to have no impact on the iron reactivity and permeability. The injections of a rapid breaking enzyme is capable of breaking iron/gel mixtures even at the highest pH. The poor performance of this PRB is believed to be due to changes in the GW flow regime resulting from the injection of granular iron into the fractured bedrock, and the slow breakdown of guar gum gel.

### 3.3.2.2 Jetting

High-pressure jetting is an established practice to inject grouting agents to improve the structural characteristics of soil for construction purposes, as shown in Figure 3.10. Recently, injection of grout has been used to construct funnel walls and in order to get the required length and thickness of the funnel (Geo-Solutions Inc). Jetting is a sound technique for installing a wall around obstructions such as boulders or a utility line. Furthermore, it has the advantage of the small equipment requirements and a lower mobilization cost.

### 3.3.2.3 Soil Mixing

In deep-soil mixing, a hollow-stem auger with special mixing paddles is lowered to the desired depth and rotated to mix the soil (see Figure 3.11). As the auger is removed, reactive slurry is injected through the drill stem. A succession of these penetrations will create a row of columns to form a

**FIGURE 3.10**
Jet grout drill with jets activated. (Reproduced with permission from Soilmec® Drilling and foundation equipments.)

**FIGURE 3.11**
Vertical soil mixing. (Reproduced with permission from Geo-Con, Geo-Solutions Inc.)

PRB. This technique is best used in soft soil. Deep-soil mixing can be used at depths up to 40 m, creates minimal spoils, and has a higher production rate with lower costs than jet grouting (Geo-Solutions Inc). Because the soil is mixed with the slurry instead of being replaced, the slurry must be more concentrated than is the case for jet grouting. Soil mixing generally achieves greater PRB uniformity compared to the injection and fracturing techniques. The mixing results in a higher hydraulic conductivity of the columns relative to the surrounding soil and consequently a slight increase in the capture zone of the wall. Mixing is also provided upstream of the barrier to maintain parallel flow lines entering the PRB.

## 3.4 Conclusion

Among the existing remediation options, PRB is a very promising technology in terms of operation and maintenance costs, as well as stability of performance. PRB installation is expensive but once installed it remains in place for decades. However, the scope for troubleshooting after installation is limited, so design selection and installation of the system are the keys for ensuring consistent performance. The type of design determines the way in which the plume is captured. The choice of the design and construction techniques is essential for the success of a PRB which is site-specific. To date, there are only about 20 years of operating data available on PRBs, although a contaminated plume may last for many decades. More research is therefore needed on the evaluations of performance and longevity of reactive barriers. Contingency plans needs to be incorporated in all PRB designs in the event of the barrier failing to perform. Existing data show that it may take many years to reach the regulatory limits on contaminant concentrations downstream of the barrier, and so an agreement with the relevant regulatory body about the decision to use a PRB and its design is crucial.

---

# References

Andromalos, K.B., and R.M. Schindler. 2005. Current state of the art installation techniques for the in-situ reactive wall groundwater treatment systems. *Geo-Con, Inc. Visited July* 27.

Birke, V., H. Burmeier, S. Jefferis, H. Gaboriau, S. Touzé, and R. Chartier. 2007. Permeable reactive barriers (PRBs) in Europe: Potentials and expectations. *Italian Journal of Engineering Geology and Environment*, 1:1–7.

Birke, V., H. Burmeier, and D. Rosenau. 2003. Design, construction, and operation of tailored permeable reactive barriers. *Practice Periodical of Hazardous, Toxic, and Radioactive Waste Management*, 7(4):264–280.

Carey, M.A., B.A. Fretwell, N.G. Mosley, and J.W.N. Smith. 2002. Guidance on the use of permeable reactive barriers for remediating contaminated groundwater. National Groundwater and Contaminated Land Centre report NC/01/51 p. 69.

Day, S.R., S.F. O'Hannesin, and L. Marsden. 1999. Geotechnical techniques for the construction of reactive barriers. *Journal of Hazardous Materials*, 67(3):285–297.

Focht, R., J. Vogan, and S. O'Hannesin. 1996. Field application of reactive iron walls for in-situ degradation of volatile organic compounds in groundwater. *Remediation Journal*, 6 (3):81–94.

Gavaskar, A., N. Gupta, B. Sass, T. Fox, and R. Jonosy. 1997. Design guidance for application of permeable barriers to remediate dissolved chlorinated solvents. DTIC Document.

Gavaskar, A., N. Gupta, B. Sass, R. Janosy, and J. Hicks. 2000. Design guidance for application of permeable reactive barriers for groundwater remediation. DTIC Document.

Gavaskar, A., B. Sass, N. Gupta, E. Drescher, and W.-S. Yoon. 2002. Cost and performance report-evaluating the longevity and hydraulic performance of permeable reactive barriers at Department of Defense Sites. DTIC Document.

Gavaskar, A.R. 1999. Design and construction techniques for permeable reactive barriers. *Journal of Hazardous Materials*, 68(1–2):41–71.

Gillham, R.W. 1996. In situ treatment of groundwater: Metal-enhanced degradation of chlorinated organic contaminants. In: M. Aral (Editor), *Advances in Groundwater Pollution Control and Remediation*. NATO ASI Series. Springer Netherlands, pp. 249–274.

Henderson, A.D., and A.H. Demond. 2007. Long-term performance of zero-valent iron permeable reactive barriers: A critical review. *Environmental Engineering Science*, 24(4):401–423.

ITRC. 1999. Regulatory Guidance for Permeable Reactive Barriers Designed to Remediate Inorganic and Radionuclide Contamination. Interstate Technology & Regulatory Council. Permeable Reactive Barriers Work Group. Available on the Internet at www.itrcweb.org.

ITRC. 2005. *Permeable Reactive Barriers: Lessons Learned/New Directions*. Interstate Technology & Regulatory Council. Washington, DC: Interstate Technology & Regulatory Council, Permeable Reactive Barriers Team. Available on the Internet at www.itrcweb.org.

ITRC. 2011. Permeable Reactive Barrier: Technology Update. The Interstate Technology & Regulatory Council PRB: Technology Update Team. Washington, DC. http://www.itrcweb.org/Guidance/GetDocument?documentID=69.

Johnson, D.B., and K.B. Hallberg. 2005. Acid mine drainage remediation options: A review. *Science of the Total Environment*, 338(1):3–14.

Johnson, T.L., M.M. Scherer, and P.G. Tratnyek. 1996. Kinetics of halogenated organic compound degradation by iron metal. *Environmental Science & Technology*, 30(8):2634–2640.

Lai, K.C.K., I.M.C. Lo, and R. Surampalli. 2006. Configuration and construction of zero-valent iron reactive barriers. In *Zero-Valent Iron Reactive Materials for Hazardous Waste and Inorganics Removal*. American Society of Civil Engineers: Virginia, pp. 224–242.

Meggyes, T. 2005. Construction methods of permeable reactive barriers. *Long-Term Performance of Permeable Reactive Barriers*. Elsevier, Amsterdam, pp. 27–52.

Naidu, R. 2013. Recent advances in contaminated site remediation. *Water, Air, & Soil Pollution*, 224(12):1–11.

O'Hannesin, S.F., and R.W. Gillham. 1998. Long-term performance of an in situ "Iron Wall" for Remediation of VOCs. *Ground Water*, 36(1):164–170.

Phillips, D.H., T. Van Nooten, L. Bastiaens, M.I. Russell, K. Dickson, S. Plant, J.M.E. Ahad, T. Newton, T. Elliot, and R.M. Kalin. 2010. Ten year performance evaluation of a field-scale zero-valent iron permeable reactive barrier installed to remediate trichloroethene contaminated groundwater. *Environmental Science & Technology*, 44(10):3861–3869.

Powell, R.M., and R.W. Puls. 1997. Proton generation by dissolution of intrinsic or augmented aluminosilicate minerals for in situ contaminant remediation by zero-valence-state iron. *Environmental Science & Technology*, 31(8):2244–2251.

Richardson, J.P., and J.W. Nicklow. 2002. In situ permeable reactive barriers for groundwater contamination. *Soil and Sediment Contamination: An International Journal*, 11(2):241–268.

Roehl, K.E., T. Meggyes, F.G. Simon, and D.I. Stewart. 2005. *Long-term Performance of Permeable Reactive Barriers*. Vol. 7: Access Online via Elsevier.

RTDF. 2001. Permeable Reactive Barrier Installation Profile. Permeable Reactive Barriers Action Team. Remediation Technologies Development Forum, Date accessed March 2014. http://www.rtdf.org/public/permbarr/PRBSUMMS/.

Starr, R.C., and J.A. Cherry. 1994. In situ remediation of contaminated ground water: The funnel-and-gate system. *Ground Water*, 32(3):465–476.

Thangavadivel, K., W.H. Wang, V. Birke, and R. Naidu. 2013. A comparative study of trichloroethylene (TCE) degradation in contaminated groundwater (GW) and TCE-spiked deionised water using zero valent iron (ZVI) under various mass transport conditions. *Water, Air, & Soil Pollution*, 224(12):1–9.

Thiruvenkatachari, R., S. Vigneswaran, and R. Naidu. 2008. Permeable reactive barrier for groundwater remediation. *Journal of Industrial and Engineering Chemistry*, 14(2):145–156.

# 4

## Groundwater Modeling Involving PRBs: General Aspects, Case Study

Sreenivasulu Chadalavada, Martin Wegner, and Ravi Naidu

**CONTENTS**

## 4.1 Introduction

Permeable reactive barrier (PRB) technology is an increasingly viable option for remediating chlorinated hydrocarbon, petroleum hydrocarbon, and dissolved heavy metals contamination (Chapters 2 and 3). The PRB is an *in situ* passive remediation technology and has certain advantages compared to other active remediation technologies such as the pump-and-treat and chemical oxidation. This technology also prevents the contamination from migrating to uncontaminated aquifer systems. About 200 PRBs have been installed worldwide (Das, 2002; ETI, 2005, see Chapter 3) for treating common contaminants like chlorinated hydrocarbons (Burris et al., 1995; Orth and Gillham, 1996; Roberts et al., 1996; McMahon et al., 1999; Vogan et al., 1999; Schlicker et al., 2000), petroleum hydrocarbons (Guerin et al., 2002) and heavy metals (Powell et al., 1995; Gu et al., 1998; Shokes and Möller, 1999). A schematic diagram demonstrating the PRB technology is shown in Figure 4.1. The most important components of the design and implementation of the PRB are a detailed understanding of the subsurface hydrogeology, the kinetics of the reactive material chosen for the barrier, and the long-term monitoring plan. The kinetics of the different reactive materials is well understood and documented. While a number of different reactive materials have been used, most of the PRBs installed worldwide utilize zerovalant iron (ZVI) as the reactive material (Rabideau et al., 2005). An overview of hydrogeological modeling for PRBs is given in Gupta and Fox (1999). The most challenging component of PRB design and implementation is the site hydraulics, and several case studies of PRBs demonstrate this aspect of the technology.

Building a reliable simulation model for the implementation of any remediation technology requires reliable data from which a conceptual site model (CSM) is developed that represents the subsurface hydrogeology of the site. Adequacy of the design in this context often conflicts with the cost involved in the characterization process. There are often various conflicting objectives in the decision-making process and this requires a robust and optimal strategy to minimize uncertainties involved with the system and maintain the accuracy of the data.

Subsurface hydrogeology modeling is implemented at various stages of the PRB technology. The general principles of PRB design and implementation have been discussed in detail by Gavaskar et al. (1998). The two basic designs of PRBs most widely applied are the funnel-and-gate and continuous-trench barriers (McMohan et al., 1999). Further information on these is provided

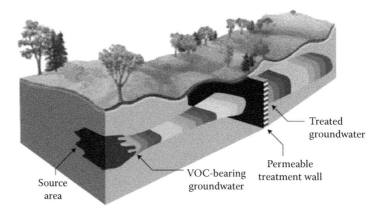

**FIGURE 4.1**
A schematic diagram demonstrating the PRB technology. (With Permission from Environmental Technologies Inc. (ETI). 2005. http://www.eti.ca/, accessed March 28, 2005, ETI, Waterloo, Ontario, Canada.)

in Chapters 2 and 3. The implementation stages of PRB technology include the initial feasibility assessment, laboratory treatability studies (including column studies), estimation of PRB design parameters, and development of a long-term monitoring network for the performance evaluation of the barrier. This chapter provides a general overview of mathematical models used for implementing the PRB technologies.

## 4.2 Design of PRBs

The most important parameter in designing the PRB is the thickness of barrier, which is a function of both hydrogeological and contaminant parameters. The contaminant concentration entering the barrier and its spatial distribution is the key for the design purposes. Since the contaminant transport depends on various hydrogeological and chemical properties it is critical to estimate the hydraulic conductivity (K), dispersion coefficient (D) of the subsurface environment and the first-order decay coefficient of the contaminant. Estimates of various parameters for designing the PRB are provided in the following sections.

A simple mathematical model (Rabideau et al., 2005) governing the transport process through the reactive barrier is represented by a one-dimensional advective–dispersive–reactive equation (ADRE). The governing equation for the single decay ADRE assuming a homogenous subsurface medium is as follows:

$$\frac{\partial c}{\partial t} = -v\frac{\partial c}{\partial x} + D\frac{\partial^2 c}{\partial x^2} - k'$$

where
$c$ = aqueous phase contaminant concentration (M/L3)
$t$ = time (T)
$x$ = distance from the entrance of the PRB (L)
$v$ = interstitial fluid velocity (L/T)
$D$ = dispersion coefficient (L2/T)
$k$ = first-order decay constant (1/T)

Application of the above equation to a PRB setting is commonly accomplished by neglecting the dispersion term and treating the PRB as an ideal plug flow reactor, which leads to the following simple design equation (e.g., Gavaskar et al., 1998; USEPA, 1998):

$$\frac{c(x = L)}{c_0} = \exp\left(-k'\frac{L}{v}\right)$$

where
  $L$ = barrier thickness (L)
  $c$ = constant contaminant concentration entering the barrier (M/L3)

Most real-world contaminated sites are heterogeneous and anisotropic in nature. To simulate the groundwater flow and pollutant transport for the actual subsurface conditions, it is desirable that the simulation model chosen is able to accommodate both these aspects of the modeling. Important concepts of groundwater modeling have been presented by Anderson and Woessner (1992) and Zheng and Bennet (2002).

### 4.2.1  Modeling of Induced Heterogeneity

The presence of a PRB induces changes in the subsurface heterogeneity, which influence the groundwater flow and contamination migration pathways. The subsurface system is viewed as a combination of adjacent flow fields of different characteristics, such as hydraulic conductivities rather than a continuous domain (Das, 2002). In this scenario, the fluid dynamics play a vital role and the contaminant transport model must take this aspect into consideration. An example of the application of this model would be former gas works sites which often contain discarded pipes and large cavities. The presence of these in the subsurface results in the zones of free flow and thus, overall contaminant transport in the subsurface is a combination of free flow and porous flow conditions. The contaminant transport in the free flow regions can be modeled using the Navier–Stokes (N–S) equations. The N–S equation of incompressible fluid flow is given as follows (Landau and Lifschitz, 1982):

$$\frac{\partial u}{\partial t} + u \cdot \nabla u = \frac{\nabla P}{\rho} + v \nabla^2 u$$

where
  $v$ is the kinematic viscosity
  u is the velocity of the fluid parcel
  $P$ is the pressure
  $\rho$ is the fluid density

The fluid mobility within the barrier can be represented by Darcy or Brinkman equations, subject to the properties of the reactive material used in the barrier. The Darcy's law is stated as follows:
  For a finite one-dimensional flow, it may be stated as

$$Q = AK \frac{\Delta h}{L}$$

where

$Q$ = volumetric flow rate (m$^3$/s or ft$^3$/s)

$A$ = flow area perpendicular to L (m$^2$ or ft$^2$)

$K$ = hydraulic conductivity (m/s or ft/s)

$l$ = flow path length (m or ft)

$h$ = hydraulic head (m or ft)

$\Delta$ = denotes the change in $h$ over the path $L$

The specific requirements of the simulation models for the PRB modeling are summarized in Gupta and Fox (1999). In view of possible underflow, overflow, and interaction between adjacent aquifers, a three-dimensional model is the most suitable option. Hanging barriers (Chapters 2 and 3) will result in significant vertical flow gradients and it is critical that the temporal distribution of vertical flow velocities should be accurately generated. Another requirement for the simulation model is that it should consider the variability of hydraulic conductivity induced by the PRB. Generally, the reactive material we use for the barrier has a higher K value than the surrounding matrix. This will result in a significant gradient in the hydraulic conductivity of the aquifer. In the case of a funnel-and-gate type of PRB, the funnel walls are very thick and highly impermeable, this is in contrast with the high-permeability barrier. For further information on funnel-and-gate type of PRBs, readers are referred to Chapters 2 and 3. The stability conditions of most numerical models will not allow us to solve the resulting high-contrast hydraulic conductivity distributions.

The finite difference-based three-dimensional groundwater flow model MODFLOW (McDonald and Harbaugh 1988) is the most widely used model for simulating groundwater flow. MT3DMS (Zheng and Wang, 1999) and RT3D (Clement et al., 1997) are also common reactive transport models and these models can be used in conjunction with MODFLOW. Hsieh and Freckleton (1993) developed a computer program to simulate the horizontal flow barriers using a finite difference model. In order to evaluate the capture zone of the PRBs, the model results should be compatible with the use of particle tracking algorithms. MODPATH (Pollock, 1989) is a widely used particle tracking code used in conjunction with MODFLOW. The finite element-based flow and transport simulation model, FEFLOW (Diersch, 2013) can also be used to model PRB systems. Finite element-based models have advantages over finite difference-based models in terms of the stability criteria, and they have the potential to address the complex nature of the conditions induced by subsurface heterogeneity.

Many sites across the world—such as Somersworth Landfill, Sunnyvale, CA, USA—have used MODFLOW in the PRB modeling. FLOWPATH developed by Waterloo Hydrogeologic has been used in Belfast, Ireland, and a DOE site in Kansas (USA) to evaluate the design of PRBs. Besides these, FLONET (Guiguer et al., 1992), FRAC3DVS (Therrien and Sudicky, 1995) also been used at some sites (Gupta and Fox, 1999).

When the longevity of the PRB is considered, one of the key concerns is mineral fouling. Mineral fouling is the reduction in pore space caused by mineral precipitation in the reactive material of the barrier (Li et al., 2006). This will result in the reduction of the porosity and hydraulic conductivity (Johnson et al., 2005) of the reactive media that will in turn influence groundwater flow and pollutant transport pathways. Li et al. (2006) conducted a study using ZVI as the reactive medium to identify and assess the most significant parameters impacting PRB hydraulics when influenced by a reduction in porosity.

## 4.2.2 Process-Based Modeling

The contaminant migration in the subsurface environment is influenced by various physical processes such as flow and nonreactive transport mechanisms, and geochemical processes. Understanding these processes holds the key for developing the reactive transport models to simulate the contaminant transport in the subsurface environment. Process-based reactive transport modeling is an important tool for building a reliable simulation model for PRBs (Amos et al., 2004). Assessment and evaluation of a PRB requires the integration of complex biogeochemical processes occurring in the heterogeneous subsurface (Mayer et al., 2006). The primary goal of using process-based models is to predict the long-term performance of PRBs. A case study involving process-based reactive transport modeling has been conducted at Nickel Rim site (Benner et al., 1999) using the MIN3P (Mayer et al., 2002) numerical simulation model. This study integrated pore water data and solid-phase data from the reactive barrier, using the reactive transport model. The conceptual model included reduction rates and secondary geochemical parameters such as alkalinity and soil-phase data. This integration process resulted in simulated sulfate reduction rates within a factor of 1.5 of the field values. The same model was successfully used by Jeen et al. (2007) for the evolution of iron reactivity and dynamic changes in geochemical conditions and remediation. Predictions under various hydrogeochemical conditions showed that trichloroethene (TCE) could be treated effectively for an extended period without significant loss of permeability. The modeling aimed to incorporate the effects of mineral precipitation on ZVI into a reactive transport model, in order to improve the prediction of long-term performance of ZVI PRBs. MIN3P was modified and tested against observed data from long-term column experiments designed to assess the extent of secondary mineral formation and its effect on the performance of the iron. The longevity of an iron PRB under various hydrogeochemical conditions was also estimated using the modified MIN3P.

Abiotic reductive dechlorination using ZVI is considered one of the most important remediation techniques for remediating chlorinated hydrocarbon compounds (Gilliam and O'Hannesin, 1992, 1994). This technique has been demonstrated at various contaminated sites across the world using PRB technology.

Prommer et al. (2008) modeled the geochemical and isotopic changes in a column experiment for degradation of TCE using ZVI. The researchers considered that incorporation of the details of the degradation pathways of the organic contaminants had not been a primary concern and had not been taken into account in models such as Moffett Air Field (Yabusaki et al., 2001) and Elizabeth City (Mayer et al., 2001). The objective of their study was to provide a more comprehensive and integrated analysis of experimental data, including isotopic data, toward the long-term goal of process-based hydrogeochemical modeling for the efficient and economic design of ZVI PRBs.

PHT3D (Prommer et al., 2003) was used in this study to simulate TCE degradation by ZVI and corresponding geochemical changes. PHT3D was developed by coupling a MT3DMS (Zheng and Wang, 1999) and PHREEQC-2 (Parkhurst and Appelo, 1999) to compute the reactive processes. The transport of TCE-contaminated groundwater through the experimental column filled with the Fe filings was simulated using PHT3D. Fitting the observed data with reaction rate constants provided by the parameter estimation tool PEST (Doherty, 2002) was coupled with PHT3D. The TCE degradation reaction network provided very good agreement between simulated and observed concentration as shown in Figure 4.2.

Contrary to this observation, poor model calibration results were achieved with alternative versions of the reaction network, such as when the production of C3–C5 hydrocarbons was omitted, and also when the pathway

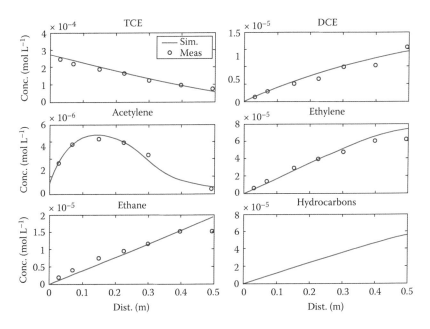

**FIGURE 4.2**
TCE degradation reaction network using PHT3D.

DCE → acetylene was excluded. It is has been observed that semiautomatic calibration is a useful mechanism to verify and/or identify reaction pathways.

A multicomponent reactive transport model was used by Yabusaki et al. (2001) to simulate mobile and nonmobile components undergoing equilibrium and kinetic reactions including TCE degradation, parallel Fe dissolution reactions, precipitation of secondary minerals, and complexation reactions. The system was modeled using 36 chemical components and 8 minerals for a systematic coupling of transport and reactions for multiple chemical components.

Sass et al. (2001) conducted a geochemical assessment at three PRB sites to evaluate performance and longevity issues. The assessment was carried out at former NAS Moffett Field (California), Dover AFB (Delaware), and former Lowry AFB (Colorado) because they all exhibited different PRB designs, hydrogeological conditions and groundwater chemistry. Geochemical modeling with PHREEQC and Geochemist's Workbench was used to simulate Fe reactivity in each of the three groundwater types and to assist in understanding precipitation sequences. Geochemical modeling results show that siderite ($FeCO_3$) and marcasite ($FeS_2$) are the initial phases to precipitate. As the reaction between iron and groundwater progresses, marcasite becomes unstable with respect to mackinawite (FeS). Authors felt that it is difficult to make a quantitative assessment of the longevity of the iron in the barriers.

## 4.3 Numerical Models

A number of numerical models have been used for the modeling of PRBs at various sites. Table 4.1 provides a summary of these models showing that they vary in input parameters.

## 4.4 Case Study

To demonstrate the modeling of a PRB, a case study demonstrating remediation of TCE at a long-term contaminated site in South Australia is summarized in the following section. At this site, PRB technology was used in combination with the pump-and-treat active remediation strategy at the source zone. The PRB technology works as a safeguard to prevent the contaminant plume from migrating off-site. The PRB technology implemented on this site consists of a network of injection and extraction wells. It is envisaged that the most appropriate option for the long-term management of the groundwater contamination at the site is some form of "funnel-and-gate"-type approach, taking advantage of the low permeability of the aquifer.

**TABLE 4.1**

Summary of Various Mathematical Models Applied to Various PRB Case Studies

| Sl No | Name of the Model Including the Developer | Important Features | Application of the Model |
|---|---|---|---|
| 1 | PHREEQC (Parkhurst and Appelo, 1999) | A reactive transport model simulating geochemical reactions Performs speciation, solubility, and reaction path calculations for aqueous, mineral, gas, surface ion-exchange solution equilibrium Powerful inverse modeling tool One dimensional advection–dispersion for dual porosity | Kowng et al. (2007), Morrison et al. (2001, 2002), Courcelleus et al. (2011), Santofirmia et al. (2009), Naftz et al. (2008), Caraballo et al. (2010), Liang et al. (2003), Komnitsas et al. (2006), Sass et al. (2001), Navarro et al. (2006) |
| 2 | Flow model MODFLOW (Mc Donald et al. 1996) along with nonreactive transport models MT3DMS (Zheng and Wang, 1999) and RT3D (Clement et al., 1997) | RT3D and MT3D models can simulate the groundwater pollutant transport in combination with the flow terms derived from modular groundwater flow model, MODFLOW | PRC (1996), Veerasekharan (2004), Jirasko and Vaníček (2009), Hemsi and Shackelford (2006), Barma (2010) |
| 3 | PHT3D:MT3DMS-PHREEQC | This model can simulate the reactive transport processes within the PRB | Prommer et al. (2008) |
| 4 | TRAFFIC (Roumane et al. (2003) | Can simulate couples groundwater flow, multispecies reactive transport and heat transport A comprehensive electro-kinetic transport | Kowang et al. (2007) |
| 5 | PHAST (Parkhurst et al. (2004) | Simulator for the groundwater flow, pollutant transport involving multi-component geochemical reactions for the three-dimensional subsurface The flow and transport processes are governed by HST3D (Kipp, 1997) and geochemical reactions are governed by PHREEQC (Parkhurst and Appelo, 1999) | Kowang et al. (2007) |
| 6 | MIN3P (Mayer et al. 2002) | A generalized formulation for kinetically controlled reactions A multicomponent reactive transport model to facilitate the investigation of a large variety of problems involving inorganic and organic chemicals in variably saturated media | Yang (2008) |

Given the depth of the groundwater and the low permeability of the aquifer unit, a number of large diameter PRB media filled boreholes have been installed with 150 mm diameter casing. This will enable both monitoring within the barrier itself and also to enable long-term extraction from the center of the PRB to assist in controlling groundwater flows through the barrier media, following the cessation of hotspot treatment. The contaminant will be removed by the media as the groundwater flows through it toward the extraction points; the extracted water will be directly reinjected into the infiltration wells. This will enable semi-active/semi-passive management of the plume over the long term.

### 4.4.1 Site Description

The site is located north of Adelaide's central business district and covers approximately 1800 ha. Several potentially contaminated areas have been identified within the precinct resulting from defense-related activities such as weapons development, army training, testing, and its use as an air force base.

From the groundwater monitoring investigations, a steady increase of TCE concentration at the site was observed, suggesting either direct disposal of TCE into the groundwater or significant leaching from contaminated soils. Historical information suggested that the contamination was likely to be due to periodic dumping of solvents and chemicals in the area. The standard practice at the time was to dispose of solvents directly onto adjacent land or to dig pits in which nonflammable and nonvolatile chemicals were buried. The sediments at the site have a very low hydraulic conductivity and represent a limited water-bearing stratum rather than a conventional aquifer.

From the present contaminant distribution, we conclude that the mean groundwater flow is toward the west as shown in Figure 4.3. However, the breadth of the contaminant plume indicates that the flow direction varies due to spatially and temporally changing recharge rates.

### 4.4.2 Model Construction

The PRB modeling calculations were carried out using the FEFLOW (Diersch, 2013), a finite element-based groundwater flow and solute transport simulation mechanism. Two separate groundwater flow and transport models were set up: one displays the whole active remediation process (model I) and the second (model II) calculates the passive impact of the extraction wells at the tip of the plume, once the active remediation ceases. This detailed model includes all the well geometry.

The area of the regional model (model I) measures 2.36 km² and includes 219,648 spatial elements. The area of site covers only 0.176 km², and the spatial elements are much smaller here (about 10 m², high spatial discretization). Model II has 159,168 elements with a model surface area of 11,400 m².

**FIGURE 4.3**
Local groundwater contours at the site in January 2010.

The simulation began with the contaminant plume, as interpolated from the chemical analyses in March 2010. The vertical limit of the plume is set uniformly at 6 m Australian Height Datum (AHD) based on the filter screen positions of the wells. The highest TCE concentrations have been measured in the wells with shallow filter screens, while concentrations in the deep screened wells are negligible.

### 4.4.3 Reactive Contaminant Transport

When referring to reactive contaminant transport, it is important to distinguish between the TCE adsorption on naturally occurring organic matter within the modeled area and TCE adsorption onto the reactive material (Remat™) in the annular space of the extraction wells.

For the reactive contaminant transport within the model area, uniform organic carbon content was defined as the mean of the measured contents. For the adsorption, a linear sorption isothermal curve was assumed (HENRY Isotherm). From the organic carbon content (0.042%) and the $K_{oc}$ value of TCE (adsorption on organic matter, 94), the distribution coefficient, $K_D$, is calculated. $K_D$ defines the ratio of dissolved TCE in the groundwater and TCE adsorbed to the organic matter in the sediment (Equation 4.1). For TCE in this aquifer it is 0.039 L/g. From the value of $K_D$, the bulk density (1.5), and the water content (0.1), the retardation factor, R, of TCE in the aquifer is calculated. It defines how much more slowly TCE is transported compared to a water molecule (Equation 4.2). The value of R for TCE at this site is 1.56. Thus, at a groundwater flow velocity of 1 m/a, TCE transported is 0.64 m/a.

$$K_D = \frac{C_{org}[100\%] \times K_{oc}}{100} \tag{4.1}$$

$$R = 1 + \left( K_D \times \frac{\rho}{\theta} \right) \tag{4.2}$$

In the second model (model II, detailed model), the annular space of the extraction well system at the tip of the plume was reproduced to scale, in order to simulate passive remediation which occurs after the pumping has ceased. For the reactive material Remat™, batch and column experiments have been carried out by CRC CARE. As a result, adsorption isotherms and breakthrough curves were provided by chemists at CERAR, University of South Australia, and the results are considered in the model. Concerning adsorption, the Freundlich isotherm gave the best correlation to the measured values and was used for the simulations as shown in Figure 4.4.

The column experimental study showed that the breakthrough curve of TCE was a straight line over the entire trial. This indicates that the adsorption process is controlled by kinetics and increases with time, and also that the material being investigated has a very high capacity for retaining TCE. The adsorption of TCE on Remat™ might be controlled by the slow diffusion of the contaminant toward the inner surfaces of the organic material. We determined the adsorption parameters from the breakthrough curve using the software Stanmod (Simunek et al., 1999). The two-site-nonequilibrium sorption model (Wagenet and Van Genuchten, 1989) fits the experimental points best. The parameters used in the numeric simulations rather underestimate

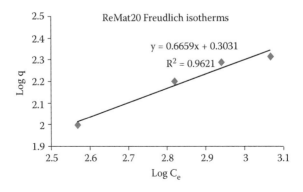

**FIGURE 4.4**
Freundlich adsorption isotherm for ReMAT™ from batch experiments.

the adsorption capacity of REMAT™ and therefore represent a worst-case scenario.

### 4.4.4 Model Parameters

The simulation model used various parameters such as hydraulic conductivity, recharge rate, porosity, and so on, as presented in Table 4.2, to simulate the contaminant concentrations over model domain at specified time intervals.

The values of the groundwater recharge and the hydraulic conductivity are the first values that have been changed during the model calibration. The parameter values were chosen according to the laboratory and field experiments. The precision of the measurements or their variance in time and space have not been taken into account, but their effect on the results is discussed in the evaluation.

### 4.4.5 Model Calibration

Calibration is achieved by varying the groundwater recharge and the hydraulic conductivity in the region. Reference points for the calibration on the groundwater recharge were the regional groundwater contour map and the TCE center line at site. In the area of the site, the hydraulic gradient of the groundwater is very low. As a result, there are no groundwater contours in the regional contour map. The center line of the TCE plume, however, deviates from the regional flow direction toward the west; the contaminants turn slightly toward the north. For this reason, the groundwater flow direction

**TABLE 4.2**

Model Parameters and Their Values

| Model Parameters | Values |
|---|---|
| Hydraulic conductivity, $k_s$ | $5.6 \times 10^{-7}$ m/s |
| Recharge rate | $-10$ mm |
| Groundwater flow | 0.1–10 m/a |
| Porosity in aquifer | 0.10 |
| Porosity REMAT™ (Modell II) | 0.45 |
| Longitudinal dispersivity $D_l$ | 0.5 m |
| Transversal dispersivity, $D_t$ | 0.05 m |
| Retardation factor aquifer, R | 1.56 |
| $K_D$ value, aquifer | 0.04 L/g |
| Solute transport TCE | 0.075 –7.5 m/a |
| FREUNDLICH parameters REMAT™ | |
| $K_f$ | $2\,\mu g/g$ |
| $N_f$ | 1.5 |
| Degradation rate | $0\,s^{-1}$ |

had to be calibrated by slightly changing the recharge rates at some points/ lines of the site: the storm water ditch south of the contaminant plume, causes a higher recharge rate which leads to the deviation of the contaminants. The groundwater recharge varies between +35 and −35 mm/a.

## 4.5 Scenarios

The remediation strategy applied for this study area is a combination of active and passive remediation approaches. The active remediation includes the pump-and-treat technology at the hot spot, that is, at the contamination source area with passive remediation which include the large diameter PRB wells coupled with extraction wells. Various scenarios have been calculated considering both remediation approaches. The scenarios simulate the remediation of the TCE groundwater contamination through the operation of large diameter wells that are filled with the reactive material REMAT™. The annular space of the wells has a large and well-defined diameter. The objective of the numeric simulations was to configure the most effective remediation plant, in order to carry out the remediation as quickly as possible and to minimize the number of extraction and injection wells, as their construction is expensive.

The principal problems of the site are the low hydraulic conductivity of the sediments, the distance of the groundwater to the soil surface (11 m) and the size of the contaminant plume (400 × 250 m). Hence, for an effective remediation, numerous large diameter wells down to depth of 20 m are needed. First simulations indicated that the maximum pumping rate of one large diameter well is only about 1 m³/d and the corresponding cone of groundwater depression has only a small diameter (see Figure 4.5).

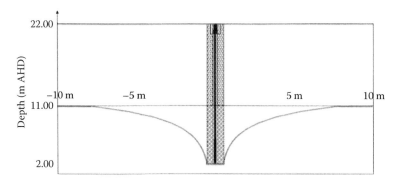

**FIGURE 4.5**
Calculated cone of groundwater depression and estimated range of depression.

## 4.6 Outcomes

The objective of the remediation was to increase the hydraulic gradient in the areas of high TCE concentration in order to achieve a significant contaminant transport toward the wells despite the low hydraulic conductivity. By infiltrating remediated groundwater at the fringe of the plume, the gradient is further increased. The simulations, however, show that an extensive increase of the hydraulic gradient is only possible with a larger number of extraction/injection wells, because even with the maximum pumping/injection rate the cone of depression/groundwater elevation of each well covers only a very small zone (see Figure 4.6).

If, on the other hand, the extraction wells are positioned close to each other, the cones of depression form a connected drawdown which enhances the remediation significantly. The evenly distributed infiltration wells at the fringe of the plume have little influence on the hydraulics.

Whether the whole contaminant plume is captured by pumping can be ascertained from particle tracking calculations. If a TCE molecule at the 500 µg/L fringe of the plume is actually transported toward an extraction well and if the injection of water at the boundary of the contamination does

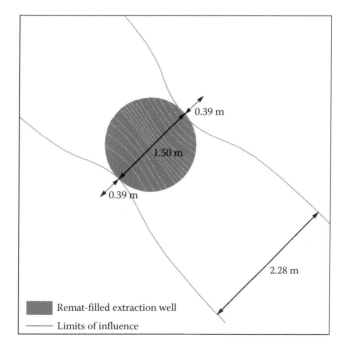

**FIGURE 4.6**
Hydraulic effect of a large diameter extraction well.

not actually broaden the plume, this provides confirmation of plume capture. The particle tracking method is therefore an important control for the correct positioning of the extraction and injection wells. The results show that in none of the scenarios the TCE plume spread during remediation. In fact, the extraction and reinjection of groundwater leads to a diminution of the plume (>500 µg/L area, see appendices).

Owing to the low hydraulic conductivity of the sediments, the hydraulic changes through extraction/injection of groundwater do not show up quickly, but need quite a long time. In order to check this, the simulations of Scenario 4.1 consider transient hydraulic conditions. The results show that stable hydraulic conditions are only reached after a duration of about 1 year. This has to be considered in the remediation design.

In the scenarios, the distribution of TCE (dissolved in the groundwater) is demonstrated after 1, 3, and 10 years. Scenario 1 describes the natural situation without any remediation (even without trial remediation).

In conclusion, the remediation of the source zone with the current wells (Scenario 2) is leading to a satisfactory clean up of TCE within the planned timeframe of 10 years. The contaminant concentration has fallen to less than 100 µg/L of TCE leading to a separation of the remaining plume from the original contaminant source. The same is true for the hot spot. Here, a small number of remediation wells are leading to a relatively rapid remediation of this spatially limited gravity center of the pollution (Scenario 3).

The remediation of the source zone and hot spot alone leads to a reduction of the TCE mass by 64%, leaving a remaining total mass of less than 20 kg (dissolved + adsorbed). The remediation at the tip of the plume has little influence on the decreasing TCE mass. However, even in Scenarios 2 and 3, the two existing extraction and injection wells at the tip of the plume contribute to TCE reduction. The effect of the additional wells at the tip of the plume in Scenario 4 has only a minor influence on the removal of total TCE mass.

The influence of the infiltration of cleaned groundwater can be seen from Scenario 4.3. Compared to Scenario 4, where water is reinjected into the aquifer at the tip of the plume, the northwestern part of the TCE plume remains much broader in Scenario 4.3 In Scenario 4.3, the extraction rate for the line of wells at the tip of the plume had to be reduced in order to avoid the wells running dry.

When building a second line of wells across the center of the plume (Scenario 5), the contaminant mass in the center is reduced even more. The reduction, however, is most significant at the tip of the plume. Moreover, by reducing the total number of wells at the tip of the plume, the north-westward spread of the TCE continues. In this scenario, the TCE mass is reduced by 70% after 10 years and therefore, the remaining contaminant mass is slightly lower than that in Scenario 4. It should be noted that with this well positioning, a large area displays a mean remaining TCE concentration after

10 years. When distributing the extraction wells in the center of the plume (Scenario 6), remediation becomes significantly more efficient.

### 4.6.1 Effectiveness of Passive Remediation

The passive remediation by the REMAT™ filled annular space of the extraction wells was evaluated in an additional model. The detailed model describes the development of the TCE concentration at the tip of the plume, once active remediation has ceased. To this, additional simulations were carried out following 10 years of remediation with a line of extraction wells at the tip of the plume. Here, the groundwater is cleaned while it passes the annular space of the extraction wells by the adsorption of TCE on the REMAT™. As the hydraulic conditions of the reactive material are better than those in the sediments, the sphere of influence of the passive wells is slightly bigger than their large diameter. Nevertheless, the groundwater is not completely captured because the distance from one well to the next is too great. The numerical simulations were carried out for a period of 30 years. The results show that the contaminant plume is not entirely captured by the line of passive wells.

## 4.7 Evaluation of the Simulations

The simulations have shown that a complete passive remediation of the site in an acceptable period of time is not possible. Despite the high number of large diameter extraction and injection wells and their great depth, each of the scenarios leads only to partial remediation. However, they do show that the TCE mass and its spread can be reduced significantly. Passive remediation approaches, on the other hand, do not lead to a significant remediation at this site. Moreover, the reactive material has only a limited adsorption capacity and the installation of large diameter wells is rather expensive.

On the basis of numerical modeling, CRC CARE demonstrated that a passive remediation would take several decades. Therefore, a remediation concept consisting of an active source remediation and passive PRB system which consists of large diameter bore wells was developed. Modeling results demonstrate that both the source area treatment and long-term monitoring of PRB shall continue in order to achieve desired remediation goals. CRC CARE has predicted the contamination scenarios at the site after 1, 2, and 10 years of treatment as shown in Figures 4.7 through 4.9. This could be compared to the status of groundwater contamination shown along with groundwater flow contours in Figure 4.2.

**FIGURE 4.7**
Simulation of TCE plume after 1 year of remediation.

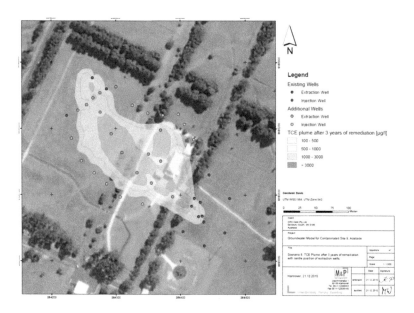

**FIGURE 4.8**
Simulation of TCE plume after 3 years of remediation.

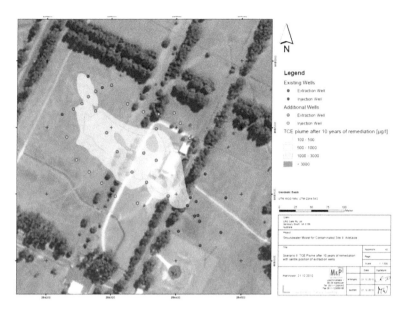

**FIGURE 4.9**
Simulation of TCE plume after 10 years of remediation.

That includes both the active source remediation and passive remediation system and the following results provide the details about the predicted outcomes. The TCE concentrations measured post commissioning of PRB are similar to the predictive modeling outcomes as the source concentrations are brought down to one-third of original concentrations. The contamination source strength could be brought down to significant levels after a minimum of 3 years operation and after 10 years of treatment, the source is completely eliminated. Considering the broader remediation interests of the site, the PRB system (source remediation+ PRB system) that we have at study areas is recommended.

## 4.8 Conclusion

Successful commissioning of a PRB depends upon the effective characterization of a subsurface environment and understanding various physical and geochemical processes that influence the groundwater flow and contaminant transport. To this end, an overview of various analytical and numerical models that have been used in designing PRBs in the world are given the chapter. An Australian case study has also been discussed to demonstrate the PRB modeling aspect. The TCE concentrations measured post commissioning of PRBs in a case study are similar to the predictive modeling outcomes, as the source concentrations are brought down to one-third of concentrations observed prior to PRB commissioning. A novel approach of coupling PRB technology in the form of large diameter bore wells with the contamination source zone is demonstrated through the case study. The potential applicability of PRBs in remediating the chlorinated hydrocarbon contamination is established.

## References

Amos R.T., Mayer K.U., Blowes D.W., and Ptacek C.J. 2004. Reactive transport modeling of column experiments for the remediation of acid mine drainage. *Environmental Science and Technology* 38: 3131–3138.

Anderson M.P. and Woessner W.W. 1992. *Applied Groundwater Modeling: Simulation of Flow and Advective Transport.* Academic Press, New York.

Barma S.D. 2010. Development of groundwater modelling linking GMS with GA-PSO based hybrid algorithm. Department of Food Processing Technology, Central Institute of Technology, Kokrajhar Assam, India.

Benner S.G., Herbert Jr., R.B., Blowes D.W., Ptacek C.J., and Gould D. 1999. Geochemistry and microbiology of a permeable reactive barrier for acid mine drainage. *Environmental Science and Technology* 33: 2793–2799.

Burris D.R., Campbell T.J., and Manoranjan V.S. 1995. Sorption of tricholoroethylene and tetracholoroethylene in a batch reactive metallic iron water system. *Environmental Science and Technology* 29(11): 2850–2855.

Clement T.P. 1997. RT3D—A modular computer code for simulating reactive multi-species reactive transport in 3-dimensional groundwater aquifers. Pacific Northwest National Laboratory Report, PNNL-11720.

Courcelles B., Modaressi-Farahmand-Razavi A., Gouvenot D., and Esnault-Filet A. 2011. Influence of precipitates on hydraulic performance of permeable reactive barrier filters. *International Journal of Geomechanics* 11(2): 142–151.

Das D.B. 2002. Hydrodynamic modelling for groundwater flow through permeable reactive barriers. *Hydrological Processes* 16: 3393–3418.

Diersch J.G. 2013. *Finite Element Modeling of Flow, Mass and Heat Transport in Porous and Fractured Media.* Springer.

Doherty J. 2002. PEST—*Model-Independent Parameter Estimation. User's Manual*, Fifth Edition. Watermark Numerical Computing, Brisbane, Australia.

Environmental Technologies Inc. (ETI). 2005. http://www.eti.ca/, accessed March 28, 2005, ETI, Waterloo, Ontario, Canada.

Gavaskar A.R., Gupta N., Sass B.M., Janosy R.J., and O'Sullivan D. 1998. *Permeable Barriers for Groundwater Remediation: Design, Construction, and Monitoring.* Battelle Press, Columbus, Ohio.

Gillham R.W. and O'Hannesin S.F. 1992. Metalcatalyzed abiotic degradation of halogenated organic compounds. IAH conference, Modern Trends in Hydrogeology, Hamilton Ontario, Canada.

Gillham R.W. and O'Hannesin S.F. 1994. Enhanced degradation of halogenated aliphatics by zero-valent iron. *Groundwater* 32: 958–967.

Gu, B. D. Watson, Goldberg W., Bogle M.A., and Allred D. 1998. Reactive Barriers for the Retention and Removal of Uranium, Technetium, and Nitrate in Groundwater. RTDF Meeting, Beaverton, OR, April 15–16.

Guiguer N.J., Molson E.O., and Frind T.F. 1992. *FLONET-Equipotential and Streamlines Simulation Package,* Waterloo Hydrogeologic Software and the Waterloo Center for Groundwater Research, Waterloo, Ontario.

Gupta N. and Fox T.C. 1999. Hydrogeologic modeling for permeable reactive barriers. *Journal of Hazardous Mater* 68: 19–39.

Hemsi P.S. and Shackelford C.D. 2006. An evaluation of the influence of aquifer heterogeneity on permeable reactive barrier design, Water Resources. Research 42, W03402.

Hsieh P.A. and Freckleton J.R. 1993. Documentation of a computer program to simulate horizontal-flow barriers using the U.S. Geological Survey modular three-dimensional finite-difference ground-water flow model, U.S. Geological Survey Open-File Report 92-477.

Jeen S.W., Mayer K.U, Gillham R.W., and Blowes D.W. 2007. Reactive transport modeling of trichloroethene treatment with declining reactivity of iron. *Environmental Science and Technology* 41: 1432–1438.

Jirásko D. and Vaníček I. 2009. The interaction of groundwater with permeable reactive barrier (PRB). *The Academia and Practice of Geotechnical Engineering.* Cairo: IFOS, pp. 2473–2478. ISBN 978-1-60750-031-5.

Johnson R.L., Tratnyek P.G., Miehr R., Thoms R.B., and Bandstra J.Z. 2005. Reduction of hydraulic conductivity and reactivity in zero-valent iron columns by oxygen and TNT. *Ground Water Monitoring and Remediation* 25(1): 129–136.

Kipp K.L. 1997. Guide to revised heat and solute transport simulator HST3D. Version 2. US geological Survey, Colorado. US.

Komnitsas K., Bartzas G., and Paspaliaris I. 2006. Modeling of reaction front progress in fly ash permeable reactive barriers. *Environmental Forensics* 7(3): 219–231.

Kwong S., Small J., and Tahir B. 2007. Modelling the remediation of contaminated groundwater using zero-valent iron barrier. WM'07 conference, Tucson, AZ.

Landau L.D. and Lifschitz E.M. 1982. *Fluid Mechanics,* 2nd edition. Pergamon Press, Oxford, England, p. 15.

Li L., Benson C.H. and Lawson E.M. 2006. Modeling porosity reductions caused by mineral fouling in continuous-wall permeable reactive barriers. *Journal of Contaminant Hydrology* 83: 89–121.

Liang L., Sullivan A.B., West O.R., Kamolpornwijit W., and Moline G.R. 2003. Predicting the precipitation of mineral phases in permeable reactive barriers. *Environmental Engineering Science* 20(6): 635–653.

Mayer K.U., Benner S.G., and Blowes D.W. 2006. Process-based reactive transport modelling of a permeable reactive barrier for the treatment of mine drainage. *Journal of Contaminant Hydrology* 85: 195–211.

Mayer K.U., Blowes D.W., and Frind E.O. 2001. Reactive transport modeling of an in situ reactive barrier for the treatment of hexavalent chromium and trichloroethylene in groundwater. *Water Resources Research* 37: 3091–3104.

Mayer K.U., Frind E.O., and Blowes D.W. 2002. Multicomponent reactive transport modeling in variably saturated porous media using a generalized formulation for kinetically controlled reactions. *Water Resources Research* 38: 1174–1195.

McDonald M.G. and Harbaugh A.W. 1988. *A Modular Three-Dimensional Finite-Difference Ground-Water Flow Model: Techniques of Water-Resources Investigations of the United States Geological Survey*, Book 6. USGS, USA. p. 586.

McMahon P.B., Dennehy K.F., and Sandstrom M.W. 1999. Hydraulic and geochemical performance of a permeable reactive barrier containing zero-valent iron, Denver Federal Center. *Groundwater* 37(3): 396–404.

Morrison S., Carpenter C., and Metzler D. 2002. Design and Performance of a Permeable reactive barrier for containment of uranium and associated contaminants at Monticello, Utah, USA. In *Groundwater Remediation of Metals, Radionuclides, and Nutrients with Permeable Reactive Barriers.*

Morrison S., Metzler D.R., and Carpenter C.E. 2001. Uranium precipitation in a Permeable Reactive Barrier by progressive irreversible dissolution of zerovalent iron. *Environmental Science and Technology* 35: 385–390.

Navarro A., Chimenos J.M., Muntaner D., and Fernández A. 2006. Permeable reactive barriers for the removal of heavy metals: Lab-scale experiments with low-grade magnesium oxide. *Ground Water Monitoring & Remediation* 26(4): 142–152.

Orth W.S. and Gillham R.W. 1996. Dechlorination of tricholoroethylene in aqueous solution using FeO. *Environmental Science and Technology* 30(1): 66–71.

Parkhurst D.L. and Appelo C.A.J. 1999. User's guide to PHREEQC (Version 2)—A computer program for speciation, batch-reaction, one-dimensional transport, and inverse geochemical calculations. US Geological Survey Water-Resources Investigations Report, 99-4259.

Parkhurst D.L., Kipp K.L., Engesgaard P., and Charlton S.R. 2004. PHAST—A program for simulating ground-water flow, solute transport, and multicomponent geochemical reactions. *U. S. Geological Survey Techniques and Methods 6—A8*, 154 pp.

Pollock D.W. 1989. Documentation of computer programs to compute and display pathlines using results from the U.S. Geological Survey Modular three-dimensional finite-difference ground-water flow model. US Geological Survey Open-File Report, 89-381.

Powell R.M., Puls R.W., Hightower S.K., and Sabatini D.A. 1995. Coupled iron corrosion and chromate reduction: Mechanisms for subsurface remediation. *Environmental Science and Technology* 29(8): 1913–1922.

PRC Environmental Management, Inc. 1996. Moffett Federal Airfield California West-Side Aquifers Treatment System Phase 2 Cost Opinion. Prepared for the Department of the Navy under Comprehensive Long-Term Environmental Action Navy (CLEAN I).

Prommer H., Aziz L.H., Bolaño N., Taubald H., and Schüth C. 2008. Modelling of geochemical and isotopic changes in a column experiment for degradation of TCE by zero-valent iron. *Journal of Contaminant Hydrology* 97: 13–26.

Prommer H., Barry D.A., and Zheng C. 2003. MODFLOW/MT3DMSbased reactive multicomponent transport modeling. *Ground Water* 41: 247–257.

Rabideau P.E. Raghavendra S., and Craig J.R. 2005. Analytical models for the design of iron-based permeable reactive barriers. *Journal of Environmental Engineering* 131(11): 1589–1597.

Roberts A.L., Totten L.A., Arnold W.A., Burris D.R., and Campbell T.J. 1996. Reductive elimination of chlorinated ethylenes by zero-valent metals. *Environmental Science and Technology* 30(9): 2654–2659.

Roumane H., Cooper N., and Bond A. 2003. Program users guide for TRAFFIC version 8.2. BNFL reports NSTS 4500 (1).

Sass B., Gavaskar A.1, Yoon W.S., Gupta N., Drescher. E., and Reeter C. 2001. Geochemical investigation of three permeable reactive barriers to assess impact of precipitation on performance and longevity. In Proceedings of the International Containment and Remediation Technology Conference and Exhibition, Orlando, FL.

Schlicker O., Ebert M., Fruth M., Weidner M., Wust W., and Dahmke A. 2000. Degradation of TCE with iron: The role of competing chromate and nitrate reduction. *Groundwater* 38(3): 409–409.

Shokes T.E. and Möller G. 1999. Removal of dissolved heavy metals from acid rock drainage using iron metal. *Environmental Science and Technology* 33: 282–287.

Therrien R. and Sudicky E. 1995. Three-dimensional analysis of variably saturated flow and solute transport in discretely fractured porous media. *Journal of Contaminant Hydrology* 23: 1–44.

United States Environmental Protection Agency USEPA. 1998. Permeable reactive barrier technologies for contaminant remediation. EPA/600/R-98/125, Office of Solid Waste and Emergency Response, USEPA, Washington, DC.

Veerasekaran B. 2004. Using MODFLOW to Predict Hydraulic Capture Zone Patterns for Arrays of Groundwater Circulation Wells. M.S. Thesis, Environmental Engineering, Texas A&M University, Kingsville.

Vogan J.L., Focht R.M., Clark D.K., and Graham S.L. 1999. Performance evaluation of a permeable reactive barrier for remediation of dissolved chlorinated solvents in groundwater. *Journal of Hazardous Materials* 68: 97–108.

Yabusaki S., Cantrell K., Sass, B., and Steefel C. 2001. Multicomponent reactive transport in an in situ zero-valent iron cell. *Environmental Science and Technology* 35: 1493–1503.

Yang C., Samper J., and Molinero J. 2008. Inverse microbial and geochemical reactive transport models in porous media. *Physics and Chemistry of the Earth* 33(2008): 1026–1034.

Zheng C. and Bennet G.D. 2002. *Applied Contaminant Transport Modelling*. Wiley, New York.

Zheng C. and Wang P.P. 1999. MT3DMS: A modular three-dimensional multispecies model for simulation of advection, dispersion, and chemical reactions of contaminants in groundwater systems; Documentation and User's Guide. Contract Reports SERDP-99-1, US Army Engineer Research and Development Center, Vicksburg, MS.

# 5

# *Impact of Trace Elements and Impurities in Technical Zero-Valent Iron Brands on Reductive Dechlorination of Chlorinated Ethenes in Groundwater*

Volker Birke, Christine Schuett, Harald Burmeier,
and Hans-Jürgen Friedrich

## CONTENTS

## 5.1  Introduction

Zero-valent iron (ZVI) applied in engineered permeable reactive barriers (PRBs) as well as application of nano and/or micro scale ZVI emulsions (NZVIs) by in situ injection are effective in situ technologies for remediating groundwater plumes or source zones, respectively, which are contaminated by chlorinated volatile organic compounds (cVOCs) or certain heavy metals.

However, ZVI or NZVI types used for these purposes so far have actually been produced chiefly for entirely different industrial applications rather than application to groundwater remediation, that is, they do not represent tailored reagents regarding their application to remediating contaminated sites. This contribution describes investigations and first results of a study which has been performed during the second operational term of the German research and development (R&D) PRB cluster "RUBIN" between

2006 and 2012 (Birke et al., 2004, 2005; RUBIN, 2014), to address this issue in order to achieve an improved quality assurance regarding proper practical field scale implementations of different ZVI and NZVI brands and production batches.

All ZVI or NZVI types, which have been applied so far to reductively dechlorinate cVOCs in contaminated groundwater, are of technical grade, especially regarding PRBs, which very often are made from scrap metals of different origins, compositions, and so on. Or regarding NZVIs, which are originally produced for application in the electronic or even food industry, therefore are very often associated with different kinds and amounts of trace elements. Thus, their reactivity regarding cVOC dehalogenation in groundwater can vary significantly, which has already been addressed and documented extensively (Johnson et al., 1996; Miehr et al., 2004; Ebert et al., 2006).

We conducted long term column and short term batch experiments as well as electrochemical investigations, to investigate the impact of different technical ZVI brands (at nano scale as well as coarser qualities to be used in PRBs) *as well as different ZVI production batches from the same supplier* on the reductive dechlorination efficacy regarding tetrachloroethene (PCE) in groundwater. It was found that trace elements and local surface elements do vary significantly regarding types and amounts, and they may have a significant impact on the degradation efficacy. It is apparent from these investigations that a useful protocol for practical applications/ implementations is required to properly and effectively select the right ZVI or NZVI type as well as its right production batch *just prior to* its particular field-scale application in a ZVI PRB, or regarding injecting NZVIs. This would be especially useful when feasibility tests have been performed significantly earlier to the actual field application, that is, wherein *different* production batches of the same ZVI type/brand might have been used/ checked earlier in the lab rather than later applied in the field. The protocol covers instructions for recording some basic chemical and physical parameters (maybe repeating a column experiment, and/or just redoing some short term, easy and inexpensive batch experiments) in order to ascertain that the very ZVI production batch (coarse ZVI or NZVI) to be loaded to a field-scale PRB or to be injected into a source zone, respectively, will not significantly differ in its reactivity compared to the batch that was tested in the column experiment earlier, thus avoiding a potential malfunction. Such a precautionary measure may prevent applicants/stakeholders (especially, vendors and site owners) from facing a potentially serious failure at the beginning of a field-scale application using ZVI or NZVI, simply due to applying the "wrong" production batch. Furthermore, these investigations may deliver a significant contribution to improved planning and help predict more precisely field-scale applications of ZVI PRBs as well as injections of NZVIs in the future.

## 5.2 Results and Discussions

In contrast to numerous other technical ZVI and NZVI brands, carbonyl micro-ZVI basically shows a low reactivity regarding the reductive dehalogenation of perchloroethene (PCE) in groundwater. This can be readily understood because it consists highly pure iron, and it is well known to be reluctant to corrosion in water at a higher degree. A cyclovoltagram (CV), as shown Figure 5.1, shows that an iron electrode made of pure iron is much nobler than technical grade iron such as iron sponge (Responge®) that has been degrading the groundwater contamination of several 1000 µg/L PCE in a pilot-scale ZVI PRB in Rheine, Northwest Germany, since 1998. It can be seen that the slope of the hysteresis loop is much steeper for Responge; moreover, the CV of pure Fe shows passivation in the potential range between −1.0 and roughly −0.7 V (related to a saturated calomel electrode [SCE]).

The CV measurements carried out with contaminated groundwater from the Rheine site showed no evidence of direct PCE reduction signals (nevertheless, they can be found through measurements in aprotic solvents, e.g., acetonitrile). Therefore, reductive dechlorination of cVOCs in water by elemental iron is, compared to anaerobic corrosion, a significantly less favorable process from an electrochemical viewpoint.

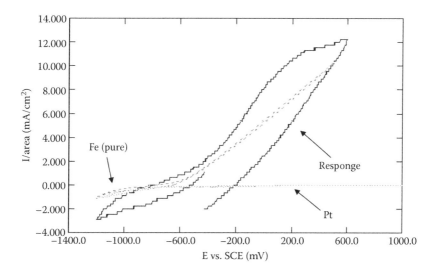

**FIGURE 5.1**
CV of a pure iron electrode in comparison to a platinum electrode and technical ZVI Responge electrode in groundwater (collected at the Rheine PRB site in Germany, being contaminated approximately by 10 mg/L PCE). SCE (saturated calomel electrode): +241 mV versus SHE (standard hydrogen electrode), mV = millivolts.

In contrast, when plated with certain metals in order to produce highly reactive local elements at the surface, carbonyl ZVI production batches become highly reactive regarding the degradation of PCE (Figure 5.2).

Checking different technical ZVI brands and production batches used for applications in PRBs, a wide range of differing trace element contents is encountered for both the average bulk composition of the particles and the surface compositions (Figures 5.3 and 5.4). To get an overview of the true (surface dependent) first-order rate constants for the reaction of PCE (10 mg/L) in groundwater from the Rheine PRB site, different technical ZVI brands and production batches as well as artificially produced combinations of pure ZVI with certain elements (produced by ion beam deposition of the trace element to the surface of a highly pure iron waver) were investigated in an electrochemical batch cell, equipped with electrodes comprising a particular metal of concern embedded into an epoxide resin (Figure 5.5).

From Figure 5.6 it can be seen that peerless iron as well as certain combinations of pure iron doped with trace elements such as sulfur, nickel, and manganese show the highest rate constants. Hence, it can be demonstrated, regarding only one groundwater type and contaminant (PCE) at a certain

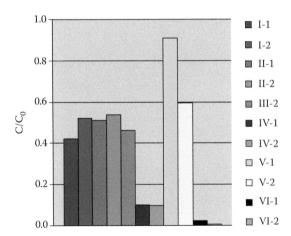

**FIGURE 5.2**
Different NZVI types/brands (they can differ in average particle sizes, their coatings with silicates, etc.) plated with around 0.2% nickel metal show different reactivities toward PCE in groundwater from the Rheine site (results of double-batch experiments at 10°C). Brand #IV and #VI show the highest turnovers of PCE within 21 days (type VI being the most effective one), only type #V showed poor turnover. As delivered/received, for pure carbonyl iron there is virtually no dehalogenation over 28 days. When plated with nickel, a virtually total degradation of PCE can be observed within 21 days. Note that a recently introduced new carbonyl micro-ZVI type shows high degradation of PCE within a few hours without any additional plating.

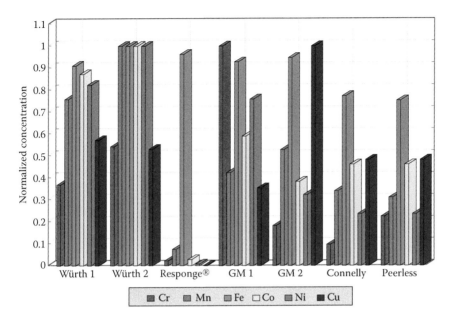

**FIGURE 5.3**
Normalized average trace element *bulk* concentrations concerning different ZVI brands (GM = Gotthart-Maier ZVI), sample workup by aqua regia dissolution and ICP-OES analyses. Great variations regarding kinds and amounts of trace elements are observed.

concentration level, that vastly differing degradation efficacies have been encountered when applying different ZVI brands or production batches.

To develop a quality assurance program to address this issue, different batch experiments using two Gotthart-Maier production batches ("GGG1" and "GGG3") were performed in comparison to the column experiment using the same production batch. It could be verified (Figure 5.7) that a good correlation between the applied amount of ZVI and the pseudo first-order rate constant k(obs) (1/h) can be attained. These findings are virtually independent of the stirring rate. In the corresponding column experiment, k(obs) could be determined at 0.05 1/h after around 20 pore volumes (the very first pore volume show adsorption and not degradation, which is a very common observation for technical ZVI brands regarding column or batch studies).

As shown in Figure 5.7, GGG1 shows higher degradation as GGG3; the reactivity of GGG3 was intentionally decreased by soaking with (noncontaminated) Rheine groundwater over 1 week ("GGG3 rusted") as well as by heating up to 600°C for 6 h in a muffle kiln ("GGG3 calcinated") in order to simulate potential failures/malicious conditions while manufacturing or storing; that is, for example heating scrap iron in a rotary kiln for too long, or storing in conditions where humidity/water may get access to the freshly produced ZVI batch.

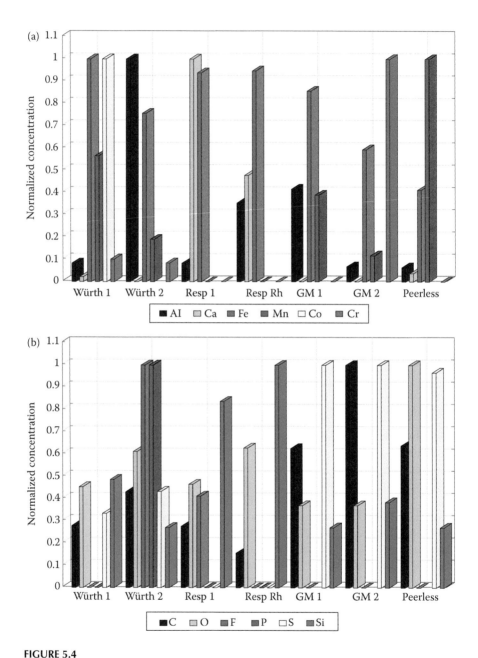

**FIGURE 5.4**

Normalized average trace element *surface* concentrations concerning different ZVI brands (GM = Gotthart-Maier), according to EDX analyses. Great variations regarding kinds and amounts of trace elements are observed.

**FIGURE 5.5**

ZVI electrodes (ZVI was embedded into a resin, cut, and polished) used for determining surface-dependent first-order rate constants regarding the degradation of PCE (10 mg/L) in groundwater of the Rheine PRB site (Germany). Left: Gotthart-Maier, right: Responge® electrodes, both no current and galvanostatically operated.

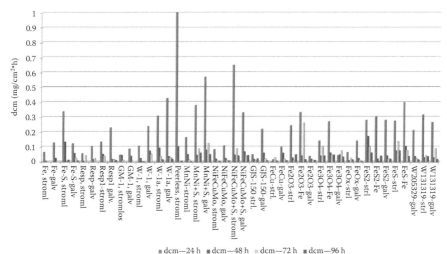

**FIGURE 5.6**

Normalized specific surface-dependent first-order rate constants $k_{SA}$ = dCm in mg PCE/cm²*h for different ZVI brands and different production batches of them as well as artificially produced combinations of pure Fe wavers spiked with certain metallic and nonmetallic elements (made by ion beam implantation) regarding degradation of PCE in groundwater from the Rheine PRB site (Germany). Sulfide containing ZVI types provide high surface normalized rate constants (dCm) after 24, 48, 72, or 96 h of treatment. "Stroml" = no current, "galv" = galvanostatical experiment.

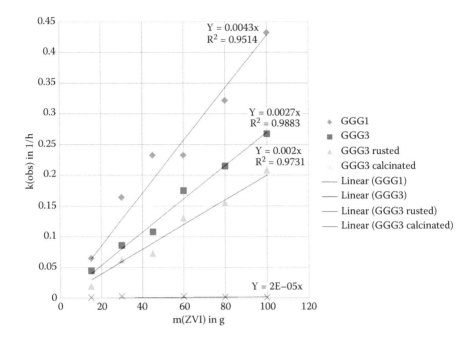

**FIGURE 5.7**

Linear relationships between k(obs) (1/h) and Gotthart-Maier ZVI ("GGG") mass employed in batch experiments using PCE in contaminated groundwater from the Rheine PRB site in Germany at 10°C. Values are virtually independent of the stirring rate/duration of stirring (using an overhead stirrer/shaker). Data points represent a single-batch experiment and the amount of two different Gotthart-Maier ZVI production batches during 1 year ("GGG1", "GGG3") per 100 mL. Calcinated GGG does not show any reactivity. Hence, conditions for degreasing scrap iron in a rotary kiln must be controlled very well if the production batch is supposed to be loaded to a PRB.

## 5.3 Conclusion (Recommendation for a Test Protocol to Check ZVI and NZVI Brands and Production Batches for Proper Field Application)

On the basis of these findings, a recommendation for a quality assurance test protocol can be outlined. Regarding the column experiment, which shows the best performance (therefore, the ZVI brand used therein will most probably be selected for the later field-scale application), batch experiments should be carried out in parallel by applying three to four different amounts of the same ZVI or NZVI brand and production batch in order to assign k(obs) of the column experiment (after 20–30 pore volumes) to a specific amount/portion of ZVI, which is required in the batch experiment to show approximately the same value. If the batch experiment is repeated again using the actual (differing) production batch of ZVI or NZVI to be charged into a field-scale PRB or to be injected into an aquifer, respectively, and it shows nearly the same value of k(obs), there would be no concern to actually apply it in the field. If k(obs) were significantly lower, one should be cautious regarding applying this production batch to the PRB in the field. Further investigations are required. In parallel or as a stand-alone solution, an electrochemical determination of the surface-dependent first-order rate constants $k_{SA}$, as outlined in Figures 5.5 and 5.6, is also a very good approach to check the ZVI or NZVI production batch for its proper reactivity towards cVOCs shortly before its field application.

In addition to checking k(obs) in a simple batch experiment or $k_{SA}$ in the electrochemical batch experiment, it could be useful to perform some further simple analyses, especially for ZVI brands produced from iron scrap, according to a test protocol of Connelly-GPM Inc., Chicago, Illinois (USA) (Klein, 2007; personal communication by the owner of Connelly-GPM Inc., Mr Stephen M. Klein, with the first author (VB) in October 2007):

- Raw borings are tested for moisture content by weight difference with oven drying.
- Borings sample is tested for oil content by weight difference with flame heating.
- Nonferrous content determined both by magnetic separation and caustic bath.
- Sample is screened to determine particle size distribution.
- Sample is ground to simulate effects of processing, then screened to estimate postmill particle size distribution.
- Based on a combination of the results of the above test protocol, a decision is made regarding the suitability of the raw material for ZVI production.
- After approval, all subsequent shipments are inspected to confirm uniformity with the approved sample.
- Borings are kept in 9–12 separate piles for proprietary blending.
- The density is checked by filling a box with product and determining the total weight.
- In the case of 3000 lbs bulk bags, samples are taken from each bag and marked "top," "middle," or "bottom" and labeled with the production code designation, so that it can be matched to the bag. These three samples are tested for water absorbency, and combined with a maximum of nine bulk bags and a screen test run. Also, there is a visual inspection of the three samples for uniformity. Any change in appearance at the beginning and at the end is noted. This is taken into consideration when samples are grouped for further testing.

Any material found to be out of specification is removed from the finished product category and dumped back to raw material, to be reprocessed.

Therefore, in the case of Connelly ZVI, the finished product must meet

- Cubic foot weight (140–160 lbs/ft$^3$ for the particular brand "ETI CC-1004" suitable for applications in PRBs)
- Water absorbency, free of oil, and grease
- Screen specifications
- Iron content at 85% minimum
- Customers packaging requirements

## Acknowledgments

The authors are grateful to the German Federal Ministry for Education and Research (BMBF) for funding their member projects of RUBIN and the scientific officer of RUBIN, Dr. Karl-Peter Knobel, for providing valuable advice over numerous years. They also thank several ZVI producers or vendors for the very good cooperation over the years: BASF Ludwigshafen (Dr. Lippert), Germany, Gotthart-Maier GmbH (Mr. Fischer), Rheinfelden, Germany, Mull-und-Partner Ing. Ges. mbH, Hannover, Germany, Mittal Steel GmbH, Hamburg, Germany, Connelly-GPM Inc. (Mr. Stephen M. Klein), Chicago, Illinois, USA, Peerless Metals Powder Inc. (Mr. David Carter), Detroit, Michigan, USA. They all provided numerous samples of ZVI materials and valuable information on them as well. Finally, tremendous support by Dr. Akmadaliev (Helmholtz-Center Dresden-Rossendorf, Institute for Ion Beam Physics) regarding the preparation of ion beam implanted iron specimens is also appreciated.

## References

Birke, V., Burmeier, H., Dahmke, A., and Ebert, M. 2005. The German Permeable Reactive Barrier (PRB) Network Rubin: Temporary Overall Results and Lessons Learned after five Years of Work, in CONSOIL 2005. 9th International FZK/TNO Conference on Contaminated Soil, Bordeaux, France, October 03–07, 2005 (computer optical disk), pp. 2865–2887.

Birke, V., Burmeier, H., Niederbacher, P., Hermanns, S.R., Koehler, S., Wegner, M., Maier, D. et al. 2004. PRBs in Germany, Austria, and Switzerland: Mainstreams, lessons learned, and new developments at 13 sites after six years. Paper 3A-14, In: A.R. Gavaskar and A.S.C. Chen (Eds.), *Remediation of Chlorinated and Recalcitrant Compounds—2004. Proceedings of the Fourth International Conference on Remediation of Chlorinated and Recalcitrant Compounds*, Monterey, CA, May 2004. ISBN 1-57477-145-0, Battelle Press, Columbus, OH.

Ebert, M., Köber, R., Parbs, A., Plagentz, V., Schäfer, D., and Dahmke, A. 2006. Assessing degradation rates of chlorinated ethylenes in column experiments with commercial iron materials used in permeable reactive barriers. *Environ. Sci. Technol.* 40(6), 2004–2010.

Johnson, T.L., Scherer, M.M., and Tratnyek, P.G. 1996. Kinetics of halogenated organic compound degradation by iron metal. *Environ. Sci. Technol.* 30(8), 2634–2640.

Miehr, R., Tratnyek, P.G., Bandstra, J.Z., Scherer, M.M., Alowitz, M.J., and Bylaska, E.J. 2004. Diversity of contaminant reduction reactions by zerovalent iron: Role of the reductate. *Environ. Sci. Technol.* 38, 139–147.

RUBIN. 2014. http://www.rubin-online.de (accessed April 1, 2014).

# 6

## Fourteen-Year Assessment of a Permeable Reactive Barrier for Treatment of Hexavalent Chromium and Trichloroethylene

Richard T. Wilkin, Tony R. Lee, Mary Sue McNeil, Chunming Su, and Cherri Adair

### CONTENTS

## 6.1 Introduction

Interest in site-specific evaluations of permeable reactive barrier (PRB) performance is high, particularly with regard to issues relating to media longevity and hydraulic performance. As compared to a large number of full-scale PRB applications around the world that have been constructed to remediate groundwater contamination, few long-term data sets are available in the literature that provide PRB performance in detail. Higgins and Olson (2009) recently conducted a life-cycle comparison of PRBs versus pump-and-treat operations for groundwater remediation. On the basis of their analysis, it was found that environmental impacts from PRBs are driven largely by material production requirements and by energy usage during construction, while for pump-and-treat systems environmental impacts are driven by operational energy demand. Higgins and Olson (2009) suggest that the minimum longevity of granular iron PRBs required to outcompete pump-and-treat systems is 10 years. Consequently, a key aspect of life-cycle analysis and cost/performance assessment is to have predictive tools that reasonably estimate long-term PRB performance using site-specific parameters such as ground-water chemistry and hydrologic conditions.

The granular iron PRB installed at the US Coast Support Center located near Elizabeth City, North Carolina, (USA) is a well-documented full-scale PRB designed and constructed for removing hexavalent chromium ($Cr^{VI}$) from groundwater. This chapter provides an update on the contaminant removal efficiency of this PRB after 14 years of operation.

## 6.2 Site Background

The US Coast Guard Support Center is located about 100 km south of Norfolk, Virginia, and 60 km inland from the Outer Banks region of North Carolina. The base is situated on the southern bank of the Pasquotank River, about 5 km southeast of Elizabeth City, North Carolina. A metal plating shop operated for more than 30 years in Hangar 79, which is about 60 m south of the river (Figure 6.1). Following its closure in 1984, soils beneath the shop were found to contain chromium concentrations up to 14,500 mg/kg.

Subsequent investigations revealed a chromate plume extending from beneath the shop to the river. At that time, the contaminant plume had high (>10 mg/L) concentrations of chromate, elevated sulfate (150 mg/L), and small amounts of volatile chlorinated organic compounds: trichloroethene (TCE), *cis*-DCE, and vinyl chloride (VC).

The groundwater flow velocity at the site is extremely variable with depth, with a highly conductive layer at roughly 4.5–6.5 m below ground surface. This layer coincides with the highest aqueous concentrations of chromate. The groundwater table ranges from about 1.5 to 2.0 m below ground surface and the average horizontal hydraulic gradient varies from 0.0011 to 0.0033. Slug tests conducted on monitoring wells with 1.5 m screened intervals between 3 and 6 m below ground surface indicate hydraulic conductivity values between 0.5 and 10 m/day. A multiple borehole tracer test showed groundwater velocities between about 0.10 and 0.20 m/day.

In June 1996, a 46 m long, 7.3 m deep, and 0.6 m wide PRB (continuous wall configuration) of zero-valent iron (Peerless Metal Powders, Inc., Detroit, MI) was installed approximately 30 m from the Pasquotank River (Figure 6.1). The reactive wall was designed to remediate hexavalent chromium-contaminated groundwater and portions of the larger overlapping plume of volatile chlorinated organic compounds. In 1999, a pilot-scale injection of sodium dithionite was conducted to evaluate the response of the source-zone hexavalent chromium (Kahn and Puls, 2003). On the basis of the success of this test, a full-scale treatment with sodium dithionite was carried out in 2001 (Malone et al., 2004). The objective of the dithionite treatment was to allow for the reduction of naturally occurring ferric iron-bearing minerals in

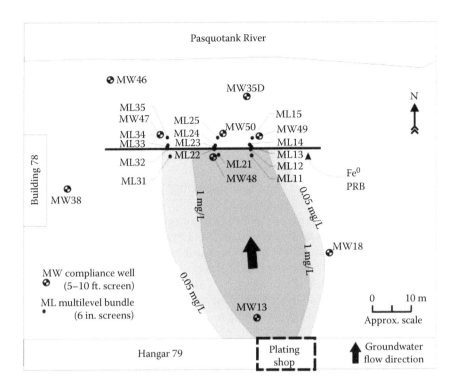

**FIGURE 6.1**
Site map showing locations of the source, PRB, monitoring wells, and Transect 2.

the aquifer matrix to a reactive ferrous iron state, which drives the reduction of mobile $Cr^{VI}$ to the insoluble trivalent ($Cr^{III}$) state.

## 6.3 Site Evaluation

The evaluation of the Elizabeth City PRB consisted of detailed groundwater sampling, hydrologic characterization, and core collection and solid-phase studies. In this contribution, we focus primarily on groundwater sampling results and contaminant distributions. A detailed monitoring network of over 130 subsurface sampling points was installed in November 1996 to provide detailed information on spatial and temporal changes in pore water geochemistry and hydrology. A series of 10 2-inch compliance wells have been sampled on quarterly, biannual, or annual basis (Figure 6.1). In addition, three detailed transects consisting of multilevel monitoring wells have been sampled, generally on an annual basis. Here, we focus on Transect 2 (ML21–ML25; Figure 6.1), which was placed in a position to coincide with the core of the $Cr^{VI}$ plume. Details of the groundwater sampling methodology have been published in Paul et al. (2003).

## 6.4 Outcomes

### 6.4.1 Chromium

Long-term trends in Cr concentrations in monitoring wells MW13 and MW48 (up gradient) and MW46, MW49, and MW50 (down gradient) are shown in Figure 6.2. After placement of the PRB, Cr concentrations above 3 ppb have never been observed in wells located on the down gradient side of PRB. Also note that beginning in 2001, Cr concentrations on the up gradient side of the PRB began to drop precipitously, a trend likely linked to the source-area dithionite treatment described above.

Concentration data collected from Transect 2 provide a snapshot series of performance of the PRB by revealing influent, interior, and effluent values of contaminant levels. The concentration data in Transect 2 over 14 years are summarized on cumulative percent diagrams to give an overall picture of performance through the lifetime of the PRB (Figure 6.3). For Cr, influent concentrations have ranged from <0.1 to 4000 ppb, with about 50% of the samples collected from the up gradient ML21 cluster above 50 ppb. The highest concentrations of chromium have been observed over the depth interval from 4 to 5 m below ground surface. Cumulative concentration data for Cr within and down gradient of the PRB show close agreement indicating that treatment of the down gradient aquifer is a consequence of groundwater transport and reaction through the reactive medium. Chromium concentrations within and down gradient of the PRB (ML24 and ML25) have ranged from <0.1 to 3 ppb; the average treatment efficiency over 14 years is >99.5%. Influent concentrations of chromium to the PRB have decreased with time (Figure 6.2), which is likely a result of the dithionite treatment of the source area and natural attenuation in the aquifer between Hangar 79 and the PRB.

Chromium treatment by the Elizabeth City PRB has been excellent over a sustained period of time. The reactive lifetime of the PRB has indeed outlasted the Cr plume. The removal mechanism of $Cr^{VI}$ has been explored in a previous publication (Wilkin et al., 2005). The sustained performance of the PRB can be linked to several key factors: (i) pH and redox conditions within the barrier have been maintained at ideal levels for Cr reduction to the trivalent state, (ii) influent groundwater chemistry is low in dissolved solids, so mineral accumulation due to carbonate precipitation has not significantly impacted reactivity and hydraulic conductivity, and (iii) the influent dissolved oxygen loading has been low, so iron corrosion reactions have not significantly impacted reactivity or hydraulic conductivity.

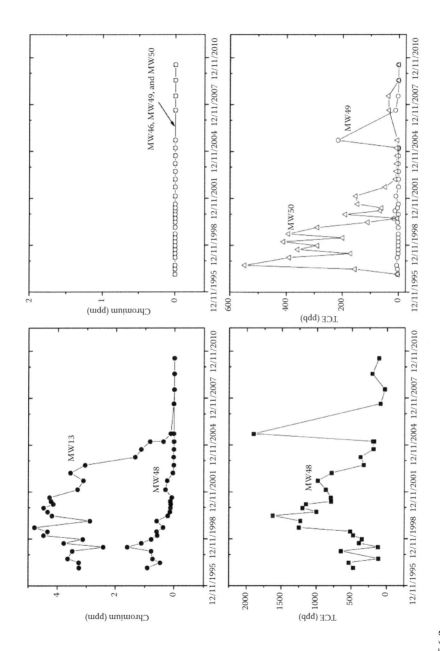

**FIGURE 6.2**
The TCE and Cr concentration in selected up gradient and down gradient wells.

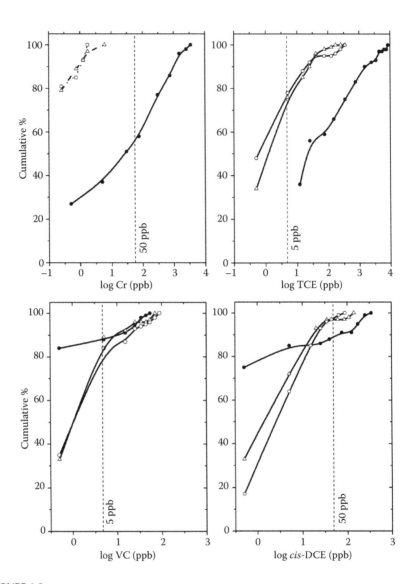

**FIGURE 6.3**
Cumulative% distribution diagrams for Cr, TCE, *cis*-DCE, and VC. Filled circles (up gradient); open triangles (down gradient) and open circles (in wall).

There are few PRB case studies which can be compared to the Elizabeth City site. Flury et al. (2009) described the 4-year performance of a granular iron PRB installed in Willisau, Switzerland, for $Cr^{VI}$ contamination. Unfortunately, the Flury et al. (2009) study does not provide any analysis of Cr uptake or removal efficiency. However, their results suggest that Fe corrosion processes and the build-up of thick deposits of Fe hydroxides on the Fe surfaces primarily limit the long-term effectiveness of the Willisau PRB.

At this site, groundwater flow velocities are comparatively high (5–6 m/day) and influent concentrations of dissolved oxygen are elevated (3.5–5 mg/L). Both of these factors are expected to limit treatment effectiveness.

## 6.4.2 Volatile Organic Compounds

Before discussing the performance characteristics of the PRB with respect to the degradation of chlorinated ethenes, it is important to point out that the Elizabeth City PRB was not originally designed to treat TCE. If it had, the PRB would have ideally been wider in some regions in order to increase residence time and it would have ideally extended deeper into the subsurface. Nevertheless, it is highly instructive to analyze the performance of the PRB with a focus on TCE treatment. As pointed out previously, the groundwater chemistry and hydrologic features of the site are ideally suited for the treatment of hexavalent chromium using granular Fe. An interesting question is whether this is also true for the treatment of chlorinated solvents.

Monitoring well data for TCE are shown in Figure 6.2 for MW48 (up gradient), MW50, and MW49 (down gradient). TCE concentration trends observed in the up gradient and down gradient regions are variable with time, with an overall trend of decreasing concentrations with time.

As shown in Figure 6.3 using a cumulative distribution diagram for data collected from Transect 2, influent TCE concentrations have ranged from ~1 to 9050 ppb, with 50% of the influent TCE concentrations above 30 ppb. In contrast to Cr, groundwater with maximum TCE values enters the PRB at its base. The cumulative concentration data for TCE within and down gradient of the PRB show close agreement, again indicating that reduced concentrations observed down gradient of the reactive medium are a consequence of groundwater transport and reaction within the PRB. About 75% of the observations of TCE concentrations within and down gradient of the PRB are below the maximum contaminant level (MCL) of 5 ppb. The remaining 25% of the observations follow a trend that corresponds to about 10% of the influent TCE. The average treatment efficiency for TCE is estimated to be >90% considering the difference of the median TCE concentrations in influent and effluent groundwater.

Interestingly, the distribution functions for transformation products, *cis*-DCE and VC, show different trends compared to TCE. Influent concentrations of *cis*-DCE and VC are low, with 90% of the observations below MCLs for these chemicals of 70 and 2 ppb, respectively. After treatment, that is, in the PRB and down gradient of the PRB, the proportion of observations with low *cis*-DCE and VC concentrations decreases, indicating that *cis*-DCE and VC are indeed products created as a consequence of TCE degradation. For *cis*-DCE, the proportion of effluent concentrations below the MCL increases to about 95% and maximum concentrations in the effluent decrease relative to maximum concentrations observed in the influent. Whereas, for VC, the proportion of effluent concentrations below the MCL drops to about 80% and maximal concentrations in the effluent actually increase relative to maximum concentrations in the influent groundwater. The TCE concentrations have been significantly reduced, yet the breakdown of daughter products is incomplete.

Results from the Elizabeth City PRB with respect to TCE remediation are comparable to other reports on PRB long-term performance (e.g., Phillips et al., 2010; Warner et al., 2005). Collectively, these studies indicate that granular iron can be effective in treating TCE contamination over for more than 10 years.

## 6.5 Conclusion

There is a continuing need for long-term performance assessments of groundwater remedial systems. The literature on PRB technology has several examples of remedial longevity that extend to 10 years and beyond. The Elizabeth City PRB is one example that has shown positive results for the treatment of hexavalent chromium and TCE. This system is perhaps the first demonstrated example of a PRB that shows reactive performance that surpasses the lifetime of the contaminant plume (in the case of chromium). The sustained performance of the PRB can be linked to several key factors: (i) pH and redox conditions within the barrier have been maintained at ideal levels for chromium reduction to the trivalent state; (ii) influent groundwater chemistry is low in dissolved solids, so mineral accumulation due to carbonate precipitation has not significantly impacted reactivity and hydraulic conductivity, and (iii) the influent dissolved oxygen loading has been low, so iron corrosion reactions have not significantly impacted reactivity or hydraulic conductivity.

## References

Flury, B., Frommer, J., Eggenberger, U., Mäder, U., Nachtegaal, M., and Kretzschmar, R. 2009. Assessment of long-term performance and chromate reduction mechanisms in a field scale permeable reactive barrier. *Environmental Science and Technology* 43, 6786–6792.

Higgins, M.R., and Olson, T.M. 2009. Life-cycle case study comparison of permeable reactive barrier versus pump-and-treat remediation. *Environmental Science and Technology* 43, 9432–9438.

Kahn, F.A., and Puls, R.W. 2003. *In situ* abiotic detoxification and immobilization of hexavalent chromium. *Ground Water Monitoring and Remediation* 23, 77–84.

Malone, D.R., Messier, J.P., Blaha, F., and Payne, F. 2004. *In situ* immobilization of hexavalent chromium in groundwater with ferrous iron: A case study. In *Remediation of Chlorinated and Recalcitrant Compounds—2004, Proceedings of the 4th International Conference*, Monterey, CA.

Paul, C., McNeil, M., Beck, F., Clark, P., Wilkin, R., and Puls, R. 2003. Capstone report on the application, monitoring, and performance of permeable reactive barriers for ground-water remediation, Vol. 2: Soil and ground water sampling, *EPA Report, EPA/600/R-03/045b*, 145pp.

Phillips, D.H., Van Nooten, T., Bastiaens, L., Russell, M.I., Dickson, K., Plant, S., Ahad, J.M.E., Newton, T., Elliot, T., and Kalin, R.M. 2010. Ten year performance evaluation of a field-scale zero-valent iron permeable reactive barrier installed to remediate trichloroethene contaminated groundwater. *Environmental Science and Technology* 44, 3861–3869.

Warner, S.D., Longino, B.L., Zhang, M., Bennett, P., Szerdy, F.S., and Hamilton, L.A. 2005. The first commercial permeable reactive barrier composed of granular iron: Hydraulic and chemical performance at 10 years of operation. In *Permeable Reactive Barriers, Proceedings of International Symposium on PRBs*, Belfast, Northern Ireland, pp. 32–42.

Wilkin, R.T., Su, C., Ford, R.G., and Paul, C.J. 2005. Chromium removal processes during groundwater remediation by a zero-valent iron permeable reactive barrier. *Environmental Science and Technology* 39, 4599–4605.

# 7

# Sequenced Permeable Reactive Barrier for the Pretreatment of Nitrate and Remediation of Trichloroethene

Keely Mundle, Janet Macmillan, and Ben McCarthy

## CONTENTS

## 7.1 Introduction

Dense nonaqueous phase liquids (DNAPLs) such as chlorinated solvents, polychlorinated biphenyl (PCB) oils, creosote, and coal tar, when released to the subsurface, will distribute themselves in the form of both pools of higher saturation distributions and disconnected blobs and ganglia of organic liquid referred to as residual. The longevity of residual and pooled DNAPL in porous media will be governed by a variety of factors, including the groundwater velocity and the aqueous solubility of the DNAPL components.

Complete source removal of DNAPL can be difficult and costly, and partial source removal may not have a significant impact on the extent of the plume but may reduce the longevity of the plume (Falta et al. 2005). Treatment of the resulting contaminant plume in order to protect downgradient sensitive receptors may be more achievable and, if required, may allow for more time to develop an effective source remediation solution.

Zero-valent iron (ZVI) permeable reactive barriers (PRBs) have been proven to be successful in removing a broad range of contaminants, including many chlorinated solvents (Roberts et al. 1996, Farrell et al. 2000). The first application of a ZVI PRB was constructed in northern California in 1994, and it continues to operate successfully to this day. The use of ZVI PRBs to remediate plumes of chlorinated solvent has become more common with hundreds of PRBs having been installed around the world (Adventus 2011).

There are two primary pathways for the dechlorination of chlorinated ethenes in ZVI PRBs: β-elimination and hydrogenolysis (IRTC 2005). β-Elimination is a reductive elimination in which halide ions are released from the molecule. Hydrogenolysis is the replacement of a halogen by a hydrogen ion. Batch experiments have indicated that the dominant degradation pathway is β-elimination (Arnold and Roberts 2000). This pathway is also the preferred pathway as it results in the chlorinated ethene degrading directly to ethane, ethene, and acetylene, thereby circumventing the production of intermediate daughter products (Eykholt 1998, Arnold and Roberts 1999).

Geochemical parameters that may affect the effectiveness of ZVI when treating chlorinated solvents have also been researched with the Interstate Technology Regulatory Council (ITRC) (2005) that summarized concentrations of nitrate, dissolved organic carbon (DOC), metals, and silica as some the most important. Of particular interest with respect to the work presented here are nitrate and DOC.

Research has found that nitrate can passivate iron (inhibit iron performance) by causing a thin layer of iron oxide to coat the iron. Some iron oxides such as goethite and maghemite have been proven to inhibit iron corrosion, resulting in iron passivation. For this reason, it is believed that as iron oxides form, the number of reactive sites available to nitrate and chlorinated solvents reduce, resulting in the advancement of both nitrate and chlorinated solvent profiles in the iron (Farrell et al. 2000).

Research has shown that certain types of DOC can also passivate the iron by coating reactive sites (ITRC 2005).

A potential solution to elevated concentrations of nitrates in the groundwater could be the use of an upgradient parallel denitrification PRB. A study by Vogan et al. (1993) concluded that sequenced PRB treatments were a potential technique to treat mixed nitrate and trichloroethene (TCE) plumes.

A literature review did not identify any field tests of denitrification and ZVI PRBs in series. However, a column experiment conducted by Vogan et al. (1993) attempted to replicate a sequenced PRB for the removal of nitrates by Nitrex™ (proprietary media incorporating waste cellulose solids) followed by the removal of TCE by ZVI. The experiment used groundwater containing 11 mg/L nitrate as N and 30 mg/L TCE. Complete removal of nitrates was observed in the Nitrex, and TCE concentrations were found to decline in the ZVI with an observed half-life of 5.7 h.

Concentrations of DOC in the Nitrex effluent were elevated, which could potentially increase passivation of ZVI. The study suggested that the increase in DOC in the effluent of the Nitrex resulted in an observed reduction in TCE degradation rates in the ZVI. However, GHD (2007) found in a pilot-scale denitrification PRB that total organic carbon (TOC) concentrations (and presumably DOC) actually decreased in the effluent due to increased biological activity.

Studies by Robertson et al. (2007, 2008) indicated that the use of highly reactive carbon sources with longer-than-necessary retention times may result in excess carbon stimulating reactions such as sulfate reduction, excess DOC leaching, and ammonium production. This suggests that to minimize concentrations of DOC in the effluent of the denitrification PRB, it is important to not overdesign the required retention times and carefully choose a less labile carbon source.

At the former waste control site (WCS), two source areas of chlorinated solvents were identified as residual DNAPL in the unsaturated zone. One of the sources has resulted in concentrations of TCE at 1000 µg/L in groundwater, which needed to be reduced to concentrations of approximately 330 µg/L prior to discharge at the downgradient Helena River. The installation of a ZVI PRB was identified as the prime candidate technology; however, further investigation identified that nitrate passivation could be a significant issue.

This chapter presents the field investigation, laboratory testing, and full-scale installation of a sequenced PRB for the purpose of pretreating the nitrate concentrations in order to remediate TCE using a ZVI PRB.

## 7.2 WCS Site Description

The WCS site is located 15 km northeast of Perth, Western Australia. The site historically operated as a chemical/oil recycling and treatment facility until a destructive fire in February 2001. Following the fire, the owners of the property went into receivership and the site has remained unoccupied since then. The site is currently an "orphan site," owned by the State of Western Australia and managed by the Department of Environment and Conservation of Western Australia. Since 2001, a series of investigations and risk assessments have been undertaken that have identified two main source areas of contamination.

During operation, the WCS had a history of accepting a wide variety of chemicals. While some records were found indicating the potential type and amount of chemicals on site at the time of the fire, it was impossible to be certain whether all potential contaminants had been identified. For this reason, initial investigations focused predominantly on characterizing the type, mass, and distribution of contaminants in the subsurface. It was found that the contamination in both soil and groundwater beneath the WCS consisted predominantly of hydrocarbons with some concentrations of chlorinated and brominated solvents. The chlorinated solvents identified at the WCS included tetrachloroethene (PCE), TCE, dichloroethene (DCE), trichloroethane (TCA), and 1,2-dichloroethane (DCA). The WCS source area is referred to as the primary source of contamination.

Groundwater investigations downgradient of the WCS in the Southwest Industrial Area (SIA) and the Damplands identified concentrations of TCE at concentrations greater than those identified at and immediately downgradient of the WCS.

In 2008, a drilling campaign using in-line sampling with a ColorTec kit was used to screen samples for chlorinated compounds. This resulted in a secondary source of contamination being identified downgradient. This source area is referred to as the secondary source and was found to consist of only TCE with no degradation products. The secondary source is located at the end of a cul-de-sac approximately 100 m to the south-east (downgradient) of the WCS site. A site plan view is presented as Figure 7.1.

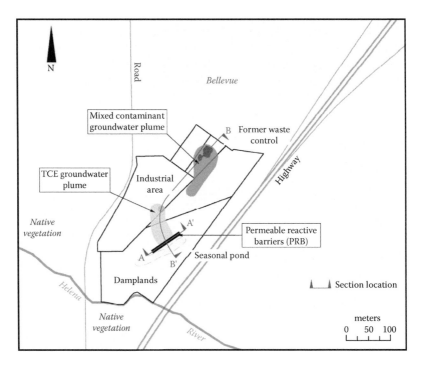

**FIGURE 7.1**
Plan view of site.

Site-specific risk-based criteria (RBC) for both human health and ecological protection were developed for each of the contaminants of concern. As the Damplands property has no onsite workers with little to no likelihood for using groundwater for irrigation purposes, the most relevant remediation criteria for TCE was the ecological criteria (330 µg/L) for the protection of the Helena River.

The two source areas are located in an area referred to as the SIA, which is north (upgradient) of the area referred to as the Damplands, a low-lying wetland area bordering the Helena River. The Damplands and the SIA are separated by an escarpment with topographical relief of approximately 10 m. The Helena River is the main drainage feature of the area and it forms the southern boundary of the study area. It is located approximately 300 m south-southwest of the WCS and approximately 120 m from the base of the river valley escarpment where the PRB system was installed.

A drain located in the northeast of the Damplands collects stormwater from the SIA and directs it southwest into the wetland depression. It is common during the winter months (May to October) for the Damplands area to be inundated and form a seasonal pond, which, when full, overflows into the Helena River. Standing water infiltrates into the alluvial aquifer and contributes base flow to the Helena River. This infiltration may dilute the concentrations of contaminants and prevent advancement of the plume during the high-rainfall autumn and winter months.

The surficial geologic unit beneath the WCS and the SIA is the Guildford Formation, which is part of the Quaternary Swan Coastal Plain sediments and extends to about 20 m depth. The Guildford Formation comprises interbedded layers and lenses of sand, silt, and clay of fluvial origin. The base of this unit is often iron-stained and/or iron-cemented and is present in the study area at about 0 to –5 m Australian Height Datum (AHD). Below the Guildford Formation is the Leederville Formation. Sediments of the Leederville Formation are alluvial in origin, and are mostly clays and siltstones.

Investigation boreholes along the base of the escarpment and a few meters upgradient of the PRB location initially encountered sand, silty sand, and clayey sand deposits. These sediments could not be readily attributed to a specific unit; rather, this was interpreted to be a transitional zone that may include a combination of colluvial deposited Guildford Formation from the escarpment interbedded with more recent deposits of the alluvial deposits from the Helena River. This transitional zone (Figure 7.2) is located where the Helena River and its predecessors have cut a valley into the Guildford and Leederville Formations. This valley has, in turn, been in-filled with younger silts, sands, and clays; the full thickness of these in-filled alluvial sediments has not been investigated.

In these same boreholes, at a depth of approximately 1 to 0 m AHD, sands and silty sands had an orange-brown color, which was visually consistent with sediments found near the base of the Guildford at other locations in the study area. The top of the Leederville Formation was encountered at

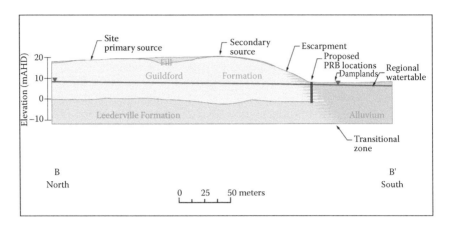

**FIGURE 7.2**
Simplified geological representation of site geology along cross-section B–B'.

approximately 0 to −0.5 m AHD. This unit was identified by the sudden appearance of feldspar and mica grains within a clay to clayey sand soil.

All the aquifers identified at site are hydraulically contiguous with one another but have variable hydraulic conductivities dependent on the lithology. The following hydrogeological units have been defined in this project:

- The alluvial unit consists of unconsolidated sediments varying in grain size from clay to gravel and with a relatively high hydraulic conductivity.

- The Regional Watertable unit comprises the uppermost part of the Guildford Formation aquifer and lithologically varies from clean sand through to silts and clays. It generally has a lower hydraulic conductivity value than the other aquifers.

- The base of the Guildford unit is the lowermost part of the Guildford Formation, which consists of unconsolidated sediments varying in grain size from sands to clay with iron-cemented sediments toward the base. It has a moderate hydraulic conductivity value, which is consistent with silty sand lithology.

- The Leederville consists of unconsolidated to cemented sediments varying from sand to clay. It has a moderate hydraulic conductivity value, which is consistent with the clayey sand lithology.

Slug tests were performed to monitor wells located at the base of the escarpment to establish the hydraulic conductivities of the lithologies present. Results from monitoring wells with 3 m screens varied between 1.3 and 7.8 m/day. In comparison, results from monitoring wells with 1 m screens varied from 0.18 to 17.9 m/day. These results indicate a highly variable and heterogeneous sand, sandy clay aquifer at the base of the escarpment.

## 7.3 ZVI Bench-Scale Tests

A bench-scale treatability study was conducted by EnviroMetals Technologies Inc. (EnviroMetals 2009) of Waterloo, Canada, to determine the feasibility of using ZVI to remediate TCE from site groundwater.

The bench-scale testing used groundwater from a monitoring well on site with the highest recorded concentrations of TCE. ETI indicated that the water provided to them had TCE concentrations of 820 µg/L and nitrate concentrations of about 3 mg/L. ZVI from two different manufacturers, Connelly GPM of Chicago, USA, and Quebec Metal Powders Ltd (QMP) of Sorel-Tracy, Canada, were used in the column experiments. A summary of the properties of the two ZVI materials is provided in Table 7.1.

The column experiments indicated that both the ZVI materials tested completely degraded TCE present in the site groundwater samples to below the remediation targets and laboratory limits of reporting without breakthrough of any chlorinated intermediate breakdown products. From these data, TCE half-lives at the 19.5°C site groundwater temperature were estimated to be 0.75 and 2.0 h for the QMP and Connelly ZVI, respectively. The residence time required to degrade TCE concentrations to below RBC was calculated to be between 2 and 9 h depending on the selected ZVI and TCE input concentrations.

An additional factor that must be considered in determining the required ZVI thickness in a PRB is the potential effect of nitrate passivation. Nitrate passivation rates of 0.44 and 0.53 $mg_{NO3-N}/g_{Fe}$ were estimated by ETI from observed column nitrate concentrations profiles for the Connelly and QMP ZVI materials, respectively.

Owing to higher reactivity and lower nitrate passivation rates, QMP ZVI was selected as the more suitable material for the construction of the ZVI PRB.

An assessment of the required thickness of ZVI to degrade TCE to concentrations below the RBC was made based on the half-life estimates from the

**TABLE 7.1**

Summary of ZVI Physical Properties

| | Grain Size | Density | Hydraulic Conductivity |
|---|---|---|---|
| QMP | <0.045 to 1.7 mm | 3.94 g/cm³ | 13 m/d |
| Connelly | 0.25 to 2.0 mm | 2.73 g/cm³ | 44 m/d |

**TABLE 7.2**

Estimated ZVI Thickness Required for 15-Year Treatment Lifetime

|          | Median TCE450 µg/L | 95% Percentile TCE 1700 µg/L | Median Nitrate 7.5 mg/L | 95% Percentile Nitrate 14 mg/L |
|----------|--------------------|------------------------------|-------------------------|--------------------------------|
| Connelly | 0.017 m            | 0.06 m                       | 0.78 m                  | 2.88 m                         |
| QMP      | 0.007 m            | 0.026 m                      | 0.45 m                  | 1.62 m                         |

laboratory results combined with the interpreted groundwater flow velocity of 0.16 m/day across the alluvial aquifer. Similarly, the thickness of ZVI required to allow for potential nitrate passivation over 15 years was assessed from the laboratory-derived passivation rates. The values for the two types of ZVI are presented in Table 7.2. The total ZVI thickness needed for an effective PRB is determined by summing the TCE degradation value with a nitrate passivation value (and adding an appropriate margin of safety).

The above analysis indicated that nitrate passivation accounts for approximately 98% of the required ZVI thickness within the PRB, while TCE treatment accounts for less than 2% (based on the QMP ZVI). Given that the ZVI material is likely to represent a significant portion of the overall project cost, it was clear that efforts to reduce nitrate influent concentrations could greatly reduce the required volume of ZVI and/or extend the longevity of the PRB.

## 7.4 Field Investigation

A detailed delineation monitoring program was undertaken in January, May, and September 2009 in order to characterize the contaminants of interest, to determine their respective concentrations along the proposed PRB location, and to better characterize the local hydrogeological conditions. This program included the installation of four multilevel monitoring wells (MWG87, 88, 89, and 90) directly upgradient of the Damplands Pond and two rounds of groundwater monitoring for volatile organic compounds (VOCs) and nitrate. The January 2009 TCE and nitrate results are presented (in cross-section) in Figures 7.3 and 7.4, respectively. The results indicate that the TCE and nitrate plumes overlap and are relatively consistent in shape over the year. The concentrations of TCE vary over time, with the maximum concentration in January 2009 being 2 mg/L in comparison with 0.82 mg/L in September 2009. Nitrate concentrations were found to be less variable with the maximum concentrations ranging from 14 mg/L in January 2009 to 19 mg/L in September 2009. It should be noted that nitrate contamination is not associated with the WCS fire and is instead likely to be the result of upgradient septic tank usage or possibly an upgradient livestock sale yard.

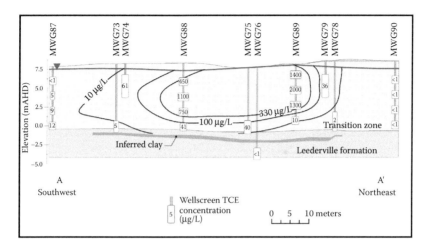

**FIGURE 7.3**
TCE concentrations along the proposed PRB alignment.

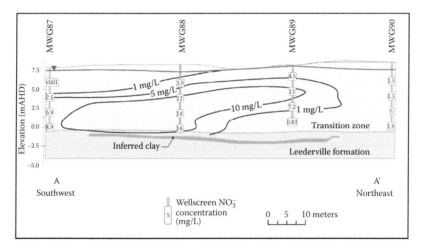

**FIGURE 7.4**
Nitrate concentrations along the proposed PRB alignment.

## 7.5 Denitrification Bench-Scale Tests

Owing to the level of nitrate concentrations detected in groundwater, an upgradient denitrification PRB to pretreat the groundwater was proposed. To minimize potential passivation of ZVI by organic carbon, it was suggested that the most suitable media for the denitrification PRB would be a relatively degraded carbon source that is not overly labile (J. Vogan, 2009, pers. comm.).

Based on this information, the most suitable media for a denitrification PRB at the Damplands site was assessed to be sawdust or woodchips as they

- Are relatively degraded carbon sources in comparison to straw, alfalfa, or other organic matter commonly used
- Have been shown to be a relatively long-lived source of carbon (Vogan 1993)
- Have been shown to be effective in removing nitrates in field trials in Western Australia (Fahrner 2002, Water Corporation 2004, GHD 2007)
- Have been shown to potentially remove DOC from groundwater (GHD 2007)

Laboratory bench-scale tests were conducted to predict how potential carbon material may perform in the proposed denitrification PRB at the WCS. The column for the apparatus was constructed of polyvinyl chloride (PVC) tubing with a length of 100 cm and a diameter of 10 cm. The column apparatus allowed for the collection of samples from the influent and effluent solution. There were also four sampling ports along the length of each column; at distances of 5, 20, 40, and 65 cm from the base of the column. These sampling ports were used to quantify denitrification rates of the solution as it flowed through the carbon media.

The porosity of the mixture was determined gravimetrically. To ensure a homogenous mixture, the materials were mixed thoroughly before placement into the column. The tests were performed at ambient room temperature in the laboratory.

Samples of groundwater were collected from a well in the vicinity of the proposed PRB, which would be expected to be representative of groundwater entering the treatment zone. During previous groundwater sampling events, this well exhibited elevated levels of nitrate and, during the experiments, the average influent nitrate concentrations was determined to be approximately 12 mg/L $NO_3$–N.

The groundwater obtained from the site was pumped upwards through the columns at a constant flow velocity using a precision variable-speed-drive peristaltic pump. A range of flow velocities were selected during the testing to mimic the flow velocity expected in the PRB.

Samples were collected from the sampling ports in appropriate containers and transported to a National Association of Testing Authorities (NATA)-accredited laboratory for analytical analyses. The samples were analyzed for nitrate as nitrogen, nitrite as nitrogen, total nitrogen, total Kjeldahl nitrogen, ammonia, DOC, and TOC.

The first round of four column experiments included four locally sourced materials:

1. 100% Sand—as a control
2. 100% Shredded aged native tree (Karri) bark

3. 100% Pine woodchips
4. 100% Pine sawdust

This initial round of experiments provided baseline denitrification rates for each material. A saline tracer test was undertaken prior to commencing the testing with site groundwater. The saline tracer testing was used to determine the particle velocity of the water flowing through the individual columns. This is governed by the porosity of the carbon material matrix within each column.

During the denitrification testing, samples were collected from the column experiments on day 1, 2, 4, and 7. The resulting effluent nitrate concentrations are presented in Figure 7.5. The results indicate that the aged Karri bark was the most efficient in consistently reducing the concentration of nitrate throughout the test. In comparison, the pine sawdust was initially ineffective; however, it improved with time, whereas the pine woodchips were effective initially but decreased during the course of the test duration.

The results from along the column (Figure 7.6) indicate that the shredded aged Karri bark also quickly denitrified the groundwater with almost 100% removal of nitrates within the first 40 cm of the column. In comparison, the pine sawdust required almost the entire 110 cm to completely denitrify the groundwater.

The concentrations of both DOC and TOC were measured (Figure 7.7). In all the columns consisting of carbon, the concentration of both TOC and DOC in the effluent decreased with respect to time. The results indicate that the TOC detected in the effluent predominantly consisted of DOC. The concentrations in the aged Karri bark effluent were about three times greater than that for sawdust and six times that for pine woodchips. In addition, the effluent from the shredded aged Karri bark column had a dark brown color

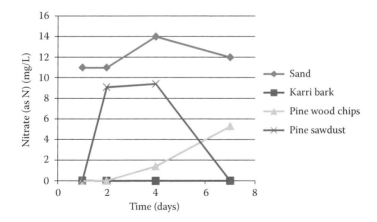

**FIGURE 7.5**
Effluent nitrate concentrations from the column experiments.

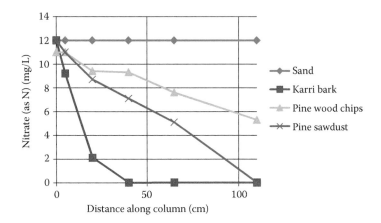

**FIGURE 7.6**
Concentration of nitrate (as N) along the length of each column experiment.

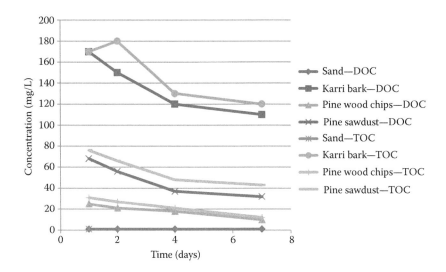

**FIGURE 7.7**
Concentration of soluble organic carbon and total organic carbon from the four columns.

as well as a strong organic odor, which corroborates the high concentrations of DOC. As the amount of DOC in the effluent of the denitrification PRB was a concern, owing to potential passivation of the ZVI, it was decided that the shredded aged Karri bark was not appropriate for the PRB system and was omitted from further testing.

As the carbon source will degrade over the life span of the PRB, it is important that some sand is used in the PRB's construction to provide long-term structural integrity of the barrier. A second round of column tests was undertaken to further investigate the optimal denitrification mix. The following column mixtures were evaluated:

1. 100% Sand
2. 80% Sand and 20% sawdust
3. 60% Sand and 40% sawdust
4. 60% Sand and 40% woodchips

As in the previous phase of testing, a saline tracer test was undertaken first to assess particle velocity. The laboratory results of this final round of testing were used to determine the relationship among denitrification rates, retention times, and carbon-to-sand ratios. The results are presented in Figure 7.8.

The results indicated that, as predicted from the results shown in Figure 7.8, the use of 100% sawdust was the most effective in nitrate removal. The results also showed that there was a high degree of variability with respect to whether sawdust or woodchips was the most effective carbon source. This is not unexpected as laboratory experiments by Carmichael (1994) that studied sawdust of varying grain size found that nitrate consumption rate was not correlated with the specific surface area of the wood. It was hypothesized by Robertson et al. (2000) that instead of being restricted to the grain surface, denitrification may be associated with reaction rims that penetrate by diffusion into the carbonaceous solid. Based on this finding, it was concluded that it was not imperative to ensure the same particle size distribution of the mixture of woodchips and sawdust. The results did, however, indicate that a mixture containing 20% sawdust was not very effective in removing nitrates, even if retention times were lengthened.

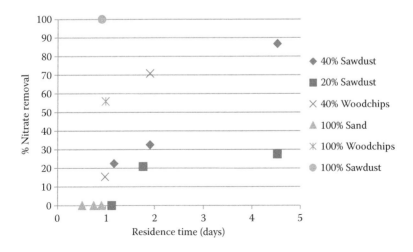

**FIGURE 7.8**
Percent nitrate removal with respect to carbon material (sawdust vs. woodchips) and amount.

As the difference between the sawdust and woodchips was not consistent, it was decided to use a carbon source consisting of a variable combination of pine sawdust and woodchips for the denitrification PRB. Owing to the production methodology, a mixture of sawdust and woodchips is much easier to source from local producers.

ETI recommended that consideration be given to retaining a section of natural aquifer between the denitrification PRB and the ZVI PRB as this could act as a buffer and reduce the transport of TOC into the ZVI PRB and thereby increase ZVI longevity. This recommendation was incorporated into the preliminary design of the remediation system.

## 7.6 Implementation

The preliminary design included material specifications and minimum sizes for both PRBs. A key preliminary design aspect of the ZVI PRB was the inclusion of three panels within the ZVI PRB (Figure 7.9). The central main panel (B) was designed to be 45 m long, which was long enough to treat the entire width of the TCE plume with concentrations greater than 330 µg/L. The two side panels (A and C) were designed to be 17 and 14 m long and have half the amount of ZVI as the central panel. These side panels were included in the design as a contingency measure in case of seasonal shifting of the plume or the permeability of the ZVI PRB decreasing and potentially resulting in flow diversion. A geotechnical investigation undertaken by Golder inferred a clay layer in the upper portion of the Leederville Formation that would be

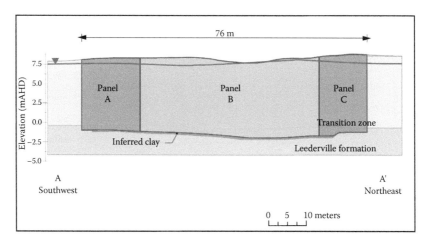

**FIGURE 7.9**
Preliminary design of the ZVI PRB.

suitable for the PRBs to be keyed into (Figure 7.9). Therefore, the preliminary design was for all panels to extend to a depth of approximately 10 m or until this upper portion of the Leederville Formation was encountered.

The preliminary design drawings, together with key performance indicators (KPIs), were included in the tender for the final design and construction contract.

The KPIs included the following criteria:

- Location—the top of the PRBs must be above the 100-year flood level and aligned as per the design specifications.
- Keyed into the silty clay at the top of the Leederville Formation
- Approximately 5 m distance between the denitrification and ZVI PRBs
- Wall thickness—continuous thickness greater than 30 cm, with no holes or gaps
- Monitoring wells—installation of three multilevel monitoring wells in the denitrification PRB and three multilevel monitoring wells in the ZVI PRB
- Appropriate hydraulic conductivity of both PRBs—five times greater than the geometric mean hydraulic conductivity along the PRB alignments
- Sawdust/woodchips and sand mixture—not less than 2:1 sawdust-to-sand ratio
- Specifications for the ZVI and sand mixtures for the central panel and side panels
- Effluent TCE concentrations downgradient of ZVI PRB to be less than 330 µg/L
- Capture contaminant plume—absence of underflow to be demonstrated by capture zone analysis using water level measurements and reference to downgradient concentrations

The successful tender was submitted by Menard Bachy and Geosolutions (the contractor) who proposed to install the sequenced PRB system using a biopolymer guar gum slurry. This method is a modified slurry trench technique that temporarily supports the trench excavation below the watertable using a biodegradable polymer instead of bentonite slurry. The slurry is made from a guar bean derivative comprising a mixture of polysaccharides (sugars, i.e., a long-chain carbon polymer), which is relatively stable in solution over a pH range of 5–7. To degrade the guar gum at the end of the PRB installation process, a slurry breaking agent is recirculated by air lifting in temporary wells.

The final design of the contract varied from the preliminary design in the following ways: the PRBs were moved away from the original alignment by approximately 4 m to the southeast (for operational reasons), the thickness of the PRBs was increased from 30 to 60 cm, and a clay cap was added to the top of the ZVI PRB.

Each PRB took approximately 7 days to install. The ZVI and sand mixture was mixed on surface, wetted and then placed in the trench using a tremmie system. The panels were installed through the use of a removable end stop. The mixture for the denitrification PRB was installed using a conventional progressive displacement technique since using the woodchips had the potential to clog the tremmie.

Some sections of the final PRBs were larger than the final design width of 60 cm. This was a result of sections of the trench collapsing due to guar gum stability issues as well as the removal of a number of unanticipated large boulders (large concretions). The geotechnical investigation that was undertaken along the original alignment did not encounter any such obstructions. The additional width resulted in additional costs associated with extra material requirements and manpower time. However, a benefit of the extra width is that the resulting retention times would likely be longer than designed. In one instance, a boulder was encountered at the base of the denitrification PRB. This boulder was left *in situ* as it was not possible to remove it and the sawdust and sand mixture emplaced around it.

## 7.7 Results

A long-term monitoring network was installed, following the completion of the PRBs. The monitoring network consisted of 48 sampling location (Figure 7.10). The network included a number of multilevel monitoring wells that were designated with the letters A, B, C, and D and which approximately represented 1 m screens at 3, 6, 9, and 12 m below ground level, respectively. The D series monitoring wells were all located in the upper portion of the Leederville Formation for the purpose of monitoring potential underflow of the system. Monitoring wells with 3 m screens across the watertable were installed along the ends of the system for the purpose of monitoring potential flow bypass.

Following the installation of the PRBs, eight quarterly monitoring rounds have been undertaken. The first occurred 3 months after the installation was completed. At each monitoring location, the following field measurements were obtained:

- Water level (depth to groundwater)
- Redox potential
- Dissolved oxygen
- pH
- Electrical conductivity

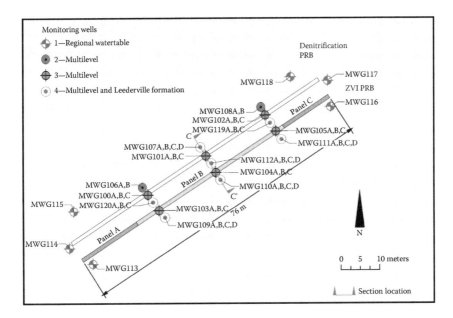

**FIGURE 7.10**
Long-term PRB monitoring network.

The first four monitoring rounds included sampling of wells located along the plume centerline and downgradient of the PRBs (locations MWG109, MWG110, MWG112, MWG107, MWG111, MWG115, MWG101, MWG106, and MWG108) for the following suite of analytes:

- VOCs: including at a minimum chlorinated ethenes, ethanes, and methanes; benzene; toluene; ethylbenzene; and xylenes
- Major ion chemistry: sulfate, chloride, bicarbonate, calcium, magnesium, sodium, potassium, and alkalinity
- Nitrate, nitrite, ammonia, and Kjeldahl nitrogen
- pH, electrical conductivity, TOC, total dissolved solids

The subsequent five monitoring rounds have focused on the B series monitoring wells, where the majority of the chlorinated solvent mass has been detected.

## 7.8 Groundwater

Groundwater levels upgradient of the remediation system have increased, resulting in larger hydraulic gradients across the PRB system. Additionally, groundwater levels within the denitrification PRB have been identified on a number of occasions as being a few centimeters higher than those upgradient, suggesting the possibility of flow bypass. However, the composition of groundwater samples collected from wells screened in the Leederville formation (D series monitoring wells) as well as at the ends of the PRBs (MWG113, MWG114, MWG115, MWG116, MAG117, and MWG118) of the system do not suggest that flow bypass is occurring.

The survey elevation of the denitrification wells was confirmed ensuring that water levels were calculated correctly. Field observations have suggested that groundwater in the denitrification and ZVI PRBs were initially slightly more viscous than background groundwater. This phenomenon was noticeably more pronounced within the denitrification PRB. The dynamic viscosity of a fluid is inversely proportional to the hydraulic conductivity of a media. Based on this fundamental understanding, a marginal increase in viscosity of a fluid could directly result in an increase of a few centimeters in water levels such as that noted within the denitrification PRB.

The increase in viscosity could be attributable to either bacteria, residual guar gum present in the groundwater, or a combination of both. Research by Rafai et al. (2010) has indicated that certain rod-shaped bacteria, which pull fluid toward themselves with their flagella and drag it along with them from behind, may increase fluid viscosity. This research has shown that increases in both living and dead cells will increase the viscosity; however, living cells result in a greater increase in viscosity. Rafai et al. (2010) found that as little as a 15% increase in concentration of living cells resulted in a doubling of the viscosity.

The TOC results indicate that despite the addition of a slurry breaking agent, residual guar gum has persisted longer than anticipated within both PRBs. TOC concentrations measured during the quarterly monitoring rounds for the B series wells along the plume centerline are presented in Figure 7.11.

Initially, concentrations of TOC were found to be significantly elevated at monitoring locations within both the denitrification and ZVI PRBs. Elevated TOC was anticipated in the denitrification PRB due to the addition of a carbon source to stimulate denitrification; however, initial field measurements were found to be 30 times greater than those measured in the laboratory column experiments.

Field observations noted during the quarterly monitoring rounds also identified that in addition to the increased viscosity, the groundwater at site had a strong organic odor, and when exposed to oxygen resulted in the formation of an unknown white precipitate.

It is thought that the initial high concentrations of TOC were associated with the guar gum and its decrease due to subsequent breakdown to shorter-chain, more labile carbon forms. With time, the concentrations of TOC in monitoring wells in the PRBs have decreased, as have the reports of odors,

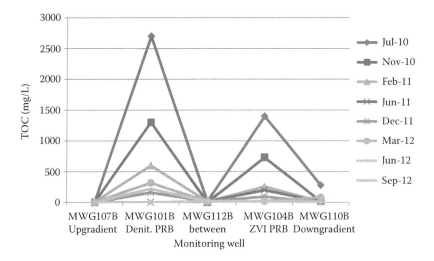

**FIGURE 7.11**
TOC concentrations measured along cross-section C–C' for the B series wells.

elevated viscosity, and precipitation. In the most recent round of monitoring, only a slight organic odor could be noted within some monitoring wells of the dentirification PRB. This may indicate that either the excess carbon has been flushed out of the system or more likely it is being consumed by bacteria within the system, with the sawdust now providing the predominant source of carbon within the system.

Yet another factor to be noted is the low concentration of TOC measured in monitoring well MWG112 at all depths, located downgradient of the denitrification PRB and upgradient of the ZVI PRB. This suggests that the natural aquifer material present between the two PRBs is acting as a buffer and is resulting in a reduction of TOC concentration in the influent groundwater to the ZVI PRB. This will likely minimize the potential passivation of ZVI due to DOC emanating from the denitrification PRB.

## 7.9 Nitrate Concentrations

Nitrate (as N) concentrations measured during the quarterly monitoring programs for B series wells along the plume centerline (Section C–C' in Figure 7.10) are presented in Figure 7.12. The results indicate that the denitrification PRB is effectively removing more than 99% of the nitrate (as N) from the influent groundwater at this depth. This ensures that passivation of the downgradient ZVI PRB by nitrate will be minimal.

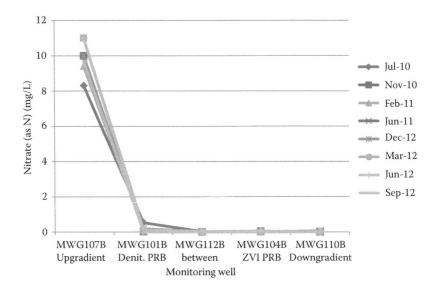

**FIGURE 7.12**
Nitrate concentrations measured along cross-section C–C′ for B series monitoring wells.

## 7.10 Chlorinated Solvent Concentrations

Concentrations of chlorinated solvents at monitoring locations at the ends of the PRBs (MWG13, MWG114, MWG115, MWG116, MWG117, and MWG118) and in the Leederville formation (D series sampling locations) indicate that the TCE plume is being captured by the remediation system. The concentration of TCE downgradient of the PRBs has been consistently below the ecological screening criteria.

Concentrations for TCE, DCE, and vinyl chloride (VC) measured during the quarterly monitoring program for the B series wells along the plume centerline (Section C–C′ shown in Figure 7.10) are presented in Figures 7.13, 7.14, and 7.15, respectively. Concentrations of TCE, DCE, and VC were converted into a molar mass and the A, B, and C series results summed to obtain a molar mass value for each location for the first four groundwater rounds and are presented in Figure 7.16. Owing to a decrease in the number of sampling location in subsequent quarterly monitoring rounds, only the total molar mass for the upgradient (MWG107) and downgradient (MWG110) sample locations are available and are presented in Figure 7.16.

Concentrations of chlorinated solvent within both the denitrification and ZVI PRBs are usually below detection limits but occasionally they are detected at levels only marginally above laboratory detection limits. Chlorinated compounds have been detected downgradient of the system. It is currently uncertain whether this is a rebound phenomenon or if samples collected from within

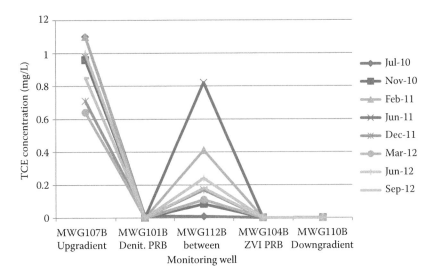

**FIGURE 7.13**
TCE concentrations along cross-section C–C′ for B series monitoring wells.

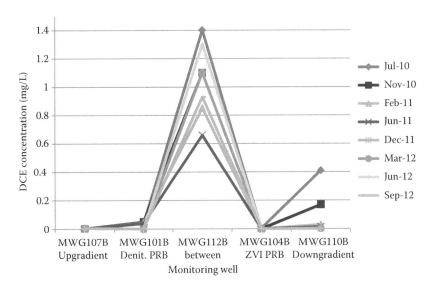

**FIGURE 7.14**
DCE concentrations along cross-section C–C′ for B series monitoring wells.

the PRBs have been underestimating the concentrations of chlorinated ethenes. Rebounding concentrations may result from desorption of chlorinated solvents from soil (Geosyntec 2007, Sale and Newell 2011). Underestimations of the concentrations of chlorinated ethenes could result from sorption to media or be related to issues with well construction or sampling techniques. In an

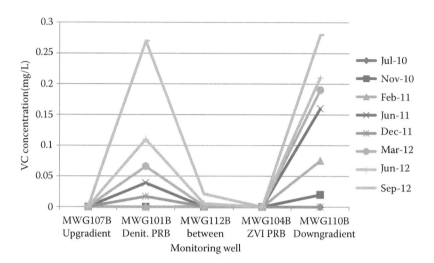

**FIGURE 7.15**
VC concentrations along cross-section C–C′ for B series monitoring wells.

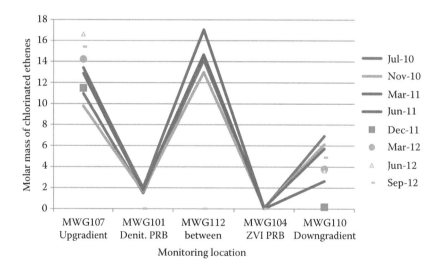

**FIGURE 7.16**
Total molar mass of chlorinated ethenes at each location along cross-section C–C′.

attempt to resolve this issue, two multilevel monitoring wells were installed in the ZVI PRB in June 2011. Subsequent monitoring results have confirmed the absence of chlorinated ethenes within the ZVI PRB.

To assess the effectiveness of the PRBs in reducing nitrate and remediating chlorinated ethenes, a comparison between the up- and downgradient monitoring data has been made (i.e., MWG107 with MWG112 for the denitrification PRB and MWG112 with MWG110 for the ZVI PRB). An examination of the data shows that concentrations of TCE are reduced and DCE concentrations are increased in the denitrification PRB (Figures 7.13 and 7.14). However, Figure 7.16 indicates that the denitrification PRB is not reducing the overall mass of chlorinated solvents, suggesting that the predominant method of TCE reduction within the denitrification PRB was through sequential dechlorination to DCE.

The detection of VC downgradient of the PRB remediation system may indicate that β-elimination may not be the preferred dechlorination process through the ZVI PRB. The most recent monitoring data (note scale change from Figures 7.13 and 7.14) indicate that concentrations of VC in MWG110B suggest that there is an upward trend with VC production higher in the second year of operation.

---

## 7.11 Eh and pH

Oxidation reduction potential (ORP) and pH were measured in the field at all monitoring locations during each monitoring round. The ORP was converted to Eh by adding 200 mV and the transformed results plotted on the Eh–pH stability diagram for iron (Figure 7.17). The figure indicates that the conditions are favorable for the formation of potentially passivating iron precipitates such as ferric hydroxide (Fe(OH)). To date, no trend in the Eh–pH relationship has been noted within the ZVI PRB.

---

## 7.12 Summary

The results from the monitoring rounds indicate that to date the remediation system has been successful. The denitrification PRB is successfully removing more than 99% of nitrates from the groundwater ensuring that the downgradient ZVI does not become passivated by nitrates. Measurements of TOC indicate that the natural aquifer strip between the two PRBs is likely to have acted as a buffer strip, minimizing the potential for ZVI to have been affected by TOC emanating from the upgradient denitrification PRB. TOC concentrations suggest that residual guar gum and its associated breakdown products remained within both the denitrification and ZVI PRBs for up to a year after installation.

While results indicate that β-elimination may not be the preferred pathway in the ZVI PRB, it is still degrading on average more than 60% of the chlorinated solvent mass. This has resulted in downgradient TCE concentrations at all monitoring locations to be below the RBC to protect the Helena River. Some elevated concentrations of VC have been detected immediately downgradient of the PRB system and further work has been undertaken to

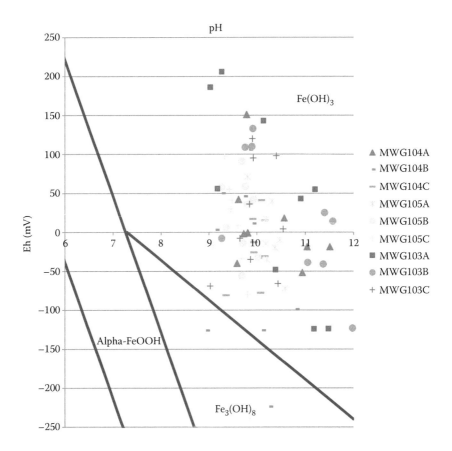

**FIGURE 7.17**
Field Eh and pH measured at ZVI PRB monitoring locations plotted on the iron Eh–pH stability diagram.

determine the cause. Monitoring during the second year of operation has included monitoring locations 50 m downgradient of the PRB, which to date have detected concentrations of VC only marginally above detection limits.

The results to date indicate that the sequencing of PRBs can be successful in targeting different chemicals of concern and should be considered when investigating potential remediation options.

# Acknowledgments

This project was conducted for LandCorp and the Department of Environment and Conservation of Western Australia. Funding for the project was provided from the Western Australian Contaminated Sites Management

Account. The authors would like to thank the many Golder employees and subcontractors who have worked on this project.

## References

Adventus. 2011. http://www.adventusgroup.com/products/prb.com.

Arnold, W.A. and A.L. Roberts. 2000. Pathways and kinetics of chlorinated ethylene and chlorinated acetylene reaction with Fe(0) particles. *Environmental Science & Technology*, 34, 1794–1805.

Carmichael, P.A. 1994. Using wood chips as a source of organic carbon in denitrification: A column experiment and field study implementing the funnel and gate design. M.Sc. thesis, Department of Earth Sciences, University of Waterloo, Waterloo, Ontario.

EnviroMetals. 2009. Bench-Scale Treatability Report in Support of a Granular Iron Permeable Reactive Barrier Installation at the Bellevue Site in Western Australia, ETI Reference: 32611.10, February 2009.

Eykholt, G.R. 1998. Analytical solution for networks of irreversible first-order reactions. *The Journal of the International Association on Water Quality*, 33(3), 814–826.

Fahrner, S. 2002. Groundwater nitrate removal using a bioremediation trench. Honours thesis, Department of Environmental Engineering, University of Western Australia.

Falta, R.W., P.S. Rao and N. Basu. 2005. Assessing the impacts of partial mass depletion in DNAPL source zones: I. Analytical modeling of source strength functions and plumes response. *Journal of Contaminant Hydrology*, 78(4), 259–280.

Farrell, J., M. Kason, N. Melitas and T. Li. 2000. Investigation of the long-term performance of zero-valent iron for reductive dechlorination of trichloroethylene. *Environmental Science & Technology*, 34, 514–521

GHD. 2007. Report for Mills Street Groundwater Treatment Trials. Final Evaluation Report Prepared for the Water Corporation, November 2007.

Geosyntec. 2007. Annual Monitoring and Demonstration of Compliance Report for 2006 Draft Volume I of II Somersworth Sanitary Landfill Superfund Site Somers Worth, New Hampshire, March 26, 2007.

The Interstate Technology & Regulatory Council (ITRC). 2005. Technical Regulatory Guidelines Permeable Reactive Barriers: Lessons Learned/New Directions, February 2005.

Rafai, S., L. Jibuti and P. Peyla. 2010. Effective viscosity of microswimmers suspensions. *Physical Review Letters*, 104(9): 098102.

Roberts, A.L., L.A. Totten, W.A. Arnold, D.R. Burris, and T.J. Campbell, 1996. Reductive elimination of chlorinated ethylenes by zero-valent metals. *Environmental Science & Technology*, 30(8): 2654–2659.

Robertson, W., D. Blowes, C. Ptacek, and J. Cherry. 2000. Long-term performance of in situ reactive barriers for nitrate remediation. *Groundwater* 38(5): 689–695.

Robertson, W.D., C.J. Ptacek, and S.J. Brown. 2007. Aquifer nitrate and perchlorate remediation using a wood particle layer. *Ground Water Monitoring and Remediation*, 27:85–95.

Robertson, W.D., C.J. Ptacek, and S.J. Brown. 2008. Rates of nitrate and perchlorate removal in a five-year-old woodparticle reactor treating agricultural nitrate. *Groundwater Monitoring and Remediation*, 29(2):87–94.

Robertson, W.D., J.L. Vogan and P.S. Lombardo. 2008. Nitrate removal rates in a 15-year-old permeable reactive barrier treating septic system nitrate. *Ground Water Monitoring and Remediation*, 28(3), 65–72.

Sale, T. and C. Newell. 2011. A Guide for Selecting Remedies for Subsurface Release of Chlorinated Solvents, ESTCP Project ER-200530, March 2011.

Vogan, J.L. 1993. The use of emplaced denitrifying layers to promote nitrate removal from septic effluent. M.Sc. thesis, Department of Earth Sciences, University of Waterloo, Waterloo, Ontario.

Water Corporation. 2004. Information Note on Groundwater Denitrification Trench, ISBN 1740432460, June 2004.

# 8

# Organic-Based Permeable Reactive Barriers for the Treatment of Heavy Metals, Arsenic, and Acidity

Ralph D. Ludwig, Richard T. Wilkin, Steven D. Acree,
Randall R. Ross, and Tony R. Lee

## CONTENTS

Three different organic-based permeable reactive barrier (PRB) systems were evaluated for their effectiveness in treating heavy metals, arsenic and acidity in groundwater. One PRB, a full-scale system consisting of composted cow manure and limestone gravel, was installed at a former battery recycling facility site located in Louisiana. The second PRB, a pilot, was installed at the same location in Louisiana. It consisted of composted cow manure, limestone gravel, and wood chips. The third PRB, also a pilot, was installed at a former phosphate fertilizer manufacturing facility site in South Carolina. It consisted of a mixture of municipal yard compost, zero-valent iron (ZVI), limestone, and granite pea gravel. In all three cases, the PRBs were designed to treat low pH, high acidity groundwater containing heavy metals and arsenic, by raising pH and promoting microbially mediated sulfate reduction.

## 8.1 Louisiana PRBs

Activities at the Louisiana battery recycling facility site included spent lead-acid battery demolition and lead smelting of battery lead plates to produce lead ingots. Initial remedial action at the site included excavation, treatment, and off-site disposal of approximately 32,400 cubic meters of impacted soils. A large low pH, high acidity, metal-laden groundwater plume remained on site and continued to discharge into a nearby creek. A full-scale PRB, consisting of 67% composted cow manure (v/v) and 33% limestone gravel, was installed at the site (by others) in May 2003 to intercept and treat the groundwater plume prior to entry of the groundwater into the creek (Figure 8.1). The dimensions of the PRB are 1.8 m wide, 4–5 m deep, and over 300 m in length. Prior to installation of the full-scale PRB, two pilot-scale PRBs were installed to evaluate two candidate organic carbon–limestone matrices. One pilot PRB consisted of the 67/33 cow manure/limestone mixture that was ultimately used in the full-scale PRB, while the other consisted of a 33/33/33 mixture of cow manure, wood chips, and limestone gravel.

Each of the two pilot PRBs measured 1.8 m wide, 4.3 m deep, and 30 m in length. The pilot PRBs and the full-scale PRB were monitored for 7 years with chemical monitoring being conducted annually. One transect through each of the two pilot PRBs and several transects through the full-scale PRB were monitored. The discussion presented herein focuses on the two pilot

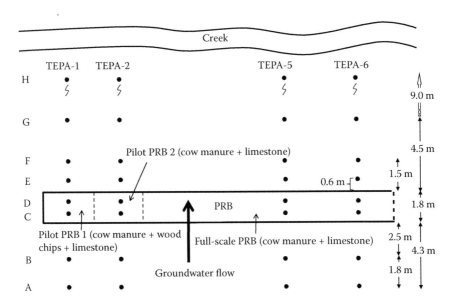

**FIGURE 8.1**
Schematic of PRB systems and transect configurations at Louisiana site.

PRB transects (TEPA-1 and TEPA-2) and two transects through the full-scale PRB (TEPA-5 and TEPA-6).

### 8.1.1 Methods

Each transect consists of 0.025-cm diameter Schedule 40 PVC screened wells over 1.5 m with two wells located within the PRB, two upgradient of the PRB, and four wells downgradient of the PRB. Wells were installed with a Geoprobe 6600 direct push unit with formation sediments being allowed to collapse around the wells following installation. The depth to the water table at the location of the PRB was measured at approximately 1.5 m. Groundwater samples were collected from the monitoring wells under low flow conditions using a peristaltic pump. Samples that were taken for cation and anion analyses were 0.45 μm filtered in the field and refrigerated until analyses. Cation samples were acidified with 12 M nitric acid following collection to pH < 2 and analyzed by ICP-OES and/or ICP-MS. Analyses for $SO_4^{2-}$ and $Cl^-$ were conducted using capillary electrophoresis. Both pH and oxidation-redox potentials (ORP) were measured in the field on unfiltered samples using combination electrodes. ORP data are reported as values relative to the standard hydrogen electrode ($E_H$). Alkalinity was measured in the field on unfiltered samples using a field titration kit (Hach® method 8221 equivalent to Standard Methods 2320 B). Sulfide measurements were made in the field on 0.45 μm filtered samples using Hach method 8131 (equivalent to Standard Methods 4500-$S^{-2}$ D). Dissolved organic carbon (DOC) samples were 0.45 μm filtered in the field, collected in 40 mL glass VOA vials with no headspace, and kept refrigerated until analyzed with a Dohrmann DC-80 carbon analyzer.

Hydrological studies at the Louisiana site included the measurement of groundwater elevations in wells within and surrounding the PRB, and the estimation of the hydraulic properties of the PRB and surrounding aquifer materials. The hydraulic conductivity structure within and adjacent to the PRB was estimated using physical slug testing techniques in approximately 70 wells. The tests were conducted based on methods proposed by Butler (1997). The methodology utilized slugs produced from PVC stock to initiate instantaneous changes in the head within the well, combined with high-frequency monitoring of the aquifer response using data loggers and pressure transducers. The aquifer response data were analyzed using the methods of Bouwer and Rice (1976) and Bouwer (1989).

### 8.1.2 Results

#### 8.1.2.1 Hydrologic Evaluation

The hydrologic data indicated that hydraulic conductivity in both the PRB and the native materials upgradient of the PRB was highly heterogeneous. The mean hydraulic conductivity of aquifer materials upgradient of the PRBs was estimated at $2.12 \times 10^{-3}$ cm/s based on measurements at 11 locations. Hydraulic conductivity measurements within the PRB varied over three orders of magnitude with a mean of $1.41 \times 10^{-3}$ cm/s. Hydraulic gradients in the aquifer immediately upgradient of the PRB were estimated using data from eight monitoring events between 2004 and 2009. The temporal head data for eight monitoring wells were evaluated using a "three point problem" approach, similar to that described by Delvin (2003). These data were used to determine the direction and magnitude of the hydraulic gradient for the centroid of six triangular elements described by the eight monitoring points. Using the available site-wide groundwater elevation data from the six monitoring events, the average magnitude of the hydraulic gradient toward the PRB was calculated at 0.0914 cm/cm. Based on these data and assuming an average effective porosity in the PRB between 0.1 and 0.2, the average seepage velocity of water moving through the PRB was 3–6 cm/d. However, there was significant uncertainty in the potential range of representative values for flow velocity through the PRB, due largely to the degree of heterogeneity observed.

### 8.1.2.2  Chemical Monitoring

Chemical monitoring data confirmed groundwater flow through the PRB, as evidenced by significant PRB-induced geochemical changes on the downgradient side of the PRB. The rate of advance of the ORP and alkalinity fronts on the downgradient side of the PRB systems (based on data collected 6 months following installation) indicated a groundwater flow rate of at least 2.5 cm/d. Selected chemical monitoring data are presented in Tables 8.1 and 8.2. The data show cumulative averages (over several years of monitoring) as well as the most recent 2010 results from the 33/33/33 cow manure/wood chip/limestone pilot PRB (TEPA-1), the 67/33 cow manure/limestone pilot PRB (TEPA-2), and the 67/33 cow manure/limestone full-scale PRB (TEPA-5 and TEPA-6).

### 8.1.2.3  pH

Data collected over a 7-year period following the installation of the pilot PRBs and full-scale PRB indicated the PRBs raised the pH of the groundwater above 6 (Table 8.1). The increase in pH within the PRBs is attributed primarily to dissolution of the limestone and production of bicarbonate associated with microbially mediated sulfate reduction reactions. The average pH values of groundwater entering the PRBs (average of upgradient [A + B] locations) ranged from 2.99 at TEPA-6 to 3.18 at TEPA-1, while pH values within the PRBs (average of [C + D] locations) ranged from 6.04 at TEPA-1 to 6.45 at TEPA-5. The most recent data (April 2010) showed pH values remained above 6 at all locations within the PRB, with the exception of location TEPA-1C in the cow manure/wood chip/limestone pilot PRB where a pH of 4.64 was measured.

**TABLE 8.1**

Data from Transects TEPA-1 and TEPA-2 through Pilot PRBs and Transects TEPA-5 and TEPA-6 through Full-Scale PRB

| | Cumulative 7-Year Average | | | | 2010 Data | | | |
|---|---|---|---|---|---|---|---|---|
| | TEPA-1 | TEPA-2 | TEPA-5 | TEPA-6 | TEPA-1 | TEPA-2 | TEPA-5 | TEPA-6 |
| pH (A/B) | 3.18 | 3.11 | 3.11 | 2.99 | 3.15 | 3.06 | 3.21 | 3.14 |
| pH (C/D) | 6.04 | 6.31 | 6.45 | 6.43 | 5.53 | 6.30 | 6.49 | 6.52 |
| Al (A/B) | 103 | 192 | 222 | 511 | 91 | 210 | 234 | 448 |
| Al (C/D) | 2.30 | 0.35 | 0.12 | 0.18 | 9.62 | 0.34 | 0.13 | 0.10 |
| % Al removed | 97.8 | 99.8 | 99.9 | 100.0 | 89.4 | 99.8 | 99.9 | 100.0 |
| mmol L$^{-1}$ yr$^{-1}$ Al removed | 34.1 | 64.7 | 75.1 | 173 | 27.5 | 70.9 | 79.0 | 151 |
| Fe (A/B) | 67.3 | 90.3 | 138 | 211 | 59.1 | 105 | 133 | 201 |
| Fe (C/D) | 35.6 | 45.2 | 8.97 | 9.46 | 16.5 | 18.3 | 23.4 | 16.4 |
| % Fe removed | 47.1 | 49.9 | 93.5 | 95.5 | 72.1 | 82.6 | 82.4 | 91.8 |
| mmol L$^{-1}$ yr$^{-1}$ Fe removed | 5.17 | 7.34 | 21.0 | 32.8 | 6.94 | 14.1 | 17.9 | 30.1 |
| Fe + Al + Mn + pH acidity (A/B) | 758 | 1302 | 1561 | 3307 | 676 | 1429 | 1618 | 2937 |
| Fe + Al + Mn + pH acidity (C/D) | 135 | 148 | 72 | 74 | 139 | 106 | 98 | 90 |
| % acidity removed | 89.0 | 93.3 | 98.9 | 99.4 | 86.4 | 97.4 | 97.2 | 99.0 |
| Alkalinity (C/D) | 533 | 814 | 981 | 1124 | 344 | 155 | 477 | 1100 |
| Net alkalinity (C/D) | 398 | 666 | 909 | 1050 | 205 | 49 | 379 | 1010 |
| H$_2$S (C/D) | 0.95 | 6.92 | 1.09 | 3.67 | 1.38 | 23.9 | 0.70 | 5.40 |
| SO$_4$ (A/B) | 1649 | 2761 | 3155 | 6874 | 1835 | 3360 | 3920 | 6435 |
| SO$_4$ (C/D) | 404 | 1245 | 299 | 692 | 770 | 2845 | 359 | 1465 |
| % SO$_4$ removed | 75.5 | 54.9 | 90.5 | 89.9 | 58.1 | 15.3 | 90.9 | 77.2 |
| mmol L$^{-1}$ a$^{-1}$ SO$_4$ removed | 118 | 144 | 271 | 587 | 101 | 49.0 | 338 | 472 |
| DOC (A/B) | 4.04 | 4.79 | 15.2 | 16.0 | 6.51 | 9.64 | 21.2 | 25.6 |
| DOC (C/D) | 19.0 | 30.6 | 55.2 | 69.9 | 9.28 | 22.9 | 16.3 | 32.9 |
| E$_H$ (A/B) in mV | 547 | 545 | 481 | 436 | 587 | 556 | 328 | 300 |
| E$_H$ (C/D) | 127 | 14 | 69 | −3 | 87 | −90 | 115 | 3 |
| Ca (A/B) | 52.1 | 93.1 | 93.2 | 129 | 64.3 | 94.3 | 124 | 153 |
| Ca (C/D) | 193 | 351 | 245 | 290 | 176 | 606 | 174 | 433 |

*Note:*  All concentrations are in mg/L; E$_H$ as mV; acidity as mg/L CaCO$_3$ equivalents; alkalinity as mg/L CaCO$_3$.

**TABLE 8.2**

Cumulative Average Yearly Concentrations (Cd, As, Pb Since 2006; Ni Since 2008) and 2010 Average Concentrations for Locations Upgradient and within PRB

| | Cumulative Yearly Average (μg/L) | | | | 2010 Data | | | |
| --- | --- | --- | --- | --- | --- | --- | --- | --- |
| | TEPA-1 | TEPA-2 | TEPA-5 | TEPA-6 | TEPA-1 | TEPA-2 | TEPA-5 | TEPA-6 |
| Pb (A/B) | 10.9 | 14.8 | 93.6 | 90.9 | 26.9 | 13.1 | 29.6 | 8.8 |
| Pb (C/D) | 0.369 | 0.560 | 0.892 | 0.329 | 0.256 | 0.832 | 0.726 | 0.170 |
| % Pb removed | 96.6 | 96.2 | 99.0 | 99.6 | 99.0 | 93.6 | 97.5 | 98.1 |
| Cd (A/B) | 302 | 461 | 254 | 136 | 384 | 470 | 113 | 17.0 |
| Cd (C/D) | 0.039 | 0.017 | 0.074 | 0.028 | 0.004 | 0.020 | 0.246 | 0.048 |
| % Cd removed | 100 | 100 | 100 | 100 | 100 | 100 | 99.8 | 99.7 |
| Ni (A/B) | 213 | 346 | 291 | 409 | 189 | 286 | 233 | 395 |
| Ni (C/D) | 26.5 | 23.7 | 9.57 | 17.3 | 24.4 | 4.89 | 2.25 | 2.94 |
| % Ni removed | 87.6 | 93.2 | 96.7 | 95.8 | 87.1 | 98.3 | 99.0 | 99.3 |
| As (A/B) | 6.33 | 13.8 | 139 | 52.3 | 7.78 | 9.59 | 58.7 | 30.8 |
| As (C/D) | 1.75 | 7.14 | 3.81 | 5.66 | 1.60 | 6.19 | 2.20 | 2.94 |
| % As removed | 72.4 | 48.1 | 97.3 | 89.2 | 79.5 | 35.5 | 96.3 | 90.5 |

*Note:* A/B denotes average of A + B concentrations upgradient of PRB; C/D denotes average of C + D concentrations within PRB.

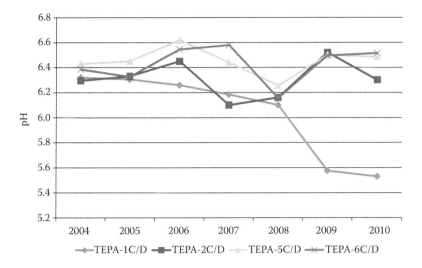

**FIGURE 8.2**
Average pH within PRB at four transect locations over time (TEPA-1C/D is through PRB with wood chips; TEPA-2, TEPA-5, and TEPA-6 are through PRB without wood chips).

The pH at TEPA-1C was as high as 6.40 in 2004 (the year following instal-lation of the PRB); however, a gradual decline in pH began in 2005 and was followed by a much sharper decline between 2008 and 2009. The decreas-ing pH trend at TEPA-1C coincided with increasing aluminum and nickel concentration trends at this location. The low pH at TEPA-1C also coincided with the lowest DOC concentration (6.4 mg/L) and highest $E_H$ (+216 mV) measured at any location within the PRB in 2010. Despite this apparent early sign of treatment failure in the upgradient half of the PRB along TEPA-1, a pH of 6.42 measured at TEPA-1D in 2010 in the downgradient half of the PRB indicated the PRB still effectively raises the pH of the groundwater prior to its discharge from the PRB (Figure 8.2).

Upon exiting the PRB, pH values remained elevated along all four transects; however, they decreased with distance from the PRB. The pH values at the furthermost H wells that were located along the shoulder of the downgradi-ent creek (see Figure 8.1) ranged from 5.93 at TEPA-2 to 6.18 at TEPA-6 in 2010.

### 8.1.2.4 Acidity

The removal of acidity (i.e., acid-producing capacity) from the groundwater is as important a component of the PRB treatment process as raising the pH of the groundwater, since acidity represents potential acid discharge to the downgradient creek. Groundwater acidity entering the PRBs was high and primarily attributable to the presence of dissolved-phase aluminum and fer-rous iron. Aluminum-based acidity accounted for approximately 80% of the total mineral acidity entering the PRB and in 2010 ranged from 506 mg/L as

CaCO$_3$ equivalents along TEPA-1 to a high of 2489 mg/L as CaCO$_3$ equivalents along TEPA-6. Table 8.1 indicates almost 100% removal of Al (presumably almost entirely as Al(OH)$_3$) along transects TEPA-2, 5, and 6 and nearly 90% removal along TEPA-1. Consistent with the decline in pH at TEPA-1C, aluminum concentrations increased to 9.62 mg/L at this location in 2010. This concentration, although still relatively low, was approximately 20 times the concentration observed within the PRB along any of the other three transects in 2010.

Iron-based acidity accounted for approximately 15% of the total mineral acidity entering the PRB and ranged in 2010 from 111 mg/L as CaCO$_3$ equivalents at TEPA-1 to 377 mg/L as CaCO$_3$ equivalents at TEPA-6. Major sinks for iron entering the PRB are expected to be ferrous sulfide (mackinawite) and ferrous carbonate (siderite) precipitation. Iron removal efficiency in 2010 ranged from 72.1% along TEPA-1 to 91.8% along TEPA-6. The removal of aluminum and iron-based acidity and corresponding precipitation of secondary minerals such as aluminum hydroxides, mackinawite, and siderite may have led to a gradual passivation of limestone surfaces. It was unlikely, however, that secondary mineral formation adversely impacted the hydraulic conductivity over time, given the relatively large sized limestone gravel used in the PRB mixtures and the gradual consumption of the organic substrate that occurs with time.

### 8.1.2.5 Alkalinity

The conversion of the impacted groundwater from an acid-producing potential to an acid-consuming potential is an important objective of the organic-based PRB systems. Alkalinity data indicated both pilot PRBs and the full-scale PRB were successful in converting the high acidity groundwater entering the PRB to an acid-consuming potential (Table 8.1). Alkalinities within the PRB based on April 2010 data ranged from 155 mg/L as CaCO$_3$ along TEPA-2 to 1100 mg/L as CaCO$_3$ along TEPA-6. Net alkalinities (total alkalinity minus [Fe + Al + Mn + pH] acidity) ranged from 49 mg/L as CaCO$_3$ along TEPA-2 to 1010 mg/L as CaCO$_3$ along TEPA-6. A gradual decline in alkalinity values within the PRB with time was observed along all four transects (Figure 8.3), which possibly indicates reduced sulfate reduction rates and/or increased passivation of limestone surfaces. Net alkalinity was maintained in groundwater downgradient of the PRB along each of the four transects although gradual reductions were observed with distance from the PRB. At H well locations near the creek, net alkalinity values in 2010 ranged from 229 mg/L as CaCO$_3$ at TEPA-2H to 524 mg/L as CaCO$_3$ at TEPA-6H.

### 8.1.2.6 Lead (Pb), Cadmium (Cd), and Nickel (Ni)

Metal sulfides are expected to precipitate out in accordance with their solubility products in the preferred order CuS > PbS > CdS> ZnS > NiS > FeS > MnS

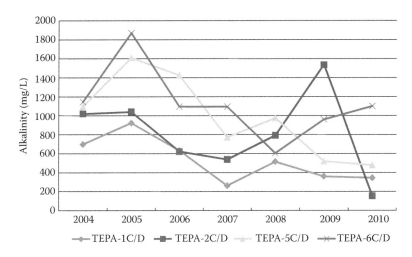

**FIGURE 8.3**
Alkalinity trends along transects within PRB.

(Hedin et al., 1994). Pb is thus expected to be one of the first metals pre-cipitated out under sulfate reducing conditions followed by Cd and Ni. Pb concentrations entering the PRB showed significant temporal and spatial variation since monitoring activities at the site were first initiated. The high-est average Pb concentrations entering the PRB (based on the average of the A + B locations along each transect between 2006 and 2010) were reported at TEPA-5, where concentrations averaged 93.6 µg/L (Table 8.2). The most recent data (2010) showed that the average of the A + B concentrations entering the PRB ranged from 8.8 µg/L at TEPA-2 to 29.6 µg/L at TEPA-5. With the excep-tion of one location within the PRB (at TEPA-2D where a Pb concentration of 1.46 µg/L was measured in 2010), Pb was consistently treated to concentra-tions less than 1.0 µg/L at all locations and to less than 0.5 µg/L at 5 of the 8 locations monitored within the PRB. The low Pb concentrations (<1 µg/L) were also maintained downgradient of the PRB along all four transects at the H locations near the creek.

The Cd concentrations entering the PRB far exceeded Pb concentrations and similar to Pb, have also shown significant temporal and spatial varia-tion. The average Cd concentrations entering the PRB ranged from 136 µg/L at TEPA-6 to 461 µg/L at TEPA-2. The most recent data (2010) indicated that cadmium concentrations entering the PRB ranged from 17.0 µg/L (TEPA-6) to 470 µg/L (TEPA-2). Within the PRB, nearly 100% treatment was achieved based on Cd concentrations being reduced to less than 0.3 µg/L along all four transects. As in the case of Pb, Cd concentrations remained at low con-centrations (<0.4 µg/L) downgradient of the PRB system.

Average Ni concentrations entering the PRB since 2008 ranged from 213 µg/L at TEPA-1 to 409 µg/L at TEPA-6, while Ni concentrations entering the PRB in

2010 ranged from 189 µg/L along TEPA-1 to 395 µg/L along TEPA-6. Within the PRB, Ni was generally treated to <5 µg/L with the exception of TEPA-1C and TEPA-2D, where concentrations of 47.1 and 7.46 µg/L were measured, respectively. The highest concentration of 47.1 µg/L measured at TEPA-1C coincided with the low pH of 4.64 also measured at this location in 2010. The increased presence of Ni relative to Pb and Cd under the declining pH conditions at TEPA-1C is consistent with the higher solubility of NiS relative to PbS and CdS. Ni in the downgradient half of the PRB at TEPA-1D was measured at 1.63 µg/L in 2010, indicating the PRB along TEPA-1 still remains effective in treating Ni. The overall Ni removal efficiency of 87.6% along TEPA-1 was significantly less than the greater than 98% removal efficiency observed along the other three transects in 2010. Ni concentrations were also higher than Pb and Cd concentrations downgradient of the PRB along all transects, with concentrations as high as 24.9 and 47.8 µg/L, respectively measured at TEPA-2H and TEPA-6E. At all other downgradient transect locations, Ni concentrations were less than 8 µg/L.

### 8.1.2.7 Arsenic

Average yearly arsenic (As) concentrations entering the PRB ranged from 6.33 µg/L at TEPA-1 to 139 µg/L at TEPA-5 (Table 8.2). Arsenate was the predominant form of arsenic entering the PRB while arsenate, arsenite, and thioarsenic forms were observed downgradient of the PRB, without any one form dominating. The highest arsenic concentration entering the PRB in April 2010 was 58.7 µg/L at TEPA-5. Within the PRB, arsenic was treated to less than 10 µg/L along all four transects. However, contrary to the heavy metals, arsenic concentrations rebounded downgradient of the PRB as shown in Figure 8.4, eventually exceeding concentrations upgradient of the PRB. This is presumed to be due to reductive dissolution of arsenic-containing iron minerals in the downgradient aquifer sediments. Arsenic concentrations at TEPA-2H near the creek were measured at 68.1 µg/L in 2010 relative to the average arsenic concentration within the PRB of only 6.2 µg/L. The value of 68.1 µg/L, however, represents a decrease from previous sampling rounds in 2009 and 2008 when arsenic concentrations at this same location were measured at 102 and 123 µg/L, respectively and in 2006 when the arsenic concentration at this location was measured at 230 µg/L. The data suggest that although arsenic was being mobilized from aquifer sediments downgradient of the PRB, the amounts being mobilized were gradually decreasing over time.

Arsenic was also monitored along a fifth transect across the full-scale PRB where concentrations within the PRB averaged 820 µg/L since 2006. This is despite sulfide concentrations within the PRB at this location having averaged 11.5 mg/L during the same period. Thioarsenic species accounted for over 25% of the dissolved-phase arsenic present within the PRB along this fifth transect. Iron concentrations within the PRB at this location, however, averaged only 1.5 mg/L since 2006, which suggests that low iron concentrations

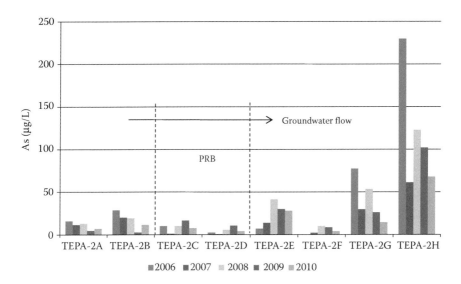

**FIGURE 8.4**
Arsenic concentration trends along TEPA-2 from 2006 to 2010.

may have been a factor in the high arsenic concentrations observed. A low iron concentration would be expected to limit arsenic removal by limiting coprecipitation reactions with iron sulfides, iron carbonates, and/or iron (oxy)hydroxides.

### 8.1.2.8 Sulfate/Sulfide

Sulfate concentrations entering the PRB ranged from 1649 mg/L at TEPA-1 to 6696 mg/L at TEPA-6, while concentrations entering the PRB in 2010 ranged from 1835 mg/L at TEPA-1 to 6435 mg/L at TEPA-6 (Table 8.1). Within the PRB, sulfate concentrations were significantly reduced, consistent with active sulfate reduction although in 2010, sulfate removal along TEPA-2 decreased significantly. Sinks for sulfate entering the PRB include formation of metal sulfides, elemental sulfur, precipitation as gypsum, and formation of organically bound sulfur (Ludwig et al., 2009). Given the generally high iron concentrations entering the PRB, mackinawite (FeS) was possibly the largest sulfate sink. Chemical equilibrium calculations indicated saturation conditions with respect to gypsum along TEPA-2. However, undersaturated conditions with respect to gypsum were indicated along the other three transects suggesting gypsum precipitation was not a significant sulfate sink. The 2010 data suggested only 15.3% of the sulfate was removed along TEPA-2, although sulfide concentrations within the PRB along TEPA-2 were consistently the highest, including a sulfide concentration of 23.9 mg/L in 2010. The sulfide data indicated that despite the apparently more limited sulfate removal occurring along TEPA-2, sulfate-reducing activity still remained sufficiently high enough to effectively remove incoming heavy metals and arsenic from the groundwater. Interestingly, contrary to the other three transects, calcium concentrations within the PRB along TEPA-2 increased steadily since 2007, despite pH remaining relatively constant. This suggests that the limestone in the PRB along TEPA-2 may have made an increasing contribution to pH buffering relative to the bicarbonate generated from microbially mediated sulfate reduction.

As might be expected, the PRB discharges a sulfide plume. The highest average sulfide concentration measured in H wells near the creek downgradient of the PRBs since 2004 was 1.2 mg/L at TEPA-6. The highest single measurement of sulfide at an H well location was 4.0 mg/L at TEPA-6H in 2007.

### 8.1.2.9 Dissolved Organic Carbon

DOC concentrations increased significantly within the PRB relative to upgradient locations as expected (Table 8.1). Since installation, the average DOC concentrations upgradient of the PRB ranged from 4.04 mg/L at TEPA-1 to 16.0 mg/L at TEPA-6. Within the PRB, average DOC concentrations since installation of the PRB ranged from 19.0 mg/L at TEPA-1 to 69.9 mg/L at TEPA-6. However, DOC concentrations within the PRB declined significantly with time (Figure 8.5), with concentrations in 2010 ranging from 9.28 mg/L at TEPA-1 to 33.0 mg/L at TEPA-6. Decreasing DOC concentrations were consistent with that observed for other organic carbon-based PRB systems and were likely a reflection of decreasing labile organic substrate content in the PRB with time. The initially high DOC concentrations may

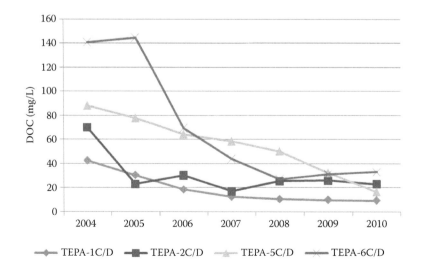

**FIGURE 8.5**
DOC trend over time within PRB along four monitored transects.

have also been the result of the biopolymer used in the construction of the PRB systems. The lowest DOC concentration measured within the PRB in 2010 was at TEPA-1C (6.4 mg/L), which coincided with the location where pH also decreased sharply over the past 2 years. DOC concentrations at all other locations monitored within the PRB in 2010 were greater than 10 mg/L and generally greater than 20 mg/L.

---

## 8.2  South Carolina PRB

Results for the pilot PRB installed at the former phosphate fertilizer production facility in South Carolina, (USA) have been previously described (Ludwig et al., 2009). The PRB at this location consisted of a mixture of granite pea gravel (45%), composted yard waste (30% v/v), ZVI filings (20% v/v), and limestone (5% v/v). The pilot PRB measured 7.9 m in length, 4.1 m in depth, and 1.8 m in width and was installed in pyrite-containing fill sediments which were disposed of at the site over decades of operation. The pilot PRB was installed near the edge of a tidal marsh located downgradient of the site. The ZVI was added to help neutralize pH, ensure removal of arsenic by promoting sorption of arsenic to ZVI surfaces, and to attempt to extend the life of the PRB through sustained production of $H_2$ for utilization by sulfate reducers. Column studies have indicated that the presence of ZVI in combination with organic carbon can enhance sulfate reduction rates by up to 15% and extend the life of an organic carbon matrix by nearly twofold (Guo and Blowes, 2009). Although the average pH of groundwater entering the pilot PRB (pH 3.68) was slightly higher than at the Louisiana site, the acidity of the groundwater entering the South Carolina pilot PRB was much higher than that of the Louisiana site. Iron acidity was greater than 25-fold while aluminum acidity was approximately twofold greater than at the Louisiana site. Vertically averaged concentrations of As, Pb, Cd, Ni, and Zn entering the South Carolina PRB (based on 48-month data) were 59.2, 0.744, 0.067, 0.481, and 215 mg/L, respectively. The PRB effectively removed heavy metals, acidity, and arsenic from the acid rock drainage-impacted groundwater, with concentrations of As, Pb, Cd, Ni, and Zn within the PRB (based on 48-month data) averaging 30, <3, <1, <1, and 37 µg/L, respectively. Iron concentrations were reduced from an average of 3384 mg/L to an average of 3.15 mg/L within the PRB, while aluminum concentrations were reduced from 520 mg/L to <17 µg/L within the PRB. Although dissolved iron concentrations entering the South Carolina PRB were high, average arsenic concentrations entering the same PRB were three orders of magnitude higher than average arsenic concentrations entering the Louisiana PRBs, thus making it conceivable that without ZVI, iron coprecipitation reactions alone would not have been sufficient to remove all of the arsenic.

Owing to pre-existing high levels of contamination on the downgradient side of the PRB at the South Carolina site, the discharge plume from the PRB could not be effectively monitored. However, as in the case of the Louisiana PRBs, the high acidity groundwater entering the PRB was effectively converted to a net-acid consuming water, with alkalinities as high as 650 mg/L as $CaCO_3$ measured within the PRB (based on 30-month data). Increases in $\delta^{34}S$ values, order of magnitude increases in sulfate reducing bacteria counts, and order of magnitude decreases in sulfate concentrations within the PRB provided strong evidence of microbially mediated sulfate reduction occurring.

The much higher total dissolved solids (TDS) concentrations (up to 35,000 mg/L) entering the South Carolina PRB relative to the Louisiana PRBs, coupled with the ZVI-induced higher average pH within the South Carolina PRB (up to pH 9.28), resulted in much larger scale precipitation of secondary minerals in the South Carolina PRB. This was reflected in decreased permeability observed within the South Carolina PRB over the course of just 1.5 years. The very high TDS concentrations and associated adverse impacts on PRB hydraulic conductivity resulted in a decision to preclude further consideration of such a PRB for full-scale use at the site. However, organic carbon combined with ZVI may offer significant benefits at other sites where TDS levels are not so unusually high.

## 8.3 Conclusions

The findings of the Louisiana PRB study indicated that a cow manure-limestone-based PRB system effectively removed heavy metals, arsenic, and acidity from acidic sulfate-containing waters; although a significant dissolved iron content entering the PRB may have been a prerequisite for removal of arsenic. The results also indicated that the cow manure-based PRBs showed persistence and performed well over a period of 7 years. The PRBs also produced an alkalinity discharge that helped to neutralize acid-impacted aquifer sediments downgradient of the PRBs. Less positive findings included the observed mobilization of arsenic downgradient of the PRB and observed groundwater mounding along the upgradient face of the PRB. Lower average hydraulic conductivity properties within the PRB were responsible for the buildup of a hydraulic head along the upgradient face of the PRB and resulted in some bypassing of groundwater around the ends of the PRB. This significant shortcoming in the design of the PRB indicated that any refurbishment efforts should include formulation of a higher permeability treatment matrix. An alternative but less preferable option would be to extend the length of the PRB to capture any potential bypass flow, as has been done by others (Jarvis et al., 2006).

Recent data have suggested that the cow manure/wood chip/limestone mixture may be beginning to fail. Reasons for this are unclear although it may be a result of the wood chips not being able to sustain microbially mediated sulfate reduction at an adequate rate compared to the cow manure. Alternatively, groundwater flow velocities may have been significantly faster in the presence of wood chips, thereby reducing the residence time in the PRB relative to the pilot PRB without the wood chips. Differences in hydraulic conductivity between the two mixtures using slug tests could not be discerned with any confidence in the study.

The addition of ZVI in an organic-based PRB system likely better ensures the removal of arsenic and also, likely enhances the longevity of organic-based PRB systems by providing a long-term source of $H_2$. However, ZVI-induced pH increases may also promote more clogging of PRB systems, particularly in the presence of highly mineralized groundwater.

---

# References

Bouwer, H. 1989. The Bouwer and Rice slug test—An update. *Ground Water* 27:304–309.

Bouwer, H. and R.C. Rice. 1976. A slug test method for determining hydraulic conductivity of unconfined aquifers with completely or partially penetrating wells. *Water Resources Research* 12:423–428.

Butler, J.J., Jr. 1997. *The Design, Performance, and Analysis of Slug Tests*. Lewis Publishers, Boca Raton, FL.

Delvin, J.F. 2003. A spreadsheet method for estimating best-fit hydraulic gradients using head data from multiple wells. *Ground Water* 41(3):316–320.

Guo, Q., and Blowes, D.W. 2009. Biogeochemistry of two types of permeable reactive barriers, organic carbon and iron-bearing organic carbon for mine drainage treatment: Column experiments. *Journal of Contaminant Hydrology* 107:128–139.

Hedin, R.S., R.W. Narin, R.L.P. Kleinmann. 1994. Passive treatment of coal mine drainage. *Bureau of Mines Information Circular* 9389, United States Department of the Interior.

Jarvis, A.P., M. Moustafa, P.H.A. Orme, P.L. Younger. 2006. Effective remediation of grossly polluted acidic, and metal-rich, spoil heap drainage using a novel, low-cost, permeable reactive barrier in Northumberland, UK. *Environmental Pollution* 143:261–268.

Ludwig, R.D., D.J.A. Smyth, D.W. Blowes, L.E. Spink, R.T. Wilkin, D.G. Jewett, C.J. Weisener. 2009. Treatment of arsenic, heavy metals, and acidity using a mixed ZVI-compost PRB. *Environmental Science and Technology* 43(6):1970–1976.

# 9

# Effective Cleanup of Groundwater Contaminated with Radionuclides Using Permeable Reactive Barriers

**Franz-Georg Simon and Tamás Meggyes**

## CONTENTS

## 9.1 Introduction

The mining of natural resources always causes environmental impacts such as land use, large quantities of waste, destruction of habitat, impairment of groundwater regime, and quite possibly contamination of soil, water, or air (Dudka and Adriano 1997). The environmental impact of mining cannot be estimated easily. The concept of total material requirement (TMR) is an attempt to quantify the environmental impact of materials. TMR is the sum of domestic and imported primary natural resources and their hidden flows (Adriaanse et al. 1997). Hidden flows are often not considered in environmental analyses because they are attributed with no cost. However, overburden from mining, earth moving for construction, and soil erosion are major sources of ecological damage. From the mining of minerals to the final products, a number of process steps take place (exploration, mine site development, extraction, milling, washing, concentration, smelting, refining, fabrication), each step being connected to other input flows such as energy or other resources. The hidden flows for metals related to metal ore mining have been investigated by Halada et al. (2001). Table 9.1 lists the data for some metals, together with the data for some other minerals and fuels. The TMR in the United States in the 1990s was between 80 and 100 tonnes per capita; in the European Union (EU-15) approximately 50 tonnes per capita (Bringezu 2002).

Although the TMR concept is widely accepted, there are no economic incentives to reduce TMR because the cost associated with the hidden flows is usually not known. However, such external costs can be calculated from the cost of the remediation for the former German uranium mining sites. In the western world uranium mining started in 1945 and reached a first maximum in 1957 and a second one in 1977, see Figure 9.1. Production of uranium from mining then decreased because fuel for nuclear power plants was partly produced by reprocessing nuclear weapons material. The cumulative global uranium production until 2004 was more than 2.5 million tonnes with Canada and the United States contributing more than 300,000 tonnes each and Germany more than 200,000 tonnes (OECD/NEA 2001, Taylor et al. 2006). The TMR of uranium is 11,000 kg/kg U (see Table 9.1).

After the end of uranium mining in Germany in 1990, remediation started with a focus on controlled flooding of underground mines, water treatment, backfilling of open pit mines, and treatment of mine tailings (Gatzweiler et al. 2002). The total cost for the remediation project was estimated to be 6.5 billion Euro (Bundesministerium für Wirtschaft und Technologie [D] 2000). So far, 4.2 billion Euro has been spent, while the project is scheduled to end sometime between 2010 and 2015. Based on a total production of 213,809 tonnes of uranium (Taylor et al. 2006) the external costs of uranium mining

**TABLE 9.1**

TMR of Some Metals, Minerals, and Fuels

| Material | TMR (t/t Material) | Global Production (t) | TMR (Mt/year) | References |
|---|---|---|---|---|
| Sand and gravel | 1.18 | 8,000,000,000 | 9440 | (2) |
| Hard coal | 2 | 3,740,000,000 | 8826 | (2) |
| Phosphate | 34 | 130,000,000 | 4420 | (2) |
| Gold | 1,800,000 | 2445 | 4401 | (1) |
| Crude oil | 1.22 | 3,485,000,000 | 4252 | (2) |
| Copper | 300 | 12,900,000 | 3870 | (1) |
| Iron | 5.1 | 571,000,000 | 2912 | (1) |
| Silver | 7500 | 160,000 | 1200 | (1) |
| Uranium | 11,000 | 45,807 | 504 | (1) |
| Lead | 95 | 2,980,000 | 283 | (1) |
| Platinum | 1,400,000 | 178 | 249 | (1) |
| Aluminum | 10 | 23,900,000 | 239 | (2) |

*Sources:* (1) Halada, K. et al. 2001. *Journal of the Japan Institute of Metals*, 65(7), 564–570. (2) Schmidt-Bleek, F. 1998. *Das MIPS-Konzept*, Droemer Knaur, München.

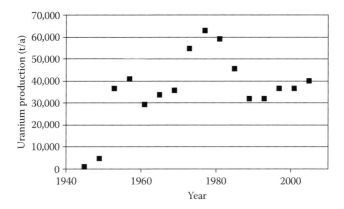

**FIGURE 9.1**
Uranium production in the Western world since 1945.

in Germany can be estimated to be 30 Euro/kg U. It should be noted that uranium concentration in German ores was much lower than those in ores exploited today (Mudd and Diesendorf 2008). The market price for uranium is far above this value, thus mining of uranium seems to be feasible in such a way that external costs can be covered by the price.

Minerals and ores are nonrenewable resources, so mining eventually leads to their exhaustion. The deposits cannot be regenerated, therefore, a remediation of the environmental damage at mining sites is a basic necessity. Polluting the environment has not been a privilege of our generation alone: even the ancient Greeks managed to pollute land and sea with the waste products of their lead mining, metallurgy, and processing some 2000 years ago. But increasing speed in development and industrial revolution multiplied the amounts and types of emissions, and kept adding more complex types of contaminant to the established ones: chemicals, toxic heavy metals, dense nonaqueous phase liquids (DNAPLs), radioactive wastes, and so on. In doing so, they have generated a steadily increasing threat to our environment, and our drinking water in particular. In the following sections remediation processes for treating groundwater contaminations arising from mining and utilization of radionuclides are described. A focus is laid on passive in situ methods. Active methods such as pump and treat are described in detail elsewhere (Simon et al. 2002).

## 9.2 Groundwater Remediation Using Permeable Reactive Barriers

Soil and groundwater remediation projects often cause high costs due to their large extent and long duration. This is the reason why a number of contaminated sites are not remediated at all. Passive in situ methods offer the advantage that after installation of the system only low operating costs occur and the environment is only disturbed to a small degree (Simon et al. 2005).

Passive in situ groundwater remediation using permeable reactive barriers (PRBs) is a relatively new and innovative technology with a high potential to significantly reduce the cost of treating contaminated shallow aquifers and therefore, contribute to the preservation of groundwater resources. A PRB is a subsurface structure situated across the groundwater flow path downstream of a contaminant source (Figure 9.2).

The barrier is a trench filled entirely or in part with a granular reactive material that is hydraulically permeable and reacts with the passing groundwater to remove or degrade the contaminants from the groundwater. Processes taking place in the reactive material of the barrier include physical, chemical, or biological contaminant retention or degradation and reactions of other groundwater constituents with the material. Suitable materials for use as reactive components in PRBs are elemental iron, activated carbon, zeolites, iron oxides/oxyhydrates, phosphates, clay minerals, and others. The most commonly used mechanisms are redox and sorption reactions. The choice of reactive materials and retention mechanisms depend upon the type of contaminant to be treated by the barrier system.

PRBs are defined by the US Environmental Protection Agency as passive in situ treatment zones of reactive material that degrades or immobilizes contaminants as groundwater flows through it. PRBs are installed as permanent, semipermanent, or replaceable units across the flow path of a contaminant plume. Natural gradients transport contaminants through strategically

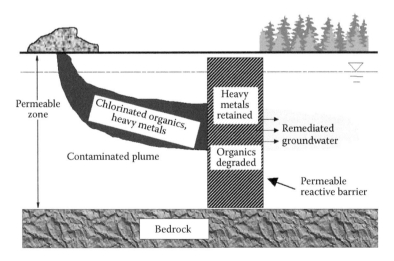

**FIGURE 9.2**
Scheme of PRBs.

placed treatment media. The media degrade, sorb, precipitate, or remove chlorinated solvents, metals, radionuclides, and other pollutants. The substantial deviation from common remediation techniques is that the contaminant plume, and not its source, is treated (Schad and Grathwohl 1998).

The concept of PRBs was first developed in North America, with pioneering work conducted at the University of Waterloo in Canada. Initially, the activities including first pilot field tests focused on "funnel-and-gate" systems and abiotic reductive dehalogenation of chlorinates and recalcitrant compounds by elemental iron (Gillham 1993, Starr and Cherry 1994, Vidic and Pohland 1996, Sivavec et al. 1997, Tratnyek et al. 2003). During the 1990s research activities on PRBs increased significantly leading to a number of new approaches in terms of PRB design, suitable reactive materials, and target contaminants.

Amongst the first and most widely studied metal compounds treated by PRBs are chromate (Blowes and Ptacek 1992, Powell et al. 1995, Blowes et al. 2000) and uranyl (Cantrell et al. 1995, Dwyer et al. 1996, Bostick et al. 2000), which are usually treated by reductive processes using, for example, elemental iron. The use of PRBs for groundwater protection or remediation has also been studied in other fields, for example, the treatment of metals-containing mine waters (Morrison and Spangler 1992, 1993, Waybrant et al. 1998, Benner et al. 1999, Naftz et al. 1999). The treatment of inorganic anions and cations can be grouped into abiotic reduction and immobilization (mostly by elemental iron), biologically mediated reduction and immobilization (bacterial sulfate reduction and precipitation of metals as sulfides), and adsorption and precipitation reactions (Blowes et al. 2000).

The selection of the reactive material to be used in a PRB depends on the type of contaminant and the remediation approach (contaminant removal mechanism). In general, contaminants can be removed from polluted water using the following processes:

- Precipitation: Immobilization of contaminants by formation of insoluble compounds (minerals), often after first reducing the contaminant to a less soluble species. The immobilized contaminants remain in the barrier material.

- Sorption: Immobilization of contaminants by adsorption or complex formation. The immobilized contaminants remain in the barrier material.

- Degradation (of organic pollutants): Application of chemical or biological reactions that lead to the decomposition of contaminants and the formation of harmless compounds which are either retained in the barrier or released downstream.

Frequently, groundwater treatment can involve a combination of these processes which cannot be individually distinguished. Nowadays, the most widely used approaches for PRBs can be grouped into two categories: reductive barriers and sorption barriers. Reductive barriers employ mechanisms that lead to the reduction of the target compound, or parts thereof, to achieve decomposition or immobilization of that compound. Barriers utilizing surface reactions that lead to immobilization of the target contaminants by adsorption, ion exchange, coprecipitation, solid-solution formation, and so on, without altering the chemical state of the contaminant are usually termed as sorption barriers.

In terms of geometry, two main types of PRB have been used in the field. These are (i) continuous reactive barriers enabling a flow through its full cross section, and (ii) "funnel-and-gate" systems (Starr and Cherry 1994) in which only special "gates" are permeable to the contaminated groundwater. The continuous PRB configuration is characterized by a single reactive zone installed across the contaminant plume, while the "funnel-and-gate" system consists of an impermeable wall that directs the contaminated plume through one or more permeable gates within the wall (Figure 9.3).

The choice between the two configurations depends on the hydrogeological characteristics of the site, the technical applicability of the barrier placement, and on the cost of the reactive material. When a high-cost reactive material is used, the "funnel-and-gate" configuration is preferable since the reactive zone requires less material. If a cheap material can be used, it is more profitable to avoid the construction of the impermeable sidewalls by employing a continuous barrier.

In PRBs, the residence time of the contaminant in the reactive material must be long enough to allow a decrease of the contaminant concentrations

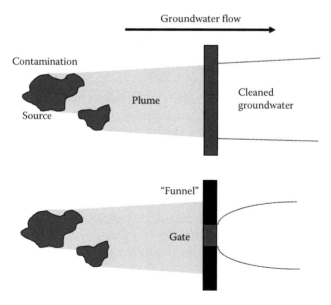

**FIGURE 9.3**
Funnel and gate versus continuous PRB.

down to an acceptable level (the remediation target). For a given contaminant and reactive material the required residence time is a function of the reaction rate and the equilibrium constant. As a PRB is basically a flow-through cell there is a continual reestablishment of equilibrium (or at least the tendency toward reestablishing equilibrium) as the groundwater passes through the barrier. For example, a given reactive material could reach equilibrium very rapidly with a contaminant, but if the initial concentration is high and the equilibrium constant is low, then a long flow path may be required to reduce the contaminant concentration to an acceptable level. Alternatively, a slow reacting material with a high equilibrium constant may reach an acceptable exit concentration in a relatively short residence time, without ever reaching equilibrium. As a range of factors may affect reaction rate, the retention time required to treat the groundwater at a particular site with a specific reactive material should always be determined in a feasibility study (e.g., by column experiments).

### 9.2.1 Removal of Metals from Groundwater

Cationic metals usually have limited mobility in soil and groundwater with high clay and organic content, high alkalinity, and low permeability (Fetter 1993). However, complexing agents such as carbonates, hydroxides, sulfates, phosphates, fluorides, and possibly silicates which are present in natural waters increase the solubility of metals (Langmuir 1978). Precipitation is a possible reaction for heavy metal contamination to lower the concentrations in the groundwater. Other possibilities are sorption, precipitation subsequent to a chemical reduction, or a combination of the different processes.

Precipitation of heavy metals is a commonly used process in wastewater treatment plants. Precipitation is used in PRBs in the same way to lower the heavy metal concentrations in groundwater. Chemical precipitation is a process by which a soluble substance is converted into an insoluble form by a reaction with the precipitant. Frequently used precipitants are hydroxides, sulfides, phosphates, and carbonates. The solubility of heavy metal hydroxides, sulfides, and carbonates is pH-dependent. Metal hydroxides exhibit amphoteric behavior, that is, solubility is high both at low pH (removal of hydroxide anions, see Reaction 9.2, with Me as a generic bivalent heavy metal) and high pH (formation of soluble hydroxo complexes, see Reaction 9.3). A minimum in solubility for most heavy metals can be observed between 9 and 11 (Chung 1989).

$$Me^{2+} + 2\ OH^- \rightarrow Me(OH)_2(s) \tag{9.1}$$

$$Me(OH)_2(s) + 2\ H^+ \rightarrow Me^{2+} + 2\ H_2O \tag{9.2}$$

$$Me(OH)_2 + OH^- \rightarrow Me(OH)_3^- \tag{9.3}$$

A cheap precipitant forming hydroxides is lime (Csövári et al. 2002). CaO was tested in lysimeter experiments with a duration of 3 years. 400 kg of uranium-containing waste from heap leaching (70 mg U/kg) was treated with a 1:20 mixture of CaO. The resulting uranium concentration was below 1 mg/L. Promising results were also achieved in subsequent field tests using horizontal barriers (1.5 kg CaO/t). Uranium concentrations lower than 1 mg/L, sometimes less than 0.1 mg/L are achievable.

The key advantage of PRBs is their low energy consumption and it is therefore important that the precipitates are not changed back by any means to soluble forms. Unlike in wastewater treatment plants, the precipitated material remains in the barrier for the whole period of operation. Spent barrier materials can be replaced by placing the reactive matrix in a double-walled structure of prefabricated elements (Beitinger and Fischer 1994). Stability of the resulting precipitates is therefore an important issue for the application of precipitation reactions in PRBs.

### 9.2.1.1 Uranium

Uranium is the heaviest naturally occurring element. All isotopes are radioactive; the half-lives of the two relevant isotopes $^{238}U$ and $^{235}U$ are $4.5 \times 10^9$ and $7.0 \times 10^8$ years, respectively (Seelmann-Eggebert et al. 1981). Uranium present in groundwater, for example, from mine tailings, is dangerous not because of its radioactivity but because of its toxicity as a heavy metal. Based on the German Radiation Protection Act, a 0.3 mg/L limit can be calculated from the radiation limit (7.0 Bq/L of the natural mixture of isotopes). Due to its toxicity an upper limit of 0.015 mg/L is being considered by the World Health Organization (WHO) (Birke et al. 2009).

Uranium occurs mainly in the oxidation states +4 and +6. The hexavalent uranium, that is, the uranyl ion $UO_2^{2+}$ and respective hydroxo- and carbonato-complexes are more mobile than U(IV) compounds, similarly to chromium where Cr(VI) has a higher mobility than Cr(III). The speciation of uranium is a complex system dependent on pH and carbonate concentration as can be seen in Figure 9.4 (calculated using MinteqA2 [Allison et al. 1991]). For more details on the complex uranium solution equilibria see Langmuir (1978).

Removal from groundwater is possible by reduction of U(VI) to U(IV) in a reducing environment, for example, by elemental iron (Cantrell et al. 1995):

$$Fe + UO_2^{2+}(aq) \rightarrow Fe^{2+} + UO_2(s) \tag{9.4}$$

The solubility of uraninite $UO_2$ is in the range of $10^{-8}$ mol/L in a pH range between 4 and 14. Below a pH of 4, uranium becomes soluble. A measurement on the dissolution of spent nuclear fuel in deionized water under non-oxidizing conditions provided results in the range of $10^{-9}$–$10^{-5}$ mol/L (Bickel et al. 1996). Under oxidizing conditions where $UO_2$ can be transformed

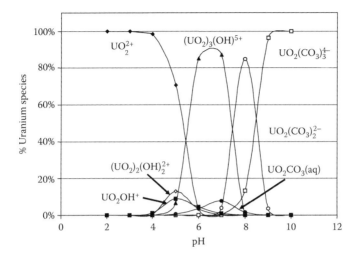

**FIGURE 9.4**

Uranium speciation in aqueous solution with atmospheric $CO_2$ and total uranium concentration of $5 \times 10^{-6}$ mol/L.

into the uranyl ion and if complexation reactions can occur, the solubility is enhanced. This can be seen from the pH–Eh diagram of uranium in the presence of carbonate in Figure 9.5 (calculated using HSC Chemistry [Roine 2007]). The reduction of U(VI) to U(IV) by elemental iron is thermodynamically possible over a wide pH range although stronger reducing conditions are needed compared to systems in the absence of or with low concentrations of carbonate. The stability fields of the U(VI)-carbonate complexes are larger than those of the respective hydroxo-complexes in the absence of or at low concentrations of carbonate (see dotted line in Figure 9.5).

Elemental iron as a reactive material for the removal of uranium has already been applied in field experiments (Naftz et al. 2002, Csövári et al. 2005a). In Hungary, uranium was mined and processed in the southern part of the country, in the Mecsek Mountains, near the city of Pécs. The mined out rock amounted to 46 million tonnes from which different wastes resulted. Mining activity was terminated at the end of 1997. Rehabilitation of the former industrial sites is in progress. Among the different remediation tasks, one of the most important issues is the restoration of the quality of groundwater contaminated by seepage from tailings, waste rock piles, and heap leaching residues. A pilot-scale PRB was built at the site near Pécs during the course of the EU project PEREBAR (Csövári et al. 2005b) (Figure 9.9). The PRB test facility has been in operation since August 2002. The PRB consists of two different zones: zone 1 with a lower content of coarse elemental iron (12% by volume or 0.39 t/m³, grain size 1–3 mm) and zone 2 with a higher content of fine elemental iron (41% by volume or 1.28 t/m³, grain size 0.2–3 mm). On both sides (upstream and downstream) sand layers were inserted to

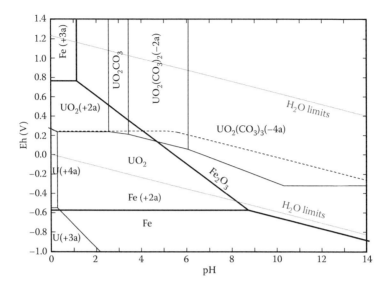

**FIGURE 9.5**

pH–Eh diagram showing the stability fields of uranium species ($c = 10^{-6}$ mol/L) in different oxidation stated in combination with stability filed of elemental iron, $Fe^{2+}$ and $Fe^{3+}$ (bold lines). The dotted line indicates the stability field of U(VI)-hydroxo complexes in the absence of carbonate. (From Roine, A. 2007. *HSC Chemistry V.6.12*. Outotec Research Oy, Finland.)

distribute water inflow and outflow. The PRB is sealed with clay and geosynthetic clay liners at the bottom and with a geomembrane (high-density polyethylene) at both sides and on the top. The design is displayed in Figure 9.6. The total mass of elemental iron installed as reactive material was 38 t, from which 5 t was coarser material.

The pilot installation has yielded good removal results for uranium (99%) and is still in operation. Table 9.2 lists average values for different parameters for the operation period. It can be seen that the pH increased in the PRB by almost 2 units, uranium concentration decreased by at least 95%, TDS was halved and calcium was almost completely removed. The impact of precipitates on the long-term performance of the installation will be investigated in further research.

Monitoring data of the site over a period of 4500 days (12 years) is shown in Figure 9.7. This includes data about the performance of the experimental PRB.

The removal mechanism of uranium in PRBs with elemental iron as reactive material is discussed controversially (Fiedor et al. 1998, Gu et al. 1998, Noubactep et al. 2003). Although reductive precipitation of U(VI) by elemental iron (see Reaction 9.4) is thermodynamically possible, evidence exists that coprecipitation on aging iron corrosion products is the main removal mechanism (Noubactep et al. 2006). On two DOE sites in the United States, U(VI) was found the prevailing species after more than 1 or 3 years of operation of a $Fe^0$ reactive barrier, respectively (Matheson et al. 2003). The presence

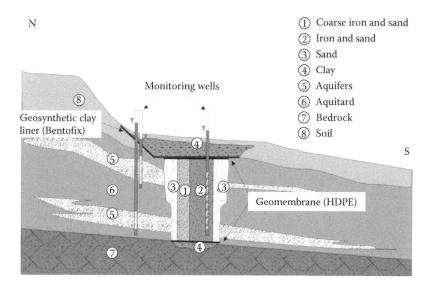

N

1. Coarse iron and sand
2. Iron and sand
3. Sand
4. Clay
5. Aquifers
6. Aquitard
7. Bedrock
8. Soil

Monitoring wells

Geosynthetic clay
liner (Bentofix)

Geomembrane (HDPE)

S

**FIGURE 9.6**
Sketch of the design of the experimental PRB near Pécs. (Reprinted from *Long-Term Performance of Permeable Reactive Barriers, Trace Materials and Other Contaminants in the Environment*, Vol. 7, Csövári, M. et al., Experimental iron barrier in Pecs, Hungary case study, 261–281, Copyright 2005b, with permission from Elsevier.)

of reduced uranium species could be the result of biogenic processes (Duff et al. 2002, Schöner et al. 2009, Kelly 2010).

Lowering the uranium concentration without changing the oxidation state is possible by the precipitation of sparingly soluble uranyl phosphates. The addition of phosphate, hydroxyapatite (HAP), or bone char (HAP with a small amount of carbon) to the water may trigger the formation of uranyl phosphate $(UO_2)_3(PO_4)_2$ (log $K_{sp} = -49.09$ [Brown et al. 1981]), autunite $Ca(UO_2)_2(PO_4)_2$ (log $K_{sp} = -47.28$, [Brown et al. 1981]) or chernikovite $H_2(UO_2)_2(PO_4)_2$ (log $K_{sp} = -45.48$, [Grenthe et al. 1992]). The mechanism of the interaction of uranium with HAP is not yet completely understood. Jeanjean et al. (1995) proposed a dissolution–precipitation mechanism. With either autunite or chernikovite as the precipitation product, the reaction may occur via the sequence displayed below.

$$Ca_5(PO_4)_3OH \rightleftharpoons 5Ca^{2+} + 3PO_4^{3-} + OH^- \tag{9.5}$$

$$H^+ + OH^- \rightleftharpoons H_2O \tag{9.6}$$

$$2UO_2^{2+} + Ca^{2+} + 2PO_4^{3-} \rightleftharpoons Ca(UO_2)_2(PO_4)_2 \tag{9.7}$$

$$2UO_2^{2+} + 2H^+ + 2PO_4^{3-} \rightleftharpoons H_2(UO_2)_2(PO_4)_2 \tag{9.8}$$

**TABLE 9.2**

Monitoring Data of the PRB as of March 22, 2005, after 31 Months of Operation in the Pécs, Hungary, Experimental Barrier

| Well | Location | pH | EC µS/cm | U(VI) µg/L | TDS mg/L | Ca mg/L | SO$_4$ mg/L | HCO$_3$ mg/L | Fe mg/L |
|---|---|---|---|---|---|---|---|---|---|
| PRB-1 | Inflow | 6.9 | 1400 | 940 | 1010 | 150 | 320 | 525 | 0.002 |
| PRB-10 | Zone 1 | 7.3 | 1330 | 37 | 937 | 125 | 300 | 275 | 5.5 |
| PRB-11 | Zone 2 | 8.7 | 865 | 10 | 550 | 10 | 185 | 299 | 0.03 |
| PRB-14 | Outflow | 8.7 | 780 | 2 | 540 | 14 | 170 | 268 | 0.03 |
| PRB-3[a] | Downstream | 8.5 | 420 | 56 | 240 | 5 | 95 | 116 | 0.017 |

*Source:* From *Long-Term Performance of Permeable Reactive Barriers, Trace Materials and Other Contaminants in the Environment*, Vol. 7, Csővári,, M. et al., Experimental iron barrier in Pecs, Hungary case study, 261–281, Copyright 2005b, with permission from Elsevier.

[a] Water is probably diluted by lateral water flow.

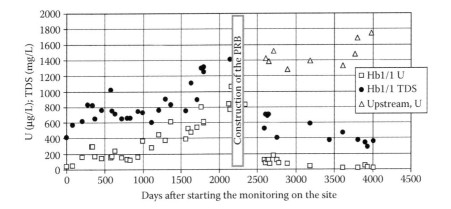

**FIGURE 9.7**
Long-term performance of the PRB near Pécs: uranium concentration and total dissolved solid (TDS) in the observation well Hb1/1 (15 m downstream of the PRB). (From Csövári, M. 2009, personal communication.)

Ion-exchange processes (see Reaction 9.9) or surface sorption using HAP (two possible surface groups, Reactions 9.10 and 9.11) are described by the following equations (Wu et al. 1991, Leyva et al. 2001).

$$\equiv Ca^{2+} + UO_2^{2+} \rightleftharpoons \equiv UO_2^{2+} + Ca^{2+} \tag{9.9}$$

$$\equiv OH + UO_2^{2+} \rightleftharpoons \equiv O\text{-}UO_2^+ + H^+ \tag{9.10}$$

$$\equiv O_3P\text{-}OH^+ + UO_2^{2+} \rightleftharpoons \equiv O_3P\text{-}O\text{-}UO_2^{2+} + H^+ \tag{9.11}$$

In the work of Fuller et al. (2002) autunite and chernikovite have been identified as solid phases after adding uranyl ions to a saturated solution of hydroxyapatite. Evidence for adsorption of U(VI) to hydroxyapatite surfaces as an inner-sphere complex was found for certain concentration ratios. Similarly uranium phosphates have been found in laboratory experiments on uranium removal from artificial groundwater (Simon et al. 2008), see Figure 9.8.

Uranium can also be removed by adsorption on surfaces. Morrison and Spangler (1992) have evaluated a range of uranium and molybdenum adsorption tests using a variety of materials. Good removal results have been obtained using lime, hematite, peat, ferric oxyhydroxide, phosphate, and TiO$_2$ while clays exhibited low sorption potential. Precipitation and adsorption onto a surface are processes that can simultaneously occur in a chemical barrier. The sorption of uranium from groundwater was studied in a series of publications (Morrison and Spangler 1992, 1993, Morrison et al.

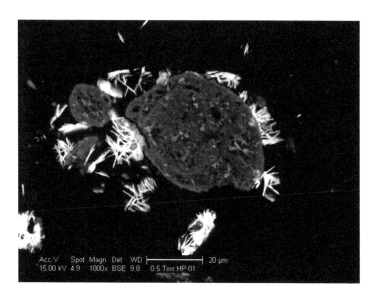

**FIGURE 9.8**
Scanning electron microscopic image recorded in backscatter mode (heavy elements appear bright) from HAP particles with uranium phosphate crystals (verified by the attached EDX system). (From Simon, F.G., Biermann, V. and Peplinski, B. 2008. *Applied Geochemistry* 23(8), 2137–2145.)

1995). Surface site complexation can be described by different models with or without electrostatic influence on charged surfaces (Allison et al. 1991). With SOH as a notation for a surface site, the adsorption reaction of uranium onto ferric oxyhydroxide as sorbent can be written as follows:

$$SOH + UO_2^{2+} + 3H_2O \rightleftarrows SOH - UO_2(OH)_3 + 2H^+ \qquad (9.12)$$

Sorption, rather than precipitation, depends strongly on pH. If several contaminants should be removed from the groundwater, an optimum pH for operation of the barrier is needed. This is difficult to achieve for uranium in the presence of molybdenum because the latter is mobile at pH values above 8 while uranium exhibits low mobility (Morrison and Spangler 1993). The distribution coefficient $K_d$ of uranium at different pH values in various soils is displayed in Figure 9.9 (Office of Radiation and Indoor Air and Office of Environmental Restoration 1999). $K_d$ is defined as the ratio of mass of adsorbate sorbed (mg/kg) to mass of adsorbate in solution (mg/L). High $K_d$ values (l/kg) were derived from adsorption experiments with ferric oxyhydroxide and kaolinite, low values from those with quartz which has low adsorptive properties. The pH dependence arises from surface charge properties of the soil and from the complex aqueous speciation of U(VI).

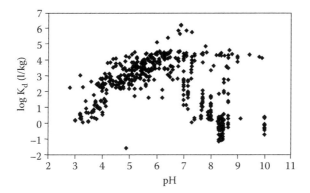

**FIGURE 9.9**
Distribution coefficient $K_d$ of uranium as a function of pH.

## 9.2.2 Remediation by Biomineralization

Reducing conditions or conditions leading to precipitation of metals or radio-nuclides from groundwater can also be established by natural processes that are, in principle, similar for all metals and radionuclides; in the following, the sequence is described for uranium (Kalin et al. 2004). In a first step, uranium is associated with organic material (e.g., plants, algae, or microbes). This can occur via adsorption, active cellular uptake, or bioprecipitation. Microorganisms are able to adsorb uranium from solution. For *Pseudomonas*, the distribution coefficient $K_d$ was found to be above 8000 (Sar and D'Souza 2001) and for *Aspergillus fumigatus* around 10,000 (Bhainsa and D'Souza 1999). This is comparable to adsorption on soils at pH values from 5 to 7 (see Figure 9.9). Uranium can also be actively taken up into the cells although the radius of uranium is comparably high (Lloyd and Macaskie 2002). Bioprecipitation may take place outside or inside the cells. Sparingly soluble metal hydroxides, carbonates, sulfides, or phosphates may be formed (Kalin et al. 2004). Then, sedimentation of the associated structures occurs. Finally, biomineralization takes place by the provision of conditions with low redox potential in the sediments with metal-reducing microbial populations. In these anaerobic sediments uranium remains stable and no redistribution occurs for millennia (Edgington et al. 1996). Possible interactions of radionuclides with microorganisms are summarized in Figure 9.10.

In natural and constructed wetlands biomineralization processes are responsible for the removal of uranium (Schöner et al. 2009) and other radionuclides like radium or heavy metals like arsenic (Groudev et al. 2008). Seasonal and hydrological fluctuation in wetlands may have an influence on the long-term performance regarding pollutant removal and could lead to remobilization. However, evidence exists that sediment aging leads to more stable immobilization (Schöner et al. 2009).

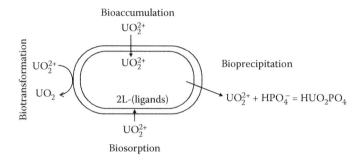

**FIGURE 9.10**
Radionuclide interactions with microorganisms. (Adapted from *Interactions of Microorganisms with Radionuclides*, Lloyd, J.R. and Macaskie, L.E., Biochemical basis of microbe-radionuclide interactions, 313–342, Copyright 2002, with permission from Elsevier.)

## 9.3 Construction Methods of PRBs

First, cut-off wall construction techniques were applied to PRBs. Single- and two-phase diaphragm walls, bored-pile walls, jet grouting, thin walls, sheet-pile walls, driven cutoff walls, injection, and frozen walls are the most common cutoff wall alternatives. To date, in addition to using cutoff wall construction methods, an increasing number of innovative techniques are being used to construct PRBs such as: drilling methods; deep-soil mixing; high-pressure jet technology; injected systems; column and well arrays; deep aquifer remediation tools (DART); and hydraulic fracturing and biobarriers. The main configurations of PRBs are

- Continuous reactive barriers
- Funnel-and-gate systems
- Arrays of wells
- Injected systems

The basic performance requirements for a reactive zone within a PRB are (Beitinger and Bütow 1997, Smyth et al. 1997, Beitinger 1998)

- Replaceability of the reactive materials
- Higher permeability than the surrounding reservoir
- Stability against fines washed into barriers from the surrounding soil
- Long life-span

The selection of the construction technique mainly depends on site characteristics (Gavaskar 1999), for example,

- Depth (the most important factor): increasing depth requires more specialized equipment, longer construction times, and is accompanied by higher costs
- Geotechnical considerations: soil/rock strength and presence of obstacles
- Soil excavation: handling and disposal of (contaminated) soil
- Health and safety during construction (entry of personnel into excavations)

The design of PRBs, as with any other technology, should meet the requirements of the best available technique (BAT). In most cases, it may be advantageous to develop a reliable conceptual site model and to perform pumping and treatability tests. The conceptual design report should include the following information (Beitinger 2002):

- The amount and type of any emissions from the remediation scheme and details of any emission control measures
- The volume and quality of any discharge or re-infiltration of treated groundwater
- The power consumption (electricity, fuels, etc.)
- A description of any waste streams generated, and details of their disposal
- Quantification of any material inputs, such as GAC, lime, and so on
- The remediation target levels
- The anticipated overall efficiency
- The anticipated maintenance requirements (manpower, parts)
- The monitoring requirements
- A detailed cost estimate (including the capital costs, construction costs, operating costs, and decommissioning costs)
- A detailed health and safety evaluation of the project

## 9.4  Electrochemical Remediation

Technologies used to remediate contaminated soils, sediments, and groundwater based on physicochemical, thermal, and biological principles are often costly, energy-intensive, ineffective, and create adverse environmental impacts (Sharma and Reddy 2004). Low permeability and heterogeneities in the layers and contaminant mixtures very often result in poor remediation results. Electrochemical remediation is a promising approach in difficult site conditions and has been extensively researched worldwide.

To implement an electrochemical technology, first wells are drilled around the contaminated region. Electrodes are installed in the wells and a low direct current or low potential gradient is applied to the electrodes (Reddy and Cameselle 2009). The generated electric field induces several transport, transfer, and transformation processes moving the contaminants to the electrodes where they can be removed. Also, the contaminants may be immobilized or degraded within the contaminated site. Other terms used for electrochemical remediation include *electrokinetics, electrokinetic remediation, electroremediation, electroreclamation,* and so on. If only water is used around the electrodes, the process is called unenhanced electrochemical remediation. Enhanced electrochemical remediation uses conditioning solutions and ion exchange membranes.

Key advantages of electrochemical remediation are

- Flexibility in terms of choosing ex situ or in situ
- Applicability to low-permeability and heterogeneous layers (clay, silt, loess, etc.)
- Both saturated and unsaturated soils can be treated
- Heavy metals, radionuclides, organic contaminants, and their mixtures can be treated
- It can be integrated into conventional systems

The complex character of the transport, transfer, and transformation processes and the influence of buffer capacity, mineralogy, organic matter content, geochemistry soil–contaminant interactions and heterogeneity on the efficacy require extensive investigations in the design phase. All this is necessary to guarantee successful implementation of electrochemical remediation technologies.

### 9.4.1 Electrochemical Processes

The fundamental transport mechanisms caused by an electric field are (Gregolec et al. 2005)

- Electromigration (transport of charged ions or ion complexes in solution)
- Electroosmosis (movement of liquid relative to a charged stationary surface)
- Electrophoresis (movement of charged particles relative to a stationary fluid)

Electromigration and electroosmosis occur in fine-grained soils, electromigration and electrophoresis are dominant in coarse-grained soils.

In addition to mass transport processes, chemical reactions take place at the electrodes. The principal electrode reaction observed is the electrolysis of water. At the anode, water is oxidized and oxygen gas and hydrogen ions are generated:

$$2H_2O \rightarrow O_{2(gas)} + 4H^+_{(aq)} + 4e^- E^0 = -1.229\,V \tag{9.13}$$

At the cathode, water is reduced and hydrogen gas and hydroxide ions are produced:

$$4H_2O + 4e^- \rightarrow 2H_{2(gas)} + 4OH^-_{(aq)} E^0 = -0.828\,V \tag{9.14}$$

This means that acid is produced at the anode; thus, pH is reduced and alkaline solution is produced at the cathode, that is, pH is increased. $H^+$ ions leaving the anode and $OH^-$ ions leaving the cathode change pH of the soil considerably (Acar and Alshawabkeh 1993).

### 9.4.2 Removal of Radionuclides

Improper handling of nuclear wastes and spent fuel, activities at nuclear fuel production plants and nuclear waste processing plants, accidents involving nuclear materials, especially those in nuclear reactors, and nuclear weapons tests have caused severe contamination to the soil and groundwater on a number of occasions (Korolev 2009). Radionuclides may enter the soil and/ or groundwater during storage, transport, and use of nuclear fuel and waste. Before being abolished, nuclear tests were a major source of nuclear contamination. The most serious contaminations have been caused by the isotopes $^{60}Co$, $^{90}Sr$, $^{90}Y$, $^{106}Ru$, $^{137}Cs$, $^{144}Ce$, $^{147}Pm$, $^{238}Pu$, $^{239}Pu$, $^{240}Pu$, and $^{226}Ra$. The severity of contamination is largely determined by the toxicity of the radioactive elements involved (Table 9.3).

Most soil decontamination technologies used are based on flushing soils with various chemicals, chemical leaching, and selective extraction of radionuclides. The main advantage of the new electrokinetic technique is that it can be applied in situ to soils of low filtration ability.

Another advantage of electrokinetic techniques is that they easily can be integrated (coupled) with other technologies promoting their advantageous features by synergistic effects and suppressing their shortcomings. Such integrated technologies include

- Electrokinetic biobarriers
- Electrolytic reactive barriers
- Electrokinetic-PRBs
- Electrokinetic-chemical oxidation/reduction

**TABLE 9.3**

Toxicity of Radioactive Elements

| Degree of Toxicity | Radioactive Element |
| --- | --- |
| Extremely high | $^{210}Pb$, $^{210}Po$, $^{226}Ra$, $^{228}Th$, $^{232}Th$, $^{232}U$, $^{237}Np$, $^{238}Pu$, $^{239}Pu$, $^{241}Am$, $^{242}Cm$ |
| High | $^{90}Sr$, $^{106}Ru$, $^{124}Sb$, $^{126}I$, $^{129}I$, $^{144}Ce$, $^{170}Tm$, $^{210}Bi$, $^{223}Ra$, $^{224}Ra$, $^{227}Th$, $^{234}Th$, $^{230}U$, $^{233}U$, $^{234}U$, $^{235}U$, $^{241}Ru$ |
| Medium | $^{22}Na$, $^{24}Na$, $^{32}P$, $^{35}S$, $^{36}Cl$, $^{54}Mn$, $^{56}Mn$, $^{59}Fe$, $^{60}Co$, $^{82}Br$, $^{89}Sr$, $^{90}Y$, $^{91}Y$, $^{95}Nb$, $^{95}Zr$, $^{105}Ru$, $^{125}Sb$, $^{132}I$, $^{133}I$, $^{134}I$, $^{134}Cs$, $^{137}Cs$, $^{141}Ce$, $^{171}Tm$, $^{203}Pb$, $^{206}Bi$, $^{231}Th$, $^{239}Np$ |
| Moderate | $^{14}C$, $^{38}Cl$, $^{55}Fe$, $^{64}Cu$, $^{69}Zn$, $^{71}Ge$, $^{97}Zr$, $^{131}Cs$, $^{136}Cs$ |
| Low | $^{3}H$ |

*Source:* Korolev, V.A.: *Electrochemical Remediation Technologies for Polluted Soils, Sediments and Groundwater.* 127–135. 2009. Copyright Wiley-VCH Verlag GmbH & Co. KGaA.

- Electrokinetic-bioremediation
- Electrokinetic-phytoremediation
- Electrokinetic-stabilization
- Electrokinetic-thermal treatment

Most of the integrated methods have been tested for the removal of organic contaminants, but a few of them, especially electrokinetic-PRBs, electrokinetic-phytoremediation, and electrokinetic-stabilization proved efficient for the remediation of sites contaminated with heavy metals. The combined techniques may enhance kinetics of the reactions, eliminate clogging due to mineral precipitation, and increase desorption, thus improving the long-term performance. Another future-oriented aspect is the use of nanomaterials. Research into these coupled technologies is at an early stage and additional work is needed to exploit their potential for remediation.

## 9.5 Outlook

Contaminations with radionuclides stem from nuclear weapons testing, nuclear accidents, and the legacy of uranium mining and processing. Contaminated sites can be found all over the world. Remediation methods comprise physical, chemical, and biological processes. The limiting factor for remediation is often not the lack of applicable technologies but economic reasons (Zhu and Chen 2009). Therefore, the development of environmentally friendly, economical, and efficient remediation processes are the future challenges for scientists and engineers. There is a need for further research into the mechanisms of passive physicochemical and biological methods.

However, the technical feasibility of these processes has already been demonstrated under field and large-scale conditions.

PRBs have shown good removal efficiencies for heavy metals and radionuclides. The cleanup of contaminated water bodies, especially river and lake beds, is still an unsolved problem after radioactive contamination. The application of passive remediation technologies like PRBs and most biological in situ processes, require reliable monitoring systems to ensure that remediation goals can be reached. The development of new and innovative sensor systems is therefore also indispensable.

# References

Acar, Y.B. and Alshawabkeh, A.N. 1993. Principles of electrokinetic remediation. *Environmental Science and Technology* 27, 2638–2647.

Adriaanse, A., Bringezu, S., Hammond, A., Moriguchi, Y., Rodenburg, E., Rogich, D., and Schütz, H. 1997. *Resource Flows: The Material Basis of Industrial Economies*. World Resource Institute (WRI), Washington, DC.

Allison, J.D., Brown, D.S., and Novo-Gradac, K.J. 1991. MinteqA2/ProdefA2, A geochemical assessment model for environmental systems. US Environmental Protection Agency, Database of computer programme, Version 3.0, Athens, GA, USA.

Beitinger, E. 1998. Permeable treatment walls—Design, construction and costs, NATO/CCMS pilot study. Evaluation of demonstrated and emerging technologies for the treatment of contaminated land and groundwater (Phase III). 1998 Special Session. *Treatment Walls and Permeable Reactive Barriers*, 229, 6–16, North Atlantic Treaty Organisation, Vienna, Austria.

Beitinger, E. 2002. Engineering and operation of groundwater treatment systems—Pump-and-treat versus PRBs, in: Simon, F.G., Meggyes, T. and McDonald, C. (eds.), *Advanced Groundwater Remediation—Active and Passive Technologies*. Thomas Telford, London.

Beitinger, E. and Bütow, E. 1997. *Design and Construction Requirements for Buried, Flow-Through Treatment Barriers for In-Situ Decontamination and Groundwater Remediation*. Berlin, Lühr, H.P. (ed.), Erich Schmidt Berlin, 342–356.

Beitinger, E. and Fischer, W. 1994. Permeable treatment bed for use in purifying contaminated ground water streams in situ. WCI Umwelttechnik GmbH.

Benner, S.G., Blowes, D.W., Gould, W.D., Herbert, R.B., and Ptacek, C.J. 1999. Geochemistry of a permeable reactive barrier for metals and acid mine drainage. *Environmental Science and Technology*, 33(16), 2793–2799.

Bhainsa, K.C. and D'Souza, S.F.D. 1999. Biosorption of uranium (VI) by Aspergillus fumigatus. *Biotechnology Techniques*, 13, 695–699.

Bickel, M., Feinauer, D., Mayer, K., Möbius, S., and Wedemeyer, H. 1996. Uranium, supplement volume C6, in: Fischer, D., Huisl, W., and Stein, F. (eds.), *Gmelin Handbook of Inorganic and Organometallic Chemistry*. Springer-Verlag, Berlin, New York.

Birke, M., Rauch, U., and Lorenz, H. 2009. Uranium in stream and mineral water of the Federal Republic of Germany. *Environmental Geochemistry and Health* 31, 693–706.

Blowes, D.W. and Ptacek, C.J. 1992. Geochemical remediation of groundwater by permeable reactive walls: Removal of chromate by reduction with iron-bearing solids, *Subsurface restoration Conference, Third International Conference on Groundwater Quality Research, Subsurface Restoration Conference,* Dallas, TX.

Blowes, D.W., Ptacek, C.J., Benner, S.G., McRae, C.W.T., Bennett, T.A., and Puls, R.W. 2000. Treatment of inorganic contaminants using permeable reactive barriers. *Journal of Contaminant Hydrology,* 45, 123–137.

Bostick, W.D., Stevenson, R.J., Jarabek, R.J., and Conca, J.L. 2000. Use of Apatite and bone char for the removal of soluble radionuclides in authentic and simulated DoE groundwater. *Advances in Environmental Research,* 3(4), 488–498.

Bringezu, S. 2002. Towards sustainable resource management in the European Union. Wuppertal-Institute for Climate, Environment, Energy, Wuppertal Paper, 121, Wuppertal.

Brown, D., Potter, P.E., and Wedemeyer, H. 1981. Uranium, supplement volume C14, in: Keim, R. (ed.), *Gmelin Handbook of Inorganic Chemistry,* Springer-Verlag, Berlin, New York.

Bundesministerium für Wirtschaft und Technologie (D). 2000. Wismut, Perspektiven durch Sanierung, Brochure.

Cantrell, K.J., Kaplan, D.I., and Wietsma, T.W. 1995. Zero-Valent Iron for the *in situ* remediation of selected metals in groundwater. *Journal of Hazardous Materials,* 42, 201–212.

Chung, N.K. 1989, Chemical precipitation, in: Freeman, H.M. (ed.), *Standard Handbook of Hazardous Waste Treatment and Disposal,* 7.21–7.32, McGraw-Hill Book Company, New York.

Csövári, M., Berta, Z., Csicsák, J., and Földing, G. 2005a. Mecsek Ore, Pécs, Hungary case study, in: Roehl, K.E., Meggyes, T., Simon, F.G., and Stewart, D.I. (eds.), *Long-Term Performance of Permeable Reactive Barriers, Trace Metals and Other Contaminants in the Environment,* Vol. 7, 211–259, Elsevier, Amsterdam.

Csövári, M., Csicsák, J., and Földing, G. 2002. Investigation into calcium-oxide based reactive barriers to attenuate uranium migration, in: Simon, F.G., Meggyes, T., and McDonald, C. (eds.), *Advanced Groundwater Remediation \*—Active and Passive Technologies* 223–235, Thomas Telford, London.

Csövári, M., Csicsák, J., Földing, G., and Simoncsics, G. 2005b. Experimental iron barrier in Pécs, Hungary case study, in: Roehl, K.E., Meggyes, T., Simon, F.G., and Stewart, D.I. (eds.), *Long-Term Performance of Permeable Reactive Barriers, Trace Metals and Other Contaminants in the Environment,* Vol. 7, 261–281, Elsevier, Amsterdam.

Dudka, S. and Adriano, D.C. 1997. Environmental impacts of metal ore mining and processing: A review. *Journal of Environmental Quality,* 26, 590–602.

Duff, M.C., Coughlin, J.U., and Hunter, D.B. 2002. Uranium co-precipitation with iron oxide minerals. *Geochimica et Cosmochimica Acta,* 66(20), 3533–3547.

Dwyer, B.P., Marozas, D.C., Cantrell, K., and Stewart, W. 1996. *Laboratory and Field Scale Demonstration of Reactive Barrier Systems.* Sandia National Laboratory, SAND96-2500—UC-2040.

Edgington, D.N., Robbins, J.A., Colman, S.M., Orlandini, K.A., and Gustin, M.P. 1996. Uranium-series disequilibrium, sedimentation, diatom frustules, and paleoclimate change in Lake Baikal. *Earth and Planetary Science Letters,* 142(1–2), 29–42.

Fetter, C.W. 1993. *Contaminant Hydrogeology*, Prentice Hall, Upper Saddle River, NJ, USA.

Fiedor, J.N., Bostick, W.D., Jarabek, R.J., and Farrell, J. 1998. Understanding the mechanism of uranium removal from groundwater by zero-valent iron using x-ray photoelectron spectroscopy. *Environmental Science and Technology*, 32(10), 1466–1473.

Fuller, C.C., Bargar, J.R., Davis, J.A., and Piana, M.J. 2002. Mechanism of uranium interactions with hydroxyapatite: Implications for groundwater remediation. *Environmental Science & Technology*, 36(2), 158–165.

Gatzweiler, R., Jakubick, A., Kiessig, G., Paul, M., and Schreyer, J. 2002. Flooding strategies for decommisioning of uranium mines—a systems approach, in: Simon, F.G., Meggyes, T., and McDonald, C. (eds.), *Advanced Groundwater Remediation—Active and Passive Technologies*, Thomas Telford, London.

Gavaskar, A.R. 1999. Design and construction techniques for permeable reactive barriers. *Journal of Hazardous Materials*, 68, 41–71.

Gillham, R.W. 1993. Cleaning halogenated contaminants from groundwater. US Patent 5266213.

Gregolec, G., Roehl, K.E., and Czurda, K. 2005, Electrokinetic techniques, in: Roehl, K.E., Meggyes, T., Simon, F.G., and Stewart, D.I. (eds.), *Long-Term Performance of Permeable Reactive Barriers, Trace Metals and Other Contaminants in the Environment*, Vol. 7, 183–209, Elsevier, Amsterdam.

Grenthe, I., Fuger, J., Konings, R.J.M., Lemire, R.J., Muller, A.B., Nguyen-Trung, C., and Wanner, H. 1992. *Chemical Thermodynamics of Uranium*. Chemical Thermodynamics, Wanner, H. and Forest, I. Vol. 1, North-Holland, Amsterdam.

Groudev, S., Georgiev, P., Spasova, I., and Nicolova, M. 2008. Bioremediation of acid mine drainage in a uranium deposit. *Hydrometallurgy*, 94(1–4), 93–99.

Gu, B., Liang, L., Dickey, M.J., Yin, X., and Dai, S. 1998. Reductive precipitation of uranium(VI) by zero-valent iron. *Environmental Science and Technology*, 21(21), 3366–3373.

Halada, K., Ijima, K., Katagiri, N., and Ohkura, T. 2001. An approximate estimation of total materials requirement of metals. *Journal of the Japan Institute of Metals*, 65(7), 564–570.

Jeanjean, J., Rouchaud, J.C., Tran, L., and Fedoroff, M. 1995. Sorption of uranium and other heavy metals on hydroxyapatite. *Journal of Radioanalytical Nuclear Chemistry Letters*, 201(6), 529–539.

Kalin, M., Wheeler, W.N., and Meinrath, G. 2004. The removal of uranium from mining waste water using algal/microbial biomass. *Journal of Environmental Radioactivity*, 78(2), 151–177.

Kelly, S.D. 2010. Chapter 14—Uranium Chemistry in Soils and Sediments, in: Singh, B. and Grafe, M. (eds.), *Synchrotron-Based Techniques in Soils and Sediments, Developments in Soil Science*, Vol. 34, 411–466, Elsevier, Amsterdam.

Korolev, V.A. 2009. Electrochemical removal of radionuclides, in: Reddy, K.R. and Cameselle, C. (eds.), *Electrochemical Remediation Technologies for Polluted Soils, Sediments and Groundwater*, 127–135, John Wiley & Sons, Hoboken, New Jersey.

Langmuir, D. 1978. Uranium solution-mineral equilibria at low temperatures with applications to sedimentary ore deposits. *Geochimica Cosmochimica Acta*, 42, 547–569.

Leyva, A.G., Mearrero, J., Smichowski, P., and Cicerone, D. 2001. Sorption of antimony onto hydroxyapatite. *Environmental Science & Technology*, 35, 3669–3675.

Lloyd, J.R. and Macaskie, L.E. 2002. Biochemical basis of microbe-radionuclide interactions, in: Keith-Roach, M.J. and Livens, F.R. (eds.), *Interactions of Microorganisms with Radionuclides*, 313–342, Elsevier Science Ltd, Amsterdam.

Matheson, L.J., Goldberg, W.C., Bostick, W.D., and Harris, L. 2003. Chapter 12—Analysis of Uranium-contaminated zero valent iron media sampled from permeable reactive barriers installed at U.S. department of energy sites in Oak Ridge, Tennessee, and Durango, Colorado, in: Naftz, D.L., Morrison, S.J., Davis, J.A., and Fuller, C.C. (eds.), *Handbook of Groundwater Remediation using Permeable Reactive Barriers*, 343–367, Academic Press, San Diego.

Morrison, S.J. and Spangler, R.R. 1992. Extraction of uranium and molybdenum from aqueous solutions: A survey of industrial materials for use in chemical barriers for uranium mill tailings. *Environmental Science and Technology*, 26(10), 1922–1931.

Morrison, S.J. and Spangler, R.R. 1993. Chemical Barriers for Controlling Groundwater Contamination. *Environmental Progress*, 12(3), 175.

Morrison, S.J., Spangler, R.R., and Tripathi, V.S. 1995. Adsorption of uranium(VI) on amorphous ferric oxyhydroxide at high concentrations of dissolved carbon(IV) and sulfur(VI). *Journal of Contaminant Hydrology*, 17, 333–346.

Mudd, G.M. and Diesendorf, M. 2008. Sustainability of uranium mining and milling: Toward quantifying resources and eco-efficiency. *Environmental Science & Technology*, 42(7), 2624–2630.

Naftz, D.L., Davis, J.A., Fuller, C.C., Morrison, S.J., Freethey, G.W., Feltcorn, E.M., Wilhelm, G., Piana, M.J., Joye, J., and Rowland, R.C. 1999. Field demonstration of permeable reactive barriers to control radionuclide and trace-element contamination in groundwater from abandoned mine lands. *Toxic Substances Hydrology Program—Technical Meeting, Charleston, South Carolina, USA, US Geological Survey, Conference Proceedings* Vol. 1 (99–4018A, Contamination from hardrock mining), 281–288.

Naftz, D.L., Fuller, C.C., Davis, J.A., Morrison, S.J., Wilkoske, C., and Piana, M.J. 2002. Field demonstration of three permeable reactive barriers to control uranium contamination in groundwater, Fry Canyon, Utah, in: Naftz, D.L., Morrison, S.J., Davis, J.A., and Fuller, C.C. (eds.), *Handbook of Groundwater Remediation Using Permeable Reactive Barriers*, 402–435, Academic Press, San Diego.

Noubactep, C., Meinrath, G., Dietrich, P., and Merkel, B. 2003. Mitigating uranium in groundwater: Prospects and Limitations. *Environmental Science and Technology*, 37(18), 4304–4308.

Noubactep, C., Schöner, A., and Meinrath, G. 2006. Mechanism of uranium removal from the aqueous solution by elemental iron. *Journal of Hazardous Materials*, 132(2–3), 202–212.

OECD/NEA 2001. Uranium 2001: Resources, Production and Demand; A Joint Report by the OECD Nuclear Energy Agency and the International Atomic Energy Agency, OECD Publications, Paris.

Office of Radiation and Indoor Air and Office of Environmental Restoration 1999. Review of Geochemistry and Available Kd Values for Cadmium, Cesium, Chromium, Lead, Plutonium, Radon, Strontium, Thorium, Tritium and Uranium. *US Environmental Protection Agency and US Department of Energy, Understanding Variation in Partition Coefficient, Kd, Values*, Volume II, EPA 402-R-99-004B, Washington, DC, USA.

Powell, R.M., Puls, R.W., Hightower, S.K., and Clark, D.A. 1995. Corrosive and Geochemical Mechanisms Influencing *in situ* Chromate Reduction by Metallic Iron, 209th American Chemical Society National Meeting, Anaheim, CA, *Division of Environmental Chemistry Conference Proceedings* Vol. 35, 784–787.

Reddy, K.R. and Cameselle, C. 2009. *Electrochemical Remediation Technologies for Polluted Soils, Sediments and Groundwater*, John Wiley &Sons, Hoboken, New Jersey, USA.

Roine, A. 2007. HSC Chemistry V.6.12. Outotec Research Oy, Finland.

Sar, P. and D'Souza, S.F.D. 2001. Biosorptive uranium uptake by a Pseudomaonas strain: characterization and equilibrium studies. *Journal of Chemical Technology and Biotechnology*, 76, 1286–1294.

Schad, H. and Grathwohl, P. 1998. Funnel and gate systems for in-situ treatment of contaminated groundwater at former manufactured gas plant sites, in: Burmeier, H. (ed.), *Treatment Walls and Permeable Reactive Barriers*, Vol. 229, 56–65, NATO CCMS, Vienna.

Schmidt-Bleek, F. 1998. *Das MIPS-Konzept*, Droemer Knaur, München.

Schöner, A., Noubactep, C., Büchel, G., and Sauter, M. 2009. Geochemistry of natural wetlands in former uranium milling sites (eastern Germany) and implications for uranium retention. *Chemie der Erde—Geochemistry* 69(Supplement 2), 91–107.

Seelmann-Eggebert, W., Pfennig, G., Münzel, H., and Klewe-Nebenius, H. 1981. Chart of Nuclides. Kernforschungszentrum Karlsruhe.

Sharma, H.D. and Reddy, K.R. 2004. *Geoenvironmental Engineering—Site remediation, Waste Containment, and Emerging Waste Management Technologies*, John Wiley & Sons, Hoboken, New Jersey, USA.

Simon, F.G., Biermann, V., and Peplinski, B. 2008. Uranium removal from groundwater using hydroxyapatite. *Applied Geochemistry* 23(8), 2137–2145.

Simon, F.G., Ludwig, S., Meggyes, T., Stewart, D.I., and Roehl, K.E. 2005. Regulatory and economic aspects, in: Roehl, K.E., Meggyes, T., Simon, F.G., and Stewart, D.I. (eds.), *Long-Term Performance of Permeable Reactive Barriers, Trace Metals and Other Contaminants in the Environment*, Vol. 7, 311–321, Elsevier, Amsterdam.

Simon, F.G., Meggyes, T., and McDonald, C. 2002. *Advanced Groundwater Remediation—Active and Passive Technologies*, Thomas Telford, London.

Sivavec, T.M., Mackenzie, P.D., Horney, D.P., and Baghel, S.S. 1997. Redox-Active Selection for Permeable Reactive Barriers, *International Conference on Containment Technology*, St. Petersburg, USA.

Smyth, D.A., Shikaze, S.G., and Cherry, J.A. 1997. Hydraulic performace of permeable barriers for the *in situ* treatment of contaminated groundwater. *Land Contamination & Reclamation*, 5(3), 131–137.

Starr, R.C. and Cherry, J.A. 1994. *In situ* remediation of contaminated ground water: The funnel-and-gate system. *Ground Water*, 32(3), 465–476.

Taylor, L.E., Brown, T.J., Benham, A.J., Lusty, P.A.J., and Minchin, D.J. 2006. *World Mineral Production 2000–04*, British Geological Survey, Keyworth, Nottingham.

Tratnyek, P.G., Scherer, M.M., Johnson, T.L., and Matheson, L.J. 2003. Permeable reactive barriers of iron and other zero-valent metals, in: Tarr, M.A. (ed.), *Chemical Degradation Methods for Wastes and Pollutants*, Marcel Dekker, New York.

Vidic, R.D. and Pohland, F.G. 1996. Treatment Walls. Ground-Water Remediation Technologies Analysis Center GWTAC, Technology Evaluation Report, TE-96-01, Pittsburgh, PA.

Waybrant, K.R., Blowes, D.W., and Ptacek, C.J. 1998. Selection of reactive mixtures for use in permeable reactive walls for treatment of mine drainage. *Environmental Science and Technology*, 32(13), 1972–1979.

Wu, L., Forsling, W., and Schindler, P.W. 1991. Surface complexation of calcium minerals in aqueous solution, 1. Surface protonation at fluorapatite-water interfaces. *Journal of Colloid and Interface Science*, 147(1), 178–185.

Zhu, Y.G. and Chen, B.D. 2009. Principles and technologies for remediation of uranium-contaminated environments, in: Voigt, G. and Fesenko, S. (eds.), *Remediation of Contaminated Environments, Radioactivity in the Environment*, Vol. 14, 357–374, Elsevier, Amsterdam.

# 10

## Reactive (Oxygen) Gas Barrier and Zone Technologies

**Ronald Giese, Frank Ingolf Engelmann, Dietrich Swaboda, Uli Uhlig, and Ludwig Luckner**

### CONTENTS

## 10.1 Introduction to Reactive Gas Barriers and Zones

Following a decade of investigation and field-scale testing, reactive gas barriers and zones (RGBZ or gas permeable reactive barriers [PRBs]) have been introduced as a state-of-the-art remediation technology for both organic and inorganic contaminants in the groundwater zone.

RGBZ technology is sustainable and can achieve long term and stable attenuation of the negative impacts of these contaminants on groundwater bodies and flow. It requires a modest initial investment and operational costs are very competitive with other alternatives. In addition, RGBZ consumes minimal resources (e.g., energy, materials, land, and manpower). Both the operational risks and risks to human health/and the environmental are low. The RGBZ technology has demonstrated a high efficiency in stimulating the intended transformation and exchange processes, while at the same time showing a low sensitivity to temporal changing geohydraulic and geobiochemical conditions.

Gas PRBs can be implemented as a stand-alone technology; they are also suitable for treatment train applications, which are used to treat complex contaminants (Figure 10.1). There are three basic application methods:

1. In situ gas reactors can operate as full-section gas PRBs (reactive walls) to prevent the breakthrough of contaminated groundwater into a sensible object that is being protected. These are typically used to limit plume propagation or to avoid juridical implications with respect to downstream land owners.

2. In situ gas reactors can operate as pre- and posttreatment zones for lumped reactive barriers (e.g., funnel and gate, grain, and gate) or treatment trains. Pretreatment is defined as the conditioning of a lumped stream of contaminated water (e.g., to remove iron) to guarantee the best technical performance of subsequent treatment steps (Kassahun et al., 2005). Posttreatment is a polishing step following the removal of the main contaminant mass. For this treatment to be optimal, downstream natural attenuation of some remaining or previously inaccessible compounds needs to be stimulated.

3. In situ gas reactors can also operate as reactive gas zones in cases where the objective is to lower the state of damage of a sensible subsurface domain (site decontamination). Reactive gas zones then act as retention or buffering regions against natural dynamic flow changes (e.g., coupled aquifers to river systems), impacts from the top soil (e.g., contaminated overburden or dumps) or from nearby applications of invasive technologies (e.g., construction or mining activities).

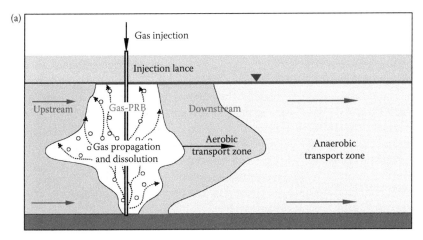

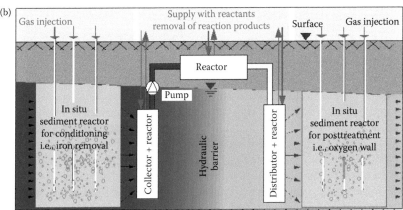

**FIGURE 10.1**
Technology application variants for RGBZ. (a) Stand-alone full-section gas PRB with sequential reactive zones (patent EP 1550519 "BIOXWAND"). (b) Pre/postreactive gas zones of a drain and gate treatment train for complex groundwater and subsurface decontamination (patent DE 10310986 "GFIadags").

The methods of RGBZ operation used in this research are direct gas injection (DGI) and application options for low-pressure (NDI) and high-pressure injections (HDI), which are discussed. It is noted that the term "sparging" is not used for RGBZ applications, as it is linked to applications that generate a gas which escapes from the groundwater zone and strips groundwater contaminants. Biodegradation is only an additional effect of sparging; a soil gas extraction and treatment system is needed.

The RGBZ technology has been approved by German Environmental authorities (ITVA, 2010) and additional applications in regard to enhanced natural attenuation (ENA) are anticipated.

Gas PRB instrumentation can be installed with minimal effort. Only a limited number of small diameter vertical perforations are needed, and sequential reactive zones can be formed in undisturbed geologic structures. In this way, the invasive effects of groundwater flow are minimized and the RGBZ operates as a hydraulically passive technology. The injection and propagation of a gaseous mixed phase in the subsurface is performed using controllable 3D gas flow networks. Reactants are temporarily stored in trapped gas clusters in the porous matrix adjacent to adsorbed contaminants and biofilms, and the delivery of gaseous reactants into the groundwater flow can be adjusted by controlling the partial pressures of gas components.

Similar to other in situ technologies, RGBZ are strongly dependent on the hydrogeological domain, described by the porous rock or sediment, groundwater flow, and migration properties. RGBZ are additionally dependent on the pneumatic or gas flow characteristics of the subsurface. Thus, the management of a complex heterogeneous multiphase multicomponent flow and migration domain demands that the engineer who is planning and applying the RGBZ displays a high level of professionalism.

RGBZ are ideally applied in horizontal multiple-layered sediment formations of nonuniform fine- to coarse-grained sands and fine gravels. Depths to 50 m below the ground surface are accessible without the use of heavy drilling techniques. Enclosed finer texture lenses or thin layers do not limit the application of RGBZ, as they are typically not continuously shaped and contain weak zones of gas-available threshold pressures. A time scale of 1–3 years is required to complete a stable formation of a gas PRB. The horizontal scale needed for a gas storage domain depends on the geological structure. In the direction of groundwater flow, it is typically in the same order of magnitude as the saturated thickness of the aquifer. Stimulation of intrinsic microorganisms can be achieved when a suitable environment is established (redox, pH) and dominant electron acceptors or donors for the biodegradation of groundwater soluble contaminants are supplied. Variable zones of redox potential can be induced by sequential reactors (Figure 10.1) or rate controlled and time-variable gas injections. Products of precipitation reactions (e.g., iron or manganese oxidation or pH-induced instability of carbonate) do not put the long-term operation of the RGBZ at risk. The well-known effects of bypassing or channelling groundwater flow due to gas clogging can also be monitored and controlled.

The most common RGBZ application uses atmospheric air and pure oxygen gas or its mixture to supply electron acceptors for aerobic biodegradation. Luckner (2001) reviewed the potentially available gaseous reactants and their impacts on biodegradation. Noble gases such as He, Ar, Ne, and $SF_6$ are used as tracers (Weber, 2007). Electron donor supply due to methane (Zittwitz and Gerhardt, 2006) and hydrogen gas (Bilek and Wagner, 2009) injection have been tested for in situ stimulation of cometabolic CHC degradation and autotrophic sulfate reduction. In situ iron removal can also be forced by oxygen and ammonia gas applications.

Due to insufficient gas storage capabilities, fissured rock domains and unconfined aquifers with a saturated thickness less than 3–5 m are less suitable for gas PRB applications. In addition to the geological domain, the type and complexity of the limiting reactants for in situ transformations, and the ability to deliver them by gas flow can also impose restrictions. A stand-alone RGBZ is unable to provide vital nutrients (e.g., available phosphorous or trace metals) where they may be deficient. A gas injection-based method to support the natural buffering capability of a subsurface domain against high proton production is still needed.

Care must be exercised when transformation of high volatility migrants (e.g., chlorinated ethenes or short-chained aliphatics) is intended. These substances can be enriched and stored in gas clusters, and even in cases where gases are not allowed to be stripped from the groundwater zone, the substances may become less accessible to biofilms. The presence of nonaqueous phase liquids (NAPLs) will lower the gas storage capability, because residual NAPL blobs occupy the same pore space portions; additional impacts are changes in the wettability or emulsifications. Toxic concentrations of contaminants, but also unfavorable environmental conditions (e.g., sulfide or pH) in the vicinity of NAPL are frequently reported.

Furthermore, the availability of sufficient time and space to achieve the given protection or remediation goals can limit the application of RGBZ.

## 10.2 Gas-Water-Dynamics in the Groundwater Environment

### 10.2.1 Basic Phenomena

Gas flow transport phenomena, capillary gas storage, or entrapment and mass transfer between the water and gas phases have been evaluated at both pore and field scales. Gas–water displacement and mass transfer due to gas injection in a water-saturated subsurface domains occur in a different manner to that in the unsaturated soil zone (Figure 10.2).

The transport of a nonwetting gas phase in groundwater environments is mainly driven by pneumatic pressure, capillary, and buoyancy forces. The pneumatic pressure gradient has to overcome the hydraulic pressure head at the injection point, an additional capillary entry pressure required to open a gas channel network, a pneumatic flow resistance that is formed by friction at nonrigid moving gas–water interfaces, and pressure-dependent gas viscosity (Geistlinger et al., 2006). With increasing distance from the injection point, gas volume portions become disconnected due to a decrease of pneumatic pressure, and pore trapping forms incoherent gas bubbles and clusters. Due to the heterogeneous layered nature of sediment domains, bubbly flow in gravel structures or channelling flow in fine-grained media cannot hold for larger distances (Brooks et al., 1999). The propagation of gas clusters is commonly

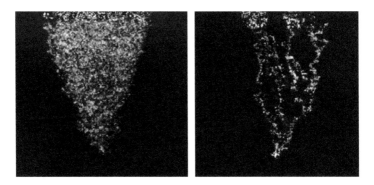

**FIGURE 10.2**
Basic gas flow types for bench scale direct gas injection in water saturated porous media. Left: incoherent bubbly flow in coarse sand (gas clusters), right: coherent channelized gas flow (viscous fingering). (Geistlinger, H. 2010: Model supported high pressure pulsed Gas Injection (HDI) for in situ Remediation of contaminated Aquifers: Laboratory scale Experiments and Computer Simulations for Optimization of the Technology. Report Nr. KF0011010SB7–2, Helmholtz-Centre for Environmental Research UFZ, 63 p. (in German).)

reported in field applications. These clusters are unstable gas-filled bodies with a magnitude in the order of several pore and solid particle diameters. Their bulk gas pressure can change due to mass transfer and they cannot equilibrate the variable capillary forces at their total water interfaces. Gas clusters are moved upward by buoyancy forces and they are laterally spread by pneumatic cluster displacement when gas is injected. This behavior is defined as pervasive gas flow. Gas propagation stops when the threshold pressure of a given sediment or rock layer cannot be overcome. Following the cessation of propagation, high local gas accumulations and highly coherent gas saturations can be present. This bulk or geological trapping can form reliable gas storage zones, and groundwater conductivity can be lowered significantly.

When applying the low-pressure NDI method to natural consolidated sediments, matrix and pore restructuring does not occur. Multiphase flow characteristics of the sediments remain stable over a large range of total mechanical stress (Giese et al., 2003). The high-pressure method HDI focuses pneumatic sediment cracking in the vicinity of the injection point. In addition, local structure reorganization is needed to generate preferential flow paths for gas pulse propagation.

## 10.2.2 Example Test Facilities

A *pore to bench scale* gravimetric-optical measurement system (Figure 10.3) was developed using coupled cameras to detect overall gas saturations (stationary camera) and local moving gas bubbles or clusters (dynamic camera) in a 2D acrylic glass chamber (0.40 m × 0.45 m × 0.01 m). The system allows for a high resolution in time and space and for simultaneous observations of

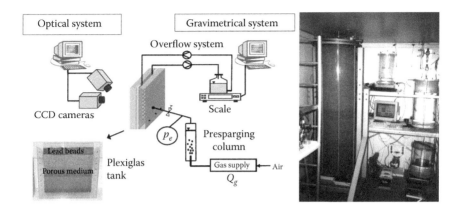

**FIGURE 10.3**
Test devices and sites for DGI studies. Upper left: gravimetric-optical system. (Geistlinger et al. 2006: Direct Gas Injection into glass beads: Transition from incoherent to coherent gas flow pattern. *Water Res. Research*, 42, paper W07403, 12 p.) Upper right: pressurized rotatable column with gravimetric balancing system, bottom image: field-scale test site BIOXWAND.

pore to local scale phenomena ($10^{-5}$–$10^{-1}$ m). Uniform sand fractions and glass beads were used as sediments and effective mechanical stress is induced by hydraulic pressure being applied to a rear-side membrane and an overburden or gravimetric load (leaden spheres). A gravimetric phase balancing system was installed for saturation measurements.

A *bench to pilot scale* research test site for gas-based remediation techniques was installed at the Dresden Groundwater Centre (Figure 10.3). Test columns and tanks (0.3 m–3.0 m in diameter) were used to evaluate the effective parameters for gas injection, storage, and dissolution for upscaling to field applications (Weber, 2007). The test devices operated under near-to-field conditions. 1D and 3D total stress and system pressures of 300 kPa (using a 20 m column of water) were applied to the columns and tanks. The temperature of the system was set at 10–15°C, which are temperatures typical of groundwater. Multiphase water and gas flow interactions were studied using parallel flow, counterflow, and cross-flow (horizontal, vertical). A phase balancing system runs under system pressure conditions. A 2D tank test device was used for imaging and the bulk estimation of trapped gas lenses in layered sediment formations. The influence of gas trapping to a stationary groundwater flow was evaluated using in situ sensor arrays and noninvasive geophysical gas monitoring systems (e.g., geoelectric-induced polarization) as discussed in Boerner et al. (1996). Bulk parameters for the field-scale NDI application were preliminary estimated by pilot scale testing.

Typical gas injection rates for the bench to pilot scale testing of the NDI were $10^{-3}$–$10^{-1}$ m³/h STP, and the flow sections ranged from $10^{-2}$ to $10^{-1}$ m².

*Field-scale* research test sites (up to 500 m² of treatment area and 5–50 m in depths) were operated by the Berlin Water Company (Figure 10.3) and the Helmholtz-Centre for Environmental Research UFZ. Several field-scale test applications were run with DGI technology (NDI and HDI methods) in sediment environments. There was also an application of NDI in fissured sandstone bedrock (Schinke, 2008). The field sites were equipped with conventional and state-of-the-art injection and monitoring techniques, and high-resolution site investigations were performed. From these field-scale tests, best available technologies and strategies for site characterization, injection, and monitoring system operation, and control of RGBZ were derived (Ehbrecht and Luckner, 2004; Beckmann et al., 2007). An integral balancing algorithm for gas injection and biodegradation and a tracer test method using noble gases was also developed.

Typical gas injection rates for the field-scale testing of NDI were $1 \times 10^{-1}$ – $2.5 \times 10^{0}$ m³/h STP and the flow sections ranged from $10^{1}$ to $10^{2}$ m².

## 10.2.3 Gas Injection and Gas-Water Displacement

There is a difference in the gas-water-displacement effects of low pressure (NDI) and high pressure (HDI) DGI methods. In particular, the effects of interest are the injection pressure gradient, gas injection rate, and apparent gas propagation velocity.

In the NDI system, the placement or mixing of a low amount of immobile (gaseous) reactants in a natural groundwater flow and their dissolution are typical of full-section PRBs. Following this, the desired in situ reactions occur in downstream aquifer regions. In addition, a stationary gas channel network is typically formed. The same gas flow paths are used multiple times, even when a pulsed injection is applied. The density of a gas channel network and the volume of gas clusters are functions of texture in homogeneous sediment regions. The coarser the material, the lower the gas network density; however, the mean dimension of moving gas clusters is higher. Typical cluster diameters in the order of <2 mm in fine- and nonuniform-grained sands and >20 mm in coarser sands bubbly clusters have been reported (Weber, 2007).

A natural groundwater domain is characterized by multiple gas transport barriers caused by the horizontal layering and compaction of sediments. The transport of gas clusters is highly sensitive to these structures and heterogeneities. Gas accumulation occurs, and regions of coherent mobile gas saturations can result. These structures must be explored during a gas-hydrogeological surveying.

There is a weak interaction between gas and water flow during NDI; pervasive and bubbly gas propagations facilitate the simultaneous use of macropore structures for water and gas flow. There is some rearrangement of the path of water flow during gas injection due to local gas accumulation in capture zones. Subsequent conductivity changes are limited to the local scale and a degree of homogenization of the water flow can be achieved by temporary clogging of coarser high-permeability zones.

An effective displacement of mobile water by mobile gas in a near flow region is induced using HDI. The displacement results from high-gradient, high-frequency pulses with injection periods in the range of seconds to minutes. HDI is applied when source zone or soil matrix decontamination is required and it has been used in combination with NDI (NDI–HDI) for local gas storage homogenization in the large scale BIOXWAND application. HDI has reported to cause more significant changes to groundwater flow in terms of flow direction, velocity, and dispersivity (Selker et al., 2007; Geistlinger et al., 2006). Applications in bedrock and other low permeable environments (e.g., sandstone structures or silt barriers) may generate gas accessible pore networks.

There is evidence from field-scale gas tracer applications that the mutual displacement of gas flow networks can occur during simultaneous injection at locally distributed lances (Uhlig, 2010 and Schinke, 2008). The effect can be explained by applying the pervasive gas flow concept of moving incoherent clusters where effective mixing of cluster flow paths is not possible. The practical outcome is that the determination of the ROI of an array of gas lances must be performed by complex lance array testing.

### 10.2.4 Gas Propagation and Storage

Gas storage in aquifers mainly appears as either mobile gas capturing or accumulation below geological barriers or the residual pore-trapping of gas clusters. Gas saturation (volume of gas per volume of pore space) is used to characterize storage.

During NDI in sandy sediments, typical gas saturations are 1%–5% for residual gas, 5%–10% for mobile gas (during injection periods), and greater than 15% for mobile gas capture zones (Weber, 2007 and Engelmann et al., 2010). Texture and mechanical stress only exert minor influences on these means of gas saturation. It has been reported that pervasive incoherent cluster flow can occupy a denser pore channel network than coherent flow over large distances, and can be maintained for hours after gas injection has ceased. It can also lead to redistribution of subsurface gas storage. Effective pervasive gas propagation is in the range of $10^{-2}$–$10^{-1}$ m/h.

When applying HDI, local increases in, as well as the homogenization of, gas saturation are induced in a near region with an ROI <3 m (Geistlinger, 2010). In this case, gases can be effectively supplied to the bottom zone of an aquifer (which is of special interest in unconfined aquifers), and when a density-driven plume propagation is under consideration (e.g., a dissolved CHC plume). With increasing distance and due to gas viscosity and compressibility characteristics, the HDI injection pressure transforms almost completely into high gas propagation velocities in coherent channelized networks. There is no additional gas saturation effect of HDI at greater distances from the injection point and the effective gas propagation of channelized flow is of $>1 \times 10^{0}$ m/h. A wide velocity range indicates the instability of this transport behavior with a few dominating macroflow paths.

Figure 10.4 and Table 10.1 summarize the current knowledge of gas storage and propagation phenomena during DGI into sediments in the groundwater zone. Assuming an injection area of $10^{-2}$ m$^2$ for bench scale testing and $10^1$ m$^2$ for field applications, observed injection pressures and gas propagation velocities during rate controlled field testing of NDI and HDI (Weber, 2007, Geistlinger, 2010 and Zittwitz et al., 2012) are very similar. The gas injection pressure difference due to the hydrostatic level typically increases during NDI from $<5 \times 10^{0}$ kPa to $3 \times 10^{1}$ kPa when the injection rate is increased from 0.5 to 2.5 m$^3$/h STP. This indicates a change to channelized flow and a

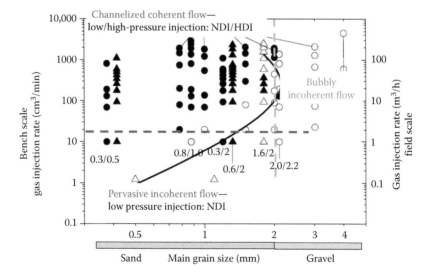

**FIGURE 10.4**

Gas flow classification scheme of bench to field-scale DGI. (Adapted from Geistlinger, H. et al. 2006: *Water Res. Research.* 42, paper W07403, 12 p.; basic scheme and data from bench scale testing.)

**TABLE 10.1**

Field Parameters for NDI and HDI Gas Injection (Sandy to Gravel Sediments)

| | Pressure Difference at Injection Point | Flow Rate (STP) at Injection Point | Mean Propagation Velocity over ROI | Flow Type/ Injection Type, Frequency |
|---|---|---|---|---|
| | $\Delta p_{IP}$ (kPa) | $Q_{g,IP}$ (m³/h) | $v_{g,ROI}$ (m/h) | – |
| NDI, low | <30 | <1.0 | 0.01–0.1 | Pervasive or bubbly/ continuous or pulsed $f < 1/d$ |
| NDI, high | 30–100 | 1.0–3.5 | 0.1–10.0 | Channelized or bubbly/ pulsed $f < 1/h$ |
| HDI | >300 | >5 | >5.0 | Channelized or bubbly/ pulsed $f > 1/min$ |

subsequent higher gas propagation is observed. It is noted that Figure 10.4 is somewhat similar to the findings of Wang et al., (1998) who analyzed the flow instability of immiscible displacement in the vadose zone during water and NAPL infiltration.

High gas saturation can be achieved using surfactant enhanced NDI (Giese and Reimann, 2003). Foam formation will lower the gas propagation velocity and the mass transfer coefficient and gas stripping can be completely avoided. There is evidence of a reliable mass transport of dispersed solid substances (e.g., bacteria, nutrients) through sediments by gas-in-water-foams pilot scale. Using surfactants, a complete local drainage of pore space can be induced, enabling up to 70% of gas saturation. Surfactant-enhanced NDI is difficult to control under field-scale conditions and is still being investigated. Potential applications of induced pH buffering and in situ gas-induced impermeable walls to optimize dewatering of construction pits are also currently being investigated.

### 10.2.5 Gas Dissolution and Degassing

*The dissolution* of gaseous components from a trapped gas phase into groundwater flow has been investigated using pore- to field-scale test facilities (Figure 10.3). Conceptually, it is understood to be a bidirectional kinetic multicomponent mass transfer of moving gas–water interfaces of multisphere gas clusters. This leads to bubbles shrinking or growing (variable volume model), and subsequently to dynamic interface areas and partial pressures of gaseous compounds. Heterogeneous gas saturation at field scales can be taken into account through coupling the multisphere distribution to a hydrogeological (e.g., water flow velocities) or geometrical (e.g., pore sizes) distribution function (Geistlinger et al., 2005).

Mass transfer is driven by partial pressure gradients of gaseous compounds in groundwater flow. A primary problem is the determination of an effective mass transfer coefficient and its scale dependency (Luckner and Schestakov, 1991). Estimation of active gas–water interface areas and water diffusion lengths are also needed. Estimates of hydraulic conductivity changes due to residual gas storage cannot easily be derived from well-known functions of vadose zone modeling due to the nature of gas-water-displacement near saturation (Giese, 2012).

A lot of experimental and modeling work to determine the mass transfer coefficients at the pore to bench scale has previously been reported; an overview of this work is presented in Geistlinger et al. (2005). Best practice scalable mass transfer calculations take into account the dimensionless numbers: the Peclet number (Pe: relates water flow velocity to diffusion), Sherwood number (Sh: relates mass transfer to Pe), and Damkoehler number (Da: relates hydraulic resistance to mass transfer times). State-of-the-art modeling techniques were tested and further developed, and field-scale modeling capabilities of multiphase multicomponent reactive transports were demonstrated for operation control of RGBZ (Horner et al., 2009, Geistlinger, 2010, Weber et al., 2013) using adapted codes of PHT3D, TOUGH2, and MIN3P. It has been reported that for the practical purposes of RGBZ control, first-order transfer functions can be applied to residual gas dissolution.

Balanced experimental data sets (Geistlinger et al., 2006; Weber, 2007; Ehbrecht and Luckner, 2004), and field-scale balance and sensing estimates (Engelmann, 2010; Ehbrecht and Luckner, 2004, Beckmann et al., 2007) are available for pure oxygen gas dissolution. Residual oxygen gas saturations of 2%–4% in sandy sediments can completely dissolve when 2–3 pore volumes of gas-free groundwater have passed. This measure is used in practice to periodically reload storage zones of the PRB BIOXWAND. Mass transfer rates decrease when inert gases are present (e.g., during air injection or in presence of high-dissolved nitrogen concentrations in natural groundwater).

*Degassing* in conjunction with DGI is defined as the reduction of gas caused by gas stripping and/or diffusive degassing from groundwater. Stripping occurs as a bulk gas escape (buoyancy and convection driven) of mobile gas clusters reaching the phreatic groundwater surface and capillary fringe. A multicomponent gas volume is injected into the coherent gas phase of the vadose zone, and the entire mass of the gas mixture is transferred. Mixing in the soil gas is only limited by gas diffusivity and the partial pressure gradients of the gaseous components. Stripping may also be generated when a gas flow network connects to unsealed technical or natural macropores (e.g., boreholes, wells, and other observation installations), or natural fissures in overlaying gas barriers. Partitioning of volatile compounds such as volatile organic compounds (VOCs) and short-chained alkanes to the gas flow and their escape to the soil gas is of concern due to safety implications.

Diffusive degassing of dissolved compounds from groundwater is a substance-specific mass transfer through the capillary fringe, and is driven by specific diffusivity and fugacity according to Raoult's law. Flux limitations typically arise from dispersivity and fluctuations in groundwater flow.

Stripping is probably the dominant degassing effect during DGI application; however it is difficult to quantify diffusive degassing due to natural soil gas fluctuations. Until recently, sensors for the direct measurement of degassing fluxes have not been available. Stripping needs to be limited by gas injection control, and should be monitored by soil gas monitoring. Best practice for flux estimation includes stationary model-based balancing of the gas injection mass, and gas tracer testing (Weber, 2007). Some light gas escape in the range of 10%–30% of the injected mass often can be tolerated to ensure a sufficient efficiency of reactant supply to the upper (near-fringe) groundwater flow region. If oxygen gas is used, aerobization of the vadose zone can be a desired additional treatment effect of immobile soil water and of leaches from the topsoil. A low-cost soil venting technique can effectively support soil gas mixing and minimize safety implications.

## 10.3 Techniques and Devices for RGBZ Formation

### 10.3.1 Set of Available Technical Tools

The first step in the technical implementation of RGBZ is a detailed gas-hydrogeological site investigation. In addition, biogeochemical and contamination information need to be obtained as part of the investigation. The best practice depth-oriented soil core sampling includes low-diameter drilling with liner sampling or percussion core probing, and direct push methods including CPT, pneumatic percussion, and Sonic® vibration sounding (e.g., Geoprobe®). A conceptual gas-hydrogeological site model is required, and can be developed using sample analysis and geophysical and hydraulic survey data. Such a model is presented in Figure 10.5. Borehole logging can include gamma (γ-γ), neutron (n-n), and electric conductivity logging and thermal and permeability flow metering. Hydraulic and immission pumping and infiltration tests can support the establishment of treatment region dimensions.

RGBZ require specialized gas injection and monitoring methods and devices; techniques to efficiently control gas dynamics and their impact on in situ transformation processes are also required. The formation and control of a homogeneous gas distribution and flow-oriented gas dissolution must be enabled, and excessive gas emission from the groundwater zone must be avoided.

A gas injection system consists of four main components: injection lances, a gas delivery and mixing station, an injection control system (pressure, flux, and time control), and a warning and safety system adapted to the expected field gas compositions.

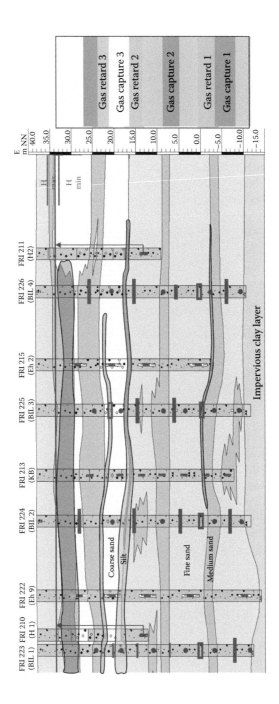

**FIGURE 10.5**
Cross section of a gas PRB domain as gas-hydrogeological structure model (BIOXWAND).

An RGBZ monitoring system is comprised of a combination of five elements. These are groundwater observation and sampling points or wells, an in situ sensing array for detection of the dissolved gas distribution domain, a detector set for estimation of pneumatic and hydraulic gas propagation, a measurement system to quantify dynamics of gas saturation in the gas storage zones, and finally, a soil gas composition control system.

Techniques that can be used to efficiently control the performance of a gas PRB and optimize the impact to in situ transformation processes are available as an integral mass balancing method for injection gases, and as an algorithm for performance optimization. Modeling techniques can be used for the planning and evaluation of gas PRB applications. Due to their reactive multiphase multicomponent nature, they are normally too complex and the uncertainty is too high, rendering them of questionable value as a decision making tool.

### 10.3.2 Gas Injection Devices

#### *10.3.2.1 Gas Lances*

Lances can be installed by drilling and sounding or direct push methods (Figure 10.6). A number of technical requirements must be met: (1) prevention of ground loosening during installation; (2) high-precision depth-oriented positioning of filter elements to 50 m below the surface, including in heavy ground penetration conditions; (3) a gas-tight vertical sealing of the injection

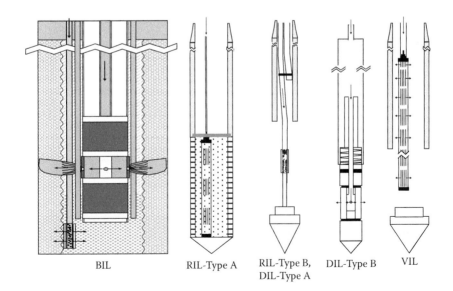

| BIL | RIL-Type A | RIL-Type B, DIL-Type A | DIL-Type B | VIL |

**FIGURE 10.6**
Variants of injection lance installations for DGI applications. BIL—drilling; RIL—percussion sounding; DIL—CPT-based penetration sounding; and VIL—vibration sounding.

filter tubing; (4) an appropriate pressure, diffusion, and reaction-resistant casing or tubing material; and (5) a gas filter backfill construction that permits homogeneous horizontal gas flow coupling to the subsurface layers.

*Drilling Injection Lances* (BIL) can be installed in heavy or variable layered sediment and bedrock environments using dry and hydraulic drilling using diameters <250 mm. There are typically no depth restrictions and multilevel injection filters can be positioned in one borehole (Schinke, 2008). The main disadvantage is that the extraction of subsurface material up to 2 m in diameter cannot be avoided (Engelmann et al., 2004). Following the installation of lances, the measurement of material extraction during the drilling process and tight grouting of the casing annulus and the loosening zone are required, even under bedrock conditions. Special grouting valve casings are available for high-pressure injection of sealing suspensions (e.g., bentonite clay). Injection volumes must be balanced and controlled, and a multistep grouting procedure has to be planned with intermittent testing of the sealing effect. Care has to be taken during grouting, as undesired clogging of the main gas transport layers or gas filters can occur. Suffusion-protected gas filter zones are built-up by gravel or coarse sand. Gas injections via multilevel filters can be performed using casing packers.

There are three types of *Sounding Injection Lances*. During direct-push installation, displacement and compaction of the rock material take place, and an autosealing effect is gained between the casing and the borehole walls. Borehole diameters are approximately 30–80 mm.

*Percussion Sounding Injection Lances* (RIL) can be installed in sediments to a depth of 10 m using 2–3 in casings (e.g., HDPE) with a filter tip and sealing packers between the casing segments (Figure 10.6—RIL-type A). The casing remains in the borehole and gas injection tubing and gravel fillings are placed into the filter zone, which is sealed by a compacted clay layer. In addition, sensors and multilevel filters can be installed. RIL are installed in medium-compacted sediments using heavy pneumatic percussion tools (e.g., Geoprobe®, up to 100 kN). Depths of 30–40 m can be reached, although care must be exercised when pushing down a cone tip with fixed injection tubing as the milling of soil material can lead to filter sealing or destruction. After reaching the planned depth, the hollow casing is drawn back and can be used multiple times. An additional hollow drilling auger can help to lower the penetration resistance of highly compacted or very coarse layers by preloosening. High pressure sealing of the borehole can be done during withdrawal of the casing (Figure 10.6—RIL-type B). It is recommended that up to 2 months consolidation time be given for the installed lances before starting lance operation, particularly for HDI applications (Engelmann, 2010).

Continuous hydraulic cone penetration tools (CPT) with up to 200 kN are used to install *Penetration Injection Lances* (DIL). Thinner casing walls allow for the installation of larger diameter tubing. The maximum depth is in the same magnitude as for RIL and predrilling or stabilizing casings are required in heavy soil layers. A filter casing injection lance (Figure 10.6—DIL-type B) can be used multiple times as the CPT is able to withdraw the complete system. There are a number of sensing, additional testing, and sampling tools available for both percussion sounding and CPT, which give the advantage of flexible multifunctional applications (Dietrich and Leven, 2006). Lances of DIL-type B allow for pressure-controlled groundwater sampling, permeability testing, and in situ groundwater screening of dissolved gases.

High-quality direct push lances can be installed up to 80 m at a moderate cost using the Sonic® sounding technology. *Vibration Injection Lances* (VIL) are good alternatives to classical dry drilling in sediment environments in terms of depth and core probing, and both multilevel and coupled sensing installations are available. Another advantage is grouting and sealing of the lances or filters is done by sonic withdrawal of the casings, which results in autocompaction and consolidation (Engelmann et al., 2009). VIL lances are preferred, even for HDI gas injections.

### 10.3.2.2 Gas Supply, Gas Mixing, and Distribution

Injection gases used in RGBZ include pure gases (e.g., oxygen), or gas mixtures. Air is typically used as a carrier gas to achieve high ROI, and partial pressure can be controlled with a few lances and oxygen. Inert trace gases (e.g., He, Ar and Ne or reactive gases such as methane and carbon dioxide) can be mixed with the injection gas.

Pure gases are economically stored in pressurized tanks, and additional gas compression is not required for injection. Oil-free compressors are used for the injection of atmospheric air mixtures, and a postdrying step for compressed air is necessary.

Mass flow controllers and flow meters are recommended for the mixing and distribution of injection gases as they allow the balancing of injected gas amount for each lance. These devices require calibration to the specific gas mixtures (Figure 10.7). Pressure meters and magnetic valves enable effective gas distribution and dynamic injection intervals.

Gas injection is performed as either low-pressure (NDI) or pulsed high-pressure injection (HDI). Continuous NDI injection is applied in the initial formation period for a gas PRB when there is a high demand for reactants (e.g., oxygen), and it results in full ROI formation, preconditioning, and preoxidation of the rock matrix. It can also lead to some emission of gas into the unsaturated zone. Constant injection pulses over a few hours are used during a regular RGBZ operation, and these pulses are interrupted by periodic break periods. HDI injection consists of high frequency, high-flow rate gas pulses in the range of seconds to minutes. Gas breakthrough to the unsaturated zone is avoided by the time limitation of coherent gas flow periods. HDI can be used for formation of local gas storage zones with higher gas saturations, and for repairing of clogged gas lances. While a gas supply system for NDI has to resist a total pressure of approximately 500 kPa, an HDI supply system (including lances) needs to be operated at >1000 kPa.

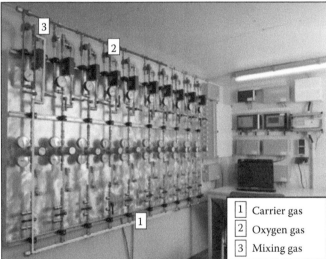

| 1 | Carrier gas |
| 2 | Oxygen gas |
| 3 | Mixing gas |

**FIGURE 10.7**
Technical equipment used for DGI (left to right): Gas compressor or blower; pressurized or liquid gas tank; and gas injection and mixing station. (From Schmolke, L.P. et al. 2007: *Proc. Dresden Groundwater Research Centre*, Nr. 31, pp. 135–146 (in German).)

### 10.3.2.3 Safety Precautions

The materials used in the gas supply system have to be chosen in accordance with the reactive gases used. Also, technical precautions for pressurized systems need to be taken into account. Leaks in the gas supply system can be automatically monitored using gas-specific sensors, pressure transducers, and smoke detectors. Limiting access, remote control systems, and an automatic off switch are needed for the gas tanks and supply system. When dealing with volatile hazardous contaminants or potentially explosive gas mixtures, gas warning devices and a soil gas venting system are required.

### 10.3.3 Monitoring Devices

The main functions of a RGBZ monitoring system are: (1) the detection of gas emissions at geological weak points, nontight boreholes, and soundings; (2) representative sampling of groundwater and soil gas; (3) detection of gas distribution and dimensions (ROI) of the RGBZ; (4) estimations of injection gas propagation and dissolution; and (5) estimation of local gas saturations in crucial regions and layers.

#### 10.3.3.1 Groundwater and Dissolved Gas Monitoring

Groundwater sampling equipment can be installed using self-grouting direct push technology. Special filters and pumps are required due to small diameters and gas-protected filter screens (Figure 10.8). Sampling using

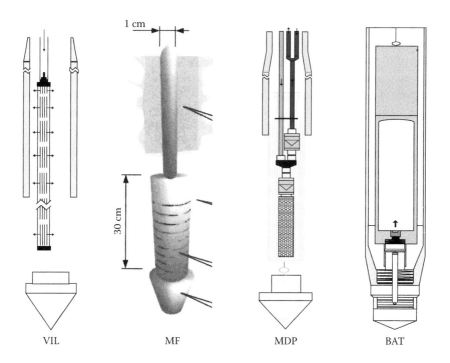

1 cm

30 cm

VIL            MF            MDP            BAT

**FIGURE 10.8**
Groundwater monitoring devices for RGBZ. (VIL—loosen sonic lance filter; MF small diameter observation well; MDP—loosen double-valve pneumatic pump; BAT pressure conserving bailer shuttle).

peristaltic (at shallow groundwater levels) or double-valve pneumatic pumps (MDP) and shuttle systems (BAT) give point information due to small sampling volumes and short filters. MDP and BAT can be used for a pressurized groundwater sampling without degassing losses.

Modified RIL-type B and DIL-type A lances can be used to install 25-mm groundwater filters. Local-scale integrated samples can be obtained using packers and either multiple MDP or button valve pumps (Uhlig, 2010 and Zittwitz et al., 2012). When using a Sonic®-system, 50-mm direct push filters can be installed. In addition to these pumps, mobile bailing systems (e.g., BAT) can be applied. Hydraulic and immission pump tests are then performed to obtain volume integrated groundwater information. In a gas injection zone, it is necessary to cover wells with a gas-tight cap.

The sampling of dissolved gases can be performed by pressure conserving bailing devices and the use of trace gases (Ehbrecht and Luckner, 2004). A headspace gas phase can be brought into equilibrium with the water sample and pre- and postsampling can be undertaken using gas chromatography. Inert gas flushing, volume and mass balancing, and multiple pressure controls are needed in order for confidence to be placed in the results obtained.

### 10.3.3.2 Gas Monitoring

It is advisable to install an array of in situ gas sensors in the gas injection zone of an RGBZ. The distribution of gases and ROI dimensions can be obtained, along with an estimation of injection gas propagation and the dissolution of the gas phase. Combinations of sensors are placed and grouted to the main gas-permeable layers using the direct push method. They can also be installed in small-diameter observation wells if packers and an automated pumping system are used. A shuttle-sensing tool MIDZ (Figure 10.9) can detect high concentrations of dissolved gases. MIDZ uses a pressurized flow chamber with integrated sensors, and is installed with CPT technology.

The interpretation of gas sensor signals is based on the gas-hydrogeological model. Currently, the best available sensors for oxygen gas are in situ redox electrodes and oxygen optodes (Engelmann, 2010). Carbon dioxide optodes are recently developed too. Flow-through monitoring systems (e.g., MIDZ or packer-sealed filters) can provide meaningful information about in situ pH and electrical conductivity conditions.

Starting a gas injection, initial gas sensor values are typically widespread. However, after matrix preoxidation and homogenization by water flow, sensor signals become meaningful. The signals can serve as a measure of the change in the heterogeneity of the reactive zone during operation of an RGBZ.

In situ redox sensors and oxygen optodes can be used to estimate gas propagation and dissolution due to their short reaction time. The travel times for coherent gases and clusters are measured as the time required for the breakthrough reactions of each sensor, and gas flow paths can be elucidated. The

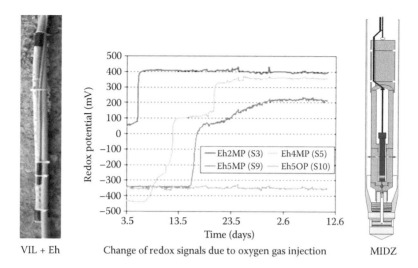

VIL + Eh     Change of redox signals due to oxygen gas injection     MIDZ

**FIGURE 10.9**
Gas monitoring devices for RGBZ and gas sensing signals (data from BIOXWAND). (VIL + Eh—loosen sonic lance filter with redox sensor; MIDZ—flow-through shuttle).

sensor value changes on the cessation of gas injections can be interpreted as the propagation of incoherent gas clusters and gas dissolution.

Another method for the estimation of gas propagation is trace gas testing; currently the best available are found to be noble gases (e.g., He, Ne, and Ar). Trace gases are mixed with a carrier gas and injected at low partial pressures. Due to a lack of interaction with soil and groundwater, environmental authorities have accepted the use of noble gases. Care is needed during trace gas sampling due to their high volatility. The use of pressurized samplers or bailers is also recommended (Uhlig, 2010 and Schinke, 2008).

### 10.3.3.3 Gas Saturation Testing

To estimate the amount of stored reactive gaseous substances, a gas saturation test that takes pressure dependency into account is required. The best available techniques for gas saturation estimations are (1) gas-hydrogeological balancing injection gas models, (2) direct gas profiling, and (3) local pumping tests in gas storage regions. Aqueous and partitioning trace gas infiltration methods are time-consuming and are currently still under evaluation.

Oxygen gas balance models can be parameterized using laboratory tests and gas monitoring. These models are suitable for the estimation of mean gas saturations in large gas storage zones, or layers during stationary operation periods (Ehbrecht and Luckner, 2004; Weber, 2007). The first step is to estimate the geometry of the storage zones, using gas-hydrogeological

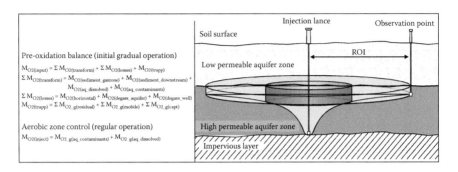

**FIGURE 10.10**
Example of a balance model scheme for NDI operation of an RGBZ. (From Weber, L. 2007: *Proc. Dresden Groundwater Research Centre*, Nr. 30, 151 p. (in German). With permission.)

surveying. Distributed gas input values; gas transfer (dissolution and consumption by groundwater and sediment), gas losses (horizontal escape from ROI and degassing) and gas storage (trapping) can be calculated or estimated (Figure 10.10). The complexity of the balance model is reduced during the regular operation of gas PRBs.

Direct gas profiling can be performed using borehole logging and sounding methods. Geophysical borehole logs are well known and evaluated (Dietrich and Leven, 2006). Calibration using a reference system (pilot or bench scales) is required. Gamma ($\gamma$-$\gamma$) logs detect the subsurface mean density distribution, while neutron (n-n) logs are sensitive to the presence and mass of hydrogen (water). The best results are achieved using the neutron method where gas saturations of 4% are significant and penetration radii are of >0.15 m.

The best resolution gas saturation data can be acquired using time domain reflectometry (TDR). TDR traces changes in the dielectric state of a domain, which is sensitive to the water content. TDR logging tubes (50 mm) can be installed by Sonic® technology. Adapted TRIME® sensors (Fundinger et al., 1996) were tested in a balanced pilot scale column device (Engelmann, 2010), and they have been used for continuous profiling. Changes in saturation of approximately 2% are significant, as are penetration radii of 0.30 m. TDR systems are recommended for identification of gas capture zones (Figure 10.11).

Hydraulic pumping tests can evaluate the impact of gas storage zones to groundwater flow. A field demonstration showed local lowering of conductivity from $4 \times 10^{-3}$ to $1 \times 10^{-3}$ m/s in gravel sediment near the tested well, and gas saturations of 7%–10% were reported.

### 10.3.3.4 Soil Gas Monitoring

Due to safety requirements, monitoring the continuous gas distribution and composition in the unsaturated zone is obligatory for RGBZ operations. A

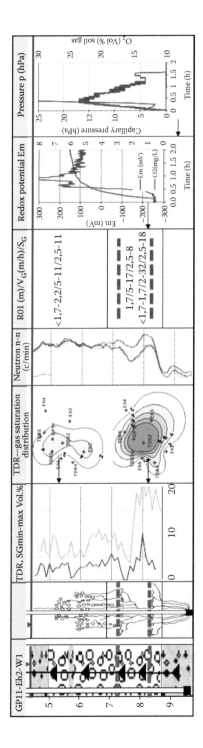

**FIGURE 10.11**

Combined gas sensing of storage zones with TDR, n-n-log, redox, O₂-optodes, and pressure detectors for gas reloading decisions. (From Engelmann, F.I. 2010: Model supported high pressure pulsed Gas Injection (HDI) for in situ remediation of contaminated aquifers: Technology for a controlled operation of gas storage zones and development of a measurement system of in situ gas saturation and a gas injection technology. Report Nr. KF0011010SB7–1, Sensatec, 46 p. (in German).)

large number of gaseous substances of interest and mixtures can be detected by sensors or on-site analyzers. Small diameter (25 mm) soil gas probes are installed at specified depths by direct push or manual electric ramming. Low flow pumping and on-site analysis is the preferred sampling method. In situ diffusive sensors are available; however the absorption rates are not as consistent or reliable.

Gas sensing and sampling near the capillary fringe can support the estimation of gas distribution and propagation in the groundwater zone. The initial signals of soil gas sensors are interpreted as breakthrough times and possible locations of gas emission from the groundwater domain. Soil gas sensors are used to check the sealing result of grouted injection lances. Additionally, gas consumption (e.g., oxygen, methane) or production (e.g., carbon dioxide) can be estimated by soil gas sampling, and the unsaturated zone can be included into an RGBZ treatment system (Uhlig, 2010).

### 10.3.4 Techniques for PRB Performance Control

Once reactive gases are dissolved into the groundwater flow, there are several measures or techniques that can be used to evaluate and control transport and the in situ transformation of dissolved compounds in the aqueous phase. These methods can also evaluate the interactions with reactive sediment surfaces (ENA). Sophisticated reactive modeling tools and techniques for site characterization and the identification of transformation process are also available. A number of these techniques require a site and contamination-specific application.

Given all of the possible methods to control the reactive zones of gas PRBs, an algorithm was constructed to take into account adaptive performance and optimization measures, and evaluated at the BIOXWAND gas PRB. The algorithm is site specific and given as an example in Table 10.2.

## 10.4 Example Applications of the RGBZ Technology

Three example applications have been chosen to demonstrate the capabilities of gas PRBs. First, the results of a full-section PRB (BIOXWAND) that has been operating for 5 years to treat an ammonium plume are presented. A homogenized nitrification effect was reached using injections of variable oxygen gas and air concentrations. The PRB was scaled up to a length of 800 m. In the second example, an oxygen gas PRB was used to naturally attenuate an organic contaminant plume containing aliphatic and aromatic hydrocarbons. The PRB technology has been accepted by the mining industry and environmental authorities as a method that can be used to prevent the future plume propagation or deviation due to mining activities using

**TABLE 10.2**

Algorithm for Optimization of the Performance of Gas PRBs

| Performance Limitations | Risk Lowering Measures | Optimization Actions |
|---|---|---|
| Distribution of reactants | Refining the injection array density by gas analysis-hydrogeological model | Horizontal ROI dimensions and overlapping<br>Injection gallery sequences in flow direction<br>Identification of gas retardants<br>Injection below treatment layer |
| | Variation of the injection rate and pulses | Injection rate (NDI) or pressure (HDI) change<br>Pulse frequency change<br>HDI–NDI combination |
| | Use of carrier gases | Hydraulic autoregulation by nitrogen clogging<br>Short-term reloading of reactants (e.g., oxygen)<br>Gas mixture supply (e.g., trace gases, methane-air) |
| | Use of a downstream reaction zone | Macrodispersion mixing of reactants<br>Amplification of reaction length and time |
| Gas dissolution | Forcing gas supply to fine to medium grained sediments | Formation of dense gas networks with high mass transfer interfaces<br>Homogenization of gas distribution |
| | Limiting gas supply to coarse sediments and capture zones | Prevention of inactive gas capture zones |
| Dissolved concentration range of reactants | Limiting reactant concentrations | Lowering gas saturation<br>Partial pressure variation of reactants |
| | Optimization of the geochemical state | Aerosol or foam injection to control, e.g., pH, Hardness and cometabolic degradation<br>Supply of higher oxidizers (e.g., $H_2O_2$) |
| | Nutrient supply | Supply of gaseous and solid nutrients (e.g., $CH_4$, $CO_2$, phosphate) |
| Clogging, permeability losses | Periodical redox state changes (aerobic/anaerobic) | Lowering the injection cycle time<br>Increasing break periods<br>Avoiding the carrier gas supply<br>Demand-oriented reactant supply |
| Bypassing of groundwater flow | Forcing autoregulation | Hydraulic forcing of bulk groundwater flow due to local pumping or drainage |

*Source:* Internal document of Sensatec GmbH, Kiel. With permission.

aerobic enhancement of biodegradation. The third example is an in situ drain and gate technology (GFIadags®) which included two RGBZs as a treatment train for a plume of a complex inorganic and organic contaminant containing ammonium, phenols, aromatics, DOC). A zone for the removal of iron by oxygen and ammonia gas and a polishing downstream oxygen gas reactor for the degradation of ammonium and DOC were formed.

### 10.4.1 The BIOXWAND Technology for Ammonium Elimination

Since the 1990s, the Berlin Water Company (BWB) has been working to safeguard a groundwater resource with a capacity of 10,000 $m^3$/d, which is used for drinking water production (reference). Approximately 200 million $m^3$ of groundwater was contaminated with 2200 tonnes of ammonium and organic trace cocontaminants including CHC (*cis*-DCE, vinyl chloride) and pesticides as a result of waste water infiltration and drainage from an unsealed sludge storage area of an upstream sewage field. A protection well gallery is being used to capture the contaminated stream, and groundwater with mean ammonium and organic trace substance concentrations of 10–20 and 0.02 mg/L respectively, are pumped out and treated at a nearby waste water plant. The extent of the contamination of the aquifer matrix is estimated to be 3000 tonnes of adsorbed ammonium, with approximately 2200 tonnes accessible to treatment using ion exchange (Ehbrecht and Luckner, 2004).

After the German Federal Ministry of Education and Research funded an evaluation of in situ cleanup approaches, the reactive gas barrier technology BIOXWAND (EP 1550519) was chosen as the best available method for the remediation and protection of the groundwater resource and therefore, the best option to replace the pump-and-treat system (Figure 10.12) (reference). Since 2007, a permeable oxygen gas barrier (length = 200 m, depth = 40 m, thickness = 25 m) has been installed approximately 500 m upstream of the drinking water well gallery A (Engelmann and Schmolke, 2014). The final length of the barrier is planned to be 800 m, and it was predicted that up to 200 kg/day of ammonium will be oxidized in situ.

Based on a mass balance approach and supported by reactive transport modeling (Horner et al., 2009), the initial annual oxygen demand for the performance of a 100 m barrier segment is approximately 64 tonnes. Approximately 28 tonnes/year of oxygen is needed to treat the inflowing groundwater (20 tonnes/year for nitrification, 8 tonnes/year for iron removal). A total of approximately 36 tonnes/year of oxygen is needed for the partial sediment matrix treatment of 22 tonnes/year of sulphide and 14 tonnes/year of adsorbed ammonium. The total oxygen demand declined with time due to gradual matrix oxidation.

The hydrogeology of the site is characterized by layered glacial sandy sediments to a depth of 50 m. Enclosed loamy lenses and sublayers in addition to sand layers which have been compacted to varying amounts act as retardants of the vertical gas propagation. In this way, there were four gas storage horizons within the unconfined aquifer (Figure 10.5).

An in-line injection gallery of sealed gas lances of types BIL, DIL, RIL, and VIL supplied the gas. The distance between the lances was 25 m and two injection filter depths (15 m and 40 m below groundwater level) were used. The low-pressure method (NDI) was applied, and gas injection rates were 0.5–2.0 $m^3$/h STP. The radii of influence (ROI) for effective horizontal gas propagation were approximately 10–25 m. In addition to the ROI and the

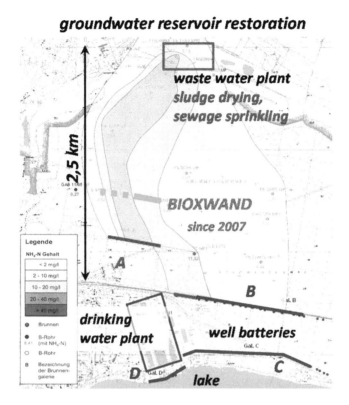

**FIGURE 10.12**
Site map of the BIOXWAND application area.

local variations in the injection regime based on hydrogeological profiling, monitoring was used to achieve a full-section PRB effect. The oxygen content of the injection gas varied between 20% (e.g., same as the atmosphere) and 100%. Gas injection cycles of 1–2 h were followed by breaks of 3–5 h, and coherent gas flow velocities >1 m/h were detected next to the injection lances. Vertical gas escape into the unsaturated soil zone was monitored as it occurred (e.g., when it exceeded the local aquifer gas storage and retardation capacities). In this case, stripping did not occur due to a lack of volatile solutes in the groundwater.

The BIOXWAND performance showed that it was impractical to aim for a quick remediation (e.g., satisfying the entire oxygen demand of 64 tonnes per 100 m) during the first year of barrier operation. This can lead to decreased operating efficiency with high gas losses mainly due to heterogeneous gas distribution, diffusion-limited gas dissolution, the variable and developing kinetics of matrix to groundwater exchange, and biochemical transformation processes. In the case of BIOXWAND, an initial 3-year operating regime was conducted, during which a total oxygen mass of 100 tonnes was supplied. The oxidation of the sediment matrix and nitrification rates were increased slowly during this time and a reliable homogenization of dissolved oxygen distribution of 5–50 mg/L were achieved. Groundwater redox potential was increased from approximately—200 mV to +500 mV after 3 years.

In situ gas storage monitoring was used to optimize the performance of the BIOXWAND. The mean gas saturation of 2%–4% was found to be an appropriate range for effective operation (Figure 10.13), however this was just the range of residual gas saturation in the sediment. The amount of time required for complete dissolution and consumption of such oxygen gas was estimated to be equal to that it required exchanging 1.5–2.0 pore volumes of groundwater. Local saturations of up to 17% were detected in some coarse sandy layers. This was linked to a localized reduction in hydraulic conductivity from $4 \times 10^{-3}$ to $1 \times 10^{-3}$ m/s. In this case, a hydraulic self-regulation and homogenization of the groundwater flow occurred. High groundwater fluxes in the coarser sections were decreased by preferential gas storage, whereas low fluxes in the finer-grained sections were increased with an increase in the local hydraulic gradients. Monitoring and control of the hydraulic flow homogenization in gas PRBs at the field scale are subjects of research, as they are important factors in the cost-effective operation of PRBs and their increasing acceptance from the point of view of the authorities.

It is reported in the literature that the supply of oxygen gas causes pyrite oxidation. Dissolved ferrous iron is predominantly precipitated as ferric iron hydroxides. Mass and volume balances for the in situ iron removal and field observations indicated that there was no significant risk of long-term pore clogging to groundwater flow or gas storage. Iron hydroxides precipitated mainly in the low-pore diameter regions (Figure 10.14).

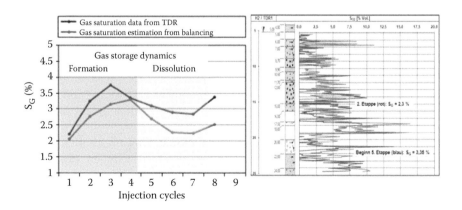

**FIGURE 10.13**
Gas storage control during BIOXWAND operation: changes of residual gas saturation (left) and local TDR gas sensing results (right).

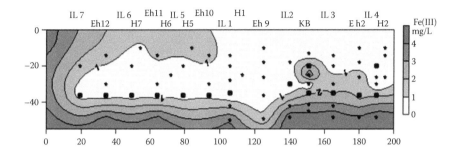

**FIGURE 10.14**
In situ iron removal in the gas storage zone after 3 years of the BIOXWAND operation.

Pyrite oxidation is accompanied by sulfate and proton production respectively. As a result of implementing the gradual barrier operation regime, production was limited and subsequent sulfate concentrations increased by 100–150 mg/L in fine-grained sands, and by 50–75 mg/L in coarser regions with lower pyrite content. The results showed that proton production due to pyrite oxidation was a reliable, but time-limited indicator of acidification potential during the initial operation period. The proton production was adjusted to the calcite buffering capacities of the aquifer matrix and the inflowing groundwater respectively, and the pH was stabilized at a mean value of 6.7, after it was decreased by 0.5–0.7 units. Calcite dissolution was accompanied by a slight hardening of the groundwater, and calcium ion exchange forced desorption of monovalent ions (e.g., sodium, potassium, and ammonium). This caused an initial increase in the ammonium concentration of 10–15 mg/L in the gas barrier zone.

Dissolved ammonium is transformed to nitrate by autochthonous microbes under aerobic conditions. The main species were *Nitrosomonas europaea, Nitrosomonas eutropha, Nitrosomonas halophila,* and *Nitrosococcus mobilis.* A lag-period of 30–50 days was needed for their activation after aerobic conditions were established. Laboratory tests indicated that an upper oxygen limit for nitrification was verified during the operation of the BIOXWAND. A significant inhibition was found when oxygen concentrations exceeded 50 mg/L. During the nitration step, no self-inhibition by nitrite was found. Proton production caused by nitrification occurred simultaneously with pyrite oxidation. It was estimated that the buffering capacity of the BIOXWAND would be lowered to approximately 90% of the initial value after 40 years of operation.

After 3 years of BIOXWAND operation, the ammonium concentration was reduced to <5 mg/L in the first 200 m-section (Figure 10.15). The nitrate was reduced to nitrogen by autotrophic denitrification under downstream anaerobic conditions. A slight lowering of DOC by approximately 1 mg/L indicated the transformation of organic compounds. CHC were degraded in the aerobic gas barrier zone.

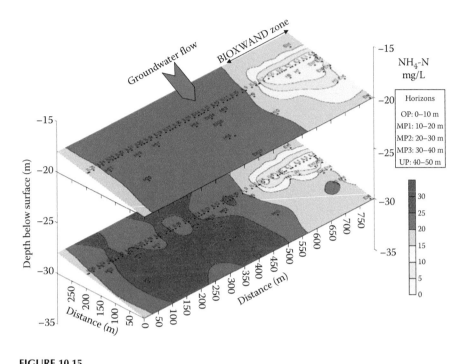

**FIGURE 10.15**
Results of ammonium degradation after 3 years of BIOXWAND operation.

## 10.4.2  Oxygen Gas PRB for Risk Coverage of MNA of an Organic Contaminant Plume

An unconfined aquifer in the vicinity of a former lignite processing site was impacted by organic pollutants (aliphatic, aromatic) and a 600 m long contamination plume had formed. Restoration of the site is conducted by a federal administration company specialized in postmining sites (the Lausitzer und Mitteldeutsche Bergbau-Verwaltungsgesellschaft mbH [LMBV]) under the supervision of the mining authority. Following a pump-and-treat decontamination, the Profen site was treated by monitored natural attenuation (MNA). The risk prognosis of the plume behavior was based on delineation of the damaged aquifer region, groundwater flow analysis and prediction, balancing of inventory and mass flow rates of contaminants, and identification of biodegradation processes. The results of the MNA showed that contaminant concentrations were reduced.

A primary protection goal was the prevention of contaminated groundwater impacts from the downstream active lignite mining. The groundwater flow is controlled by mining area drainage and a postmining pit lake. In this way, the cooperation with the mining company was needed in order to ensure an efficient long-term site restoration (Giese et al., 2012).

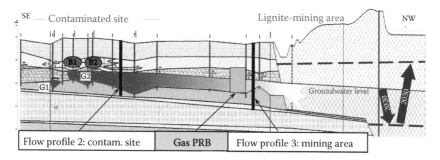

**FIGURE 10.16**
Cross section of the Profen site (in flow direction): plume propagation from the damage zones (B and G) is toward an active lignite mine pit.

The risk coverage of long-term MNA behavior was demanded by the mining authority. A technical measure was required in order to ensure risk prevention in the event MNA would fail, and to support and stabilize the accorded MNA prognoses. An oxygen gas PRB was identified as the best available technology for this purpose. A schematic representation of the PRB application is presented in Figure 10.16.

A field-scale demonstration and optimization of the gas PRB was performed over 12 months. The goals of the optimization were

- Determine the ability of the DGI to naturally attenuate the contaminants (e.g., forcing the aerobic biodegradation rates of contaminants in the plume)
- Plan a full-scale gas PRB technology that could be implemented as a risk coverage measure in the future (e.g., including owner and authority permission, evaluation of costs and time)

A 450 m² test site that included part of the plume center and a lateral inflow region was chosen. The average aquifer thickness was a 5-m saturated and a 4-m vadose zone (Figure 10.17). The upper aquifer was formed by highly permeable layers of gravel with sands, and the average groundwater velocity in the plume center was 0.8 m/day. A loamy top-layer acted as a gastight sealing. Typical BTEX, naphthalene, and petroleum hydrocarbon concentrations in the plume center were 3, 0.7, and 3 mg/L, respectively (Figure 10.18).

The groundwater level of 1.5 m fluctuated, and subsequent variations in the flow rate occurred due to a high recharge in summer and autumn 2010. Despite these fluctuations, the flow direction did not change. Soil core analysis during the site investigation indicated that there were still high concentrations of adsorbed contaminant in the plume center sediments, and in the

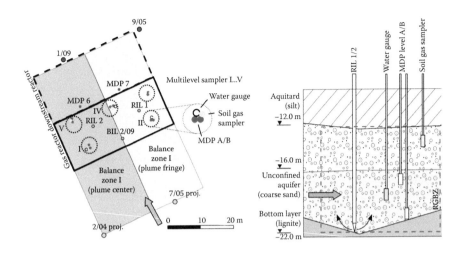

**FIGURE 10.17**
Test site scheme (left) and cross section (right) of the Profen gas PRB.

underlying impervious lignite layer. In this case, mass balancing of contaminants was limited.

The construction of the test site is presented in Figure 10.17. Packer separated double-valve pneumatic pumps (MDP 6/7) were used for the semi-integrated groundwater sampling of 25-mm mini filters (MF) (Figure 10.8).

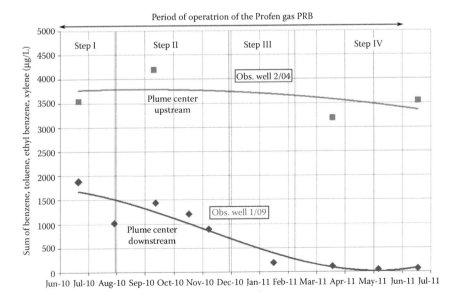

**FIGURE 10.18**
Impact of the Profen gas PRB to the groundwater load of balance zone I (plume center). The downstream trend was maintained about 6 months after PRB operation stopped (not shown).

During pump testing, the radii of influence were found to be 0.3–0.5 m. Upstream and downstream groundwater monitoring wells were used for an integrated mass flow evaluation using a 3D groundwater flow model. More wells in the vicinity and their long-term monitoring data were taken into account to determine the influence of the gas PRB on ongoing natural attenuation processes.

Neon and helium trace gases were used to determine an initial ROI of the coherent gas flow and gas escape to the vadose zone. These tests were repeated during a subsequent stationary period. An average ROI of gas flow of 5–8 m was detected in the groundwater zone; oxygen gas storage and dissolution efficiency of oxygen at approximately 60% was found using mass balance modeling. The slight aeration of the vadose zone was anticipated during the short-term testing of the gas PRB, and the effects of stripping and the safety implications were monitored in the vadose zone.

The operation of the Profen oxygen gas PRB consisted of three stages. Gas injection rates at lances were in the range of 0.25–0.5 $m^3/h$ STP.

| | |
|---|---|
| Period I (116 days): | Pre-oxidation by continuous injection of 26 kg $O_2$/day |
| Period II (77 days): | Forced sediment conditioning and initiation of biodegradation by continuous injection of 37 kg $O_2$/day |
| Period III (128 days): | Stabilization of biodegradation by pulsed injection of 21 kg $O_2$/day |

In periods I and II, almost the entire dissolved oxygen mass was needed for pyrite oxidation. Initial high sulfate production rates led to a temporary decrease in downstream pH and increased iron and manganese dissolution. During period III, the conditions for an optimized biodegradation of pollutants were established and pH >6.5 were found. Approximately, 20% of the dissolved oxygen was consumed in the transformation of contaminants. Additional details are reported in Zittwitz et al. (2012).

Aerobic biodegradation rates were estimated from mass balances, and were proven by laboratory testing and field-scale transport modeling. First-order rate coefficients of $0.07/day^{-1}$ for benzene and 0.04/day for naphthalene were found. Degradation ratios for the total mass flows of benzene and naphthalene were approximately 96% and 80% respectively.

In summary, the implementation of the oxygen gas PRB at the Profen site was performed during an ongoing MNA application. The ability to enhance natural attenuation potentials in a plume of dissolved aliphatic and aromatic hydrocarbons was demonstrated. No meaningful interference to the accorded MNA prognoses outside the PRB zone was found, due to miminal impacts on the groundwater flow. However, the efficiency of oxygen gas storage and dissolution should be increased significantly in order to optimize the cost effectiveness of a full-scale application.

The gas PRB technology was found to be suitable for the technical risk coverage of MNA. The full-scale application was based on groundwater flow and transport modeling of potential failing scenarios of MNA, due to advancing lignite mining. The mining company supported the planning by providing data regarding the anticipated water management of the lignite mine pit and by facilitating access to a PRB reservation area. The costs and safety issues were also evaluated. The mining and environmental authorities confirmed the treatment targets and the operation chart, and the gas PRB technology became part of a long-term operating closure plan for the Profen site.

### 10.4.3 Reactive Gas Zones as Part of the GFIadags®-Technology

Reactive gas zones were integrated in a drain and gate technology for the sequential plume treatment of deep aquifers. The treatment train technology was demonstrated at the Schwarze Pumpe site, a former gasification plant. A plume containing high concentrations of phenols (30 mg/L), DOC (100 mg/L), and ammonium (150 mg/L) required treatment in a 37 m deep multilayered aquifer. The thickness of the saturated zone was 20 m, and average groundwater velocity was approximately 0.12 m/day. The gate treatment (zone B) consisted of stripping and chemical oxidation of groundwater contaminants in collector and distributor well reactors (Kassahun et al., 2005). Gas injection zones were established to perform iron removal (zone A: treatment area of 900 m²) and posttreatment of ammonium and DOC (zone C: treatment area of 1.800 m²). The treatment train is presented in Figure 10.19. The construction of the gas injection zones followed the principles discussed in Section 10.3. Additional details are reported in Uhlig (2010). Due to high contamination, partial decontamination of the soil matrix was addressed to form in situ buffer zones against breakthrough of fluctuating contaminant streams.

In zone A, in situ iron removal was first induced by oxygen gas injection. As seen in Figure 10.19, the competitive effects of matrix oxidation limited the success. Carbonate precipitation of dissolved iron by ammonia gas injection was shown to be more efficient, as matrix oxidation did not exert an influence. Ammonia demand depended mainly on the buffering capacity of the groundwater flow. A conditioning pH of >7.5 was required, and injection rates were controlled by mixing ammonia gas to a nitrogen carrier gas flow of 0.5–1.0 m³/h STP. The ammonia injection approach was found to play a part in contaminated site restoration; however, in situ processes require further investigation.

An oxygen gas PRB for bio-oxidation was established in zone C, and was operated over a period of 550 days. Gas lances and observation elements of the types MDP and MF were installed by CPT (see Chapter 3). A 3D gas-hydrogeological model was constructed for groundwater flow and reactive zone balance modeling. Gas injection rates of 0.6–1.2 m³/h STP were applied, and ROI of single lances were identified using noble trace gas (Ne, He) in the range of 15 m. A downstream reactor was not monitored.

Due to high contamination and the natural pyrite content of the matrix, almost all of the 20 tonnes of injected oxygen gas was consumed by matrix

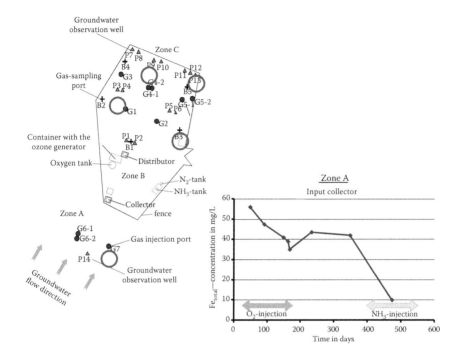

**FIGURE 10.19**
Left: map of the Schwarze Pumpe drain and gate test site with the reactive zones A to C; right: evidence of iron removal by gas injection in zone A at the collector well.

oxidation (36%), or transferred and consumed in the vadose zone (51%) without provoking a dominant stripping of VOC. With a 150-day lag time, enhanced aerobic biodegradation of the complex organic and inorganic contaminant plume was initiated and reached up to 11% of consumption of the total oxygen gas supply. Simultaneous heterotrophic and autotrophic biodegradation was found (Figure 10.20) and chemical oxidation from the initial excess supply of oxygen gas was also found. However, nitrification in zone C remained limited as the required chemical preoxidation of the hydrocarbon mass by ozone in the gate reactor of zone B was not turned on during the test period. Degradation rates were found to be 0.05–0.1/day for benzene and short-chained alkyl phenols (Uhlig, 2010).

The Schwarze Pumpe site example demonstrated the suitability of gas PRBs in treatment train applications, particularly when a complex contamination situation is present. Often, site restoration and impact reduction targets cannot be achieved or conducted economically by stand-alone applications of a main treatment technology (e.g., a pump and treat). The application of gas PRBs offers a wide range of aerobic and anaerobic conditioning. Posttreatment or polishing measures are required (e.g., for final degradation of CHC, hydrophilic alcohols [in situ flushing] or phenols [MPPE] extraction).

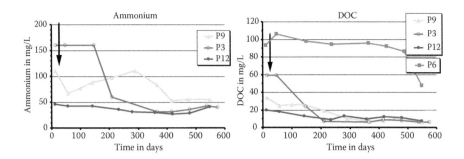

**FIGURE 10.20**
Posttreatment effect of the reactive gas zone C at Schwarze Pumpe site.

# References

Beckmann, A., M. Zittwitz, M. Gerhardt, M. Martienssen, R. Krieg, H. Geistlinger, and M. Schirmer 2007: The OXYWALL project: Application of a method of direct oxygen gas injection for remediation of contaminated by organic pollutants groundwater. Altlasten Spektrum 04/2007 (in German).

Bilek, F. and S. Wagner 2009: Testing in situ sulphate reduction by hydrogen gas injection in a bench scale column experiment. _Water Air Soil Pollution_, 203, pp. 109–122.

Boerner, F., J.R. Schopper, and A. Weller 1996: Evaluation of transport and storage properties in the soil and groundwater zone from induced polarization measurements. _Geophys. Prospecting_, 44, pp. 583–601.

Brooks, M.C., W.R. Wise, and M.D. Annable 1999: Fundamental changes in in situ air sparging flow patterns. _Ground Water Mon. Rem._, 19(2), pp. 105–113.

Dietrich, P. and C. Leven 2006: Direct push technologies. In: R. Kirsch (ed.): _Groundwater Geophysics._ pp. 321–340. Berlin, Springer.

Ehbrecht, H. and L. Luckner 2004: BIOXWAND: Development and test of a Bio-Oxidation Wall in an ammonium contaminated Aquifer. Report Nr. 02 WT 0091, Berlin Water Company, Berlin, 165 p. (in German).

Engelmann, F.I. 2010: Model supported high pressure pulsed Gas Injection (HDI) for in situ remediation of ontaminated aquifers: Technology for a controlled operation of gas storage zones and development of a measurement system of in situ gas saturation and a gas injection technology. Report Nr. KF0011010SB7–1, Sensatec, 46 p. (in German).

Engelmann, F.I., H. Ehbrecht, C. Roscher, and R. Giese 2004: Technologies for the formation of permeable reactive walls in aquifers by injection of reactive gases. _Proc. Dresden Groundwater Research Centre_, Nr. 24, pp. 115–126 (in German).

Engelmann, F.I., P. Hopp, L.P. Schmolke, and H. Ehbrecht 2009: Vibration Sounding Gas Lances and in situ Measurement Arrays. Formation of Gas Storage Zones and their monitoring for Groundwater Remediation. TerraTech, 11–12/2009 (in German).

Engelmann, F.I., P. Hopp, L.P. Schmolke, and H. Ehbrecht 2010: Operation of Gas Storage Zones. Control due to Gas Saturation Measurements. TerraTech 2011 (in German).

Engelmann, F.I. and L.P. Schmolke 2014: Treatment of a groundwater stream to the Lake Mueggelsee. wwt-wasserwirtschaft Wassertechnik, Nr. 1-2/2014, pp. 43–46 (in German).

Fundinger, R., K. Köhler, and M. Stacheder, 1996: Measurement of Material and Soil Moisture by TRIME method. In: VDI/VDE (Hrsg.): Sensors and measurement systems. Report VDI 1255, pp. 417–422 (in German).

Geistlinger, H. 2010: Model supported high pressure pulsed Gas Injection (HDI) for in situ Remediation of contaminated Aquifers: Laboratory scale Experiments and Computer Simulations for Optimization of the Technology. Report Nr. KF0011010SB7–2, Helmholtz-Centre for Environmental Research UFZ, 63 p. (in German).

Geistlinger, H., A. Beckmann, and D. Lazik 2005: Mass transfer between a multicomponent trapped gas phase and a mobile water phase: Experiment and theory. *Water Res. Research*, 41, paper W11408, 15 p.

Geistlinger, H., G. Krauss, D. Lazik, and L. Luckner 2006: Direct Gas Injection into glass beads: Transition from incoherent to coherent gas flow pattern. *Water Res. Research*, 42, paper W07403, 12 p.

Giese, R. 2012: Development of a process-oriented Model for Permeable Reactive Barriers as Gas-Water-Solid-Reactors. An Instrument for Design, Operation and Maintenance. In: Birke, V. and Burmeier, H.: Application of Permeable Reactive Walls for Brownfield Remediation. Vol. 2, pp. 65–72 (in German).

Giese, R. and T. Reimann 2003: Dispergation, transport and immobilization of reactive solids by gas injection in packed porous media. *Proc. Freiberg Research Forum*, Vol. 54/3, p. 65–73 (in German).

Giese, R., A. Thomas, J. Schmidt, M. Uhlig, and P. Tropp 2012: Safeguarding of a MNA concept under mining law conditions. 1—control and further development of accepted by authorities prognoses of MNA. TerraTech 03/2012 (in German).

Giese, R., L. Weber, and S. Baidatsch 2003: Dynamics of Soil Structure during Gas Sparging. Experiments under 3D mechanical Stress. Reports of the German Society of Soil Science, Vol. 101. (in German).

Horner, Ch., F.I. Engelmann, and G. Nuetzmann 2009: Model based Verification and Prognosis of Acidification and Sulphate releasing Processes downstream of a former Sewage Field in Berlin (Germany). *J. Contam. Hydrology*, 106, pp. 83–98.

ITVA 2010: Innovative in situ Remediation Methods. Practice guide H 1–13., German Scientific-Technical Association for Environmental remediation and Brownfield Development ITVA, (in German).

Kassahun, A., U. Uhlig, D. Burghardt, J. Masnica, and L. Luckner 2005: Reactive zone design for in situ chemical and biological oxidation of contaminated groundwater. *Proc. 2nd PRB/RZ-Symposium*, 14–16 Nov. 2005, Antwerpen, pp. 44–48.

Luckner, L. 2001: Effect of Gases to Reactive Transport in the Subsurface. Grundwasser, 04/2001, pp. 155–162 (in German).

Luckner, L. and W.M. Schestakov 1991: Migration Processes in the Soil and Groundwater Zone. Lewis publishers, 485 p.

Schinke, R. 2008: Investigation of Direct Gas Injection in damaged by Dumps and Landfills Aquifers with regard to Natural and Enhanced Attenuation Processes. *Proc. Dresden Groundwater Research Centre*, Nr. 34, 121 p. (in German).

Schmolke, L.P., F.I. Engelmann, C. Horner, P. Hopp, S. Nützmann, and S. Hüttmann 2007: Pilot application BIOXWAND and its control. *Proc. Dresden Groundwater Research Centre*, Nr. 31, pp. 135–146 (in German).

Selker, J.S., M. Niemet, N.G. McDuffy, S.M. Gorelick, and J.Y. Parlange 2007: The local geometry of gas injection into saturated homogeneous porous media. *Transport in Porous Media*, 68: 107–127; doi: 10.1007/s11242-006-0005-0.

Uhlig, U. 2010: Investigation of multi-step in situ groundwater remediation by pilot scale testing. *Proc. Dresden Groundwater Research Centre*, Nr. 42, 115 p. (in German).

Wang, Z., J. Feyen, and D.A. Elrick 1998: Prediction of fingering in porous media. *Water Res. Research*, 34/9, pp. 2183–2190.

Weber, A., A.S. Ruhl, and R.T. Amos 2013: Investigating dominant processes in ZVI permeable reactive barriers using reactive transport modelling. *J. Cont. Hydrology*, 2013 Aug, 68–82; doi: 10.1016.j.jconhyd.2013.05.001.

Weber, L. 2007: Investigation of Direct Gas Injection in Aquifers in the bench and field Scales. *Proc. Dresden Groundwater Research Centre*, Nr. 30, 151 p. (in German).

Zittwitz, M. and M. Gerhardt 2006: The method of Methane Biostimulation. TerraTech 10/2006 (in German).

Zittwitz, M., D. Swaboda, J. Schmidt, and A. Thomas 2012: Safeguarding of a MNA Concept under mining law Conditions. 2—Technology Preparation for supporting a Monitored Natural Attenuation Concept. TerraTech 03/2012 (in German).

# 11

## Remediation of PAHs, NSO-Heterocycles, and Related Aromatic Compounds in Permeable Reactive Barriers Using Activated Carbon

**Wolf-Ulrich Palm, Jan Sebastian Mänz, and Wolfgang Ruck**

### CONTENTS

## 11.1  Introduction

Permeable reactive barriers (PRBs) are alternatives to common active groundwater remediation technologies including those based on pump and treat [1–5]. PRB is a passive in situ groundwater remediation technique that avoids several inherent technical drawbacks of active systems a priori. A PRB is defined as an in situ method for remediating contaminated groundwater which combines a passive chemical or biological treatment zone with subsurface fluid flow management [6]. PRBs that were first installed in the United States in the early 1990s used zero-valent iron (e.g., elementary iron). Due to the inability of iron (Fe) to efficiently remediate polycyclic aromatic hydrocarbons (PAHs), for the past 15 years, activated carbon has been used in PRBs as an additional adsorbent for PAHs and other related organic compounds [7,8].

Commencing in 2000, the Federal Ministry of Education and Research in Germany (BMBF) funded the project RUBIN ("Reaktionswände und -barrieren im Netzwerkverbund" (German Permeable Reactive Barrier Network)). The second stage of the project RUBIN II, was launched in 2006 [9]. At the same time, a different view and philosophy of remediation techniques excluding human intervention led to the development and trial of the concept of natural attenuation. The application of natural attenuation to the remediation of NSO-heterocycles (for instance in tar oil-contaminated areas) was investigated in Germany within the project KORA ("Kontrollierter natürlicher Rückhalt und Abbau von Schadstoffen bei der Sanierung kontaminierter Grundwässer und Böden" (Retention and Degradation Processes to Reduce Contaminants in Groundwater and Soil)) [10] that was also funded by the BMBF.

Basic concepts for the analysis of PAHs and NSO-heterocycles were developed within the framework of the KORA project. However, most of the results presented here that relate to the efficiency of activated carbons in PRBs to remediate NSO-heterocycles were obtained within Subproject 3 of RUBIN II.

### 11.1.1  Solubility of NSO-Heterocycles and Their Parent Aromatics

The derivation of innovative methods for the analysis of hydrophobic compounds such as the classical 16 EPA-PAHs or phenols and highly polar compounds such as N-heterocycles (N-HETs) is necessary due to the heterogeneous spectrum of compounds typically present in contaminated groundwater. In addition, $pK_a$-values of many N-HETs and especially hydroxylated N-HETs in the range of $3 < pK_a < 7$ suggest pH-dependent properties. This is an important factor that needs to be taken into account in designing analytical methods and adsorption experiments. Due to pH-dependent adsorption parameters, the adsorption equilibrium could be different for cations (e.g., N-HETs at pH $\ll pK_a$) and anions (e.g., phenols or carboxylated compounds at pH $\gg pK_a$) compared to the neutral molecules [11–15]. Therefore, pH-values at the surface of activated carbons (for PRBs the pH of the corresponding groundwater is assumed), and $pK_a$-values of the adsorbents are important to assess the efficiencies of PRBs. A comparison of the solubilities of NSO-heterocycles and their parent hydrocarbons and PAHs is shown in Figure 11.1.

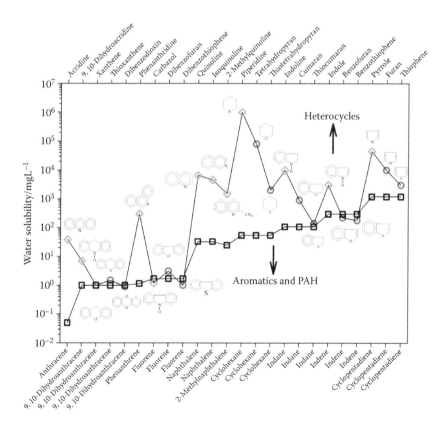

**FIGURE 11.1**
A comparison of water solubilities of representative NSO-heterocycles (neutral molecules) and their corresponding parent aromatic compounds. (Experimental data from EPI-Suite U.S. Environmental Protection Agency. 2012. Estimation Programs Interface Suite™ for Microsoft® Windows, v 4.11.)

The solubilities of short-chained alkylated phenols at pH ≪ pK$_a$ in general are approximately a factor of 50 higher than that of their parent aromatic compounds. In contrast, a comparison of NSO-heterocycles to their parent hydrocarbons and PAHs presents a different view. The solubilities of S-heterocycles are comparable to their analog hydrocarbons; a comparable and in some cases higher solubility is found for O-heterocycles. However, the solubilities of N-HETs are usually much higher (for instance the solubility of acridine is about a factor of 1000 higher compared to that of anthracene). Only a relatively few measurements are available for pH-dependent solubilities. However, the Henderson–Hasselbalch Equation 11.1 is often used to predict solubilities ($S$ = solubility, $S_0$ = intrinsic solubility of the neutral molecule) [17]:

$$S = S_0 \cdot (1 + 10^{(pK_a - pH)}) \tag{11.1}$$

2-Methylquinoline and 9-methylacridine should be used as instructive examples. For 2-methylquinoline, the pK$_a$ = 5.86 was obtained by ultraviolet (UV) spectroscopy in this study in agreement with a literature value [18]. An estimated intrinsic solubility $S_0$ = 3.6 g/L for 2-methylquinoline is available [16]. The solubilities of 2-methylacridine between pH = 5–9 are shown in Figure 11.2. Extrapolating these figures, the predicted solubility at pH = 4 is approximately 300 g/L. In contrast to 2-methylquinoline, the solubility of

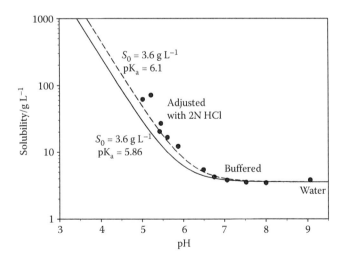

**FIGURE 11.2**
The dependence of experimentally determined water solubilities of 2-methylquinoline from pH (filled circles). Lines shown are calculated solubilities using the Henderson–Hasselbalch equation for pK$_a$ = 5.86 (experimental value, solid line) and for a slightly modified pK$_a$ – 6.10 (broken line).

9-methylacridine ($S_0 = 9.3$ mg/L, determined in this study) is assumed to be almost independent from environmentally relevant pH-values due to the low p$K_a = 3.9$ [19].

---

## 11.2 Experimental

### 11.2.1 Instrumentation and Materials

Numerous methods were applied to analyze samples of different origins. In addition to compounds such as the EPA-PAHs and the short-chained alkylated phenols, alkylated PAHs and NSO-heterocycles usually found in groundwater were incorporated into the compound list. This list of compounds was verified by analyzing contaminated groundwaters using gas chromatography–mass spectrometry (GC–MS) (full scan, m/z = 100–500) and was expanded to account for additional compounds and numerous isomers. A list of compounds analyzed in this study can be found in Table 11.7.

Compounds were purchased from a number of suppliers, typically as reference standards. In addition to the reference standards, single compounds with purities between 95% and 99% were used as received. 2-Methylquinoline was distilled before use.

Concentrations of single compounds and mixtures for the determination of adsorption isotherms were obtained using high-performance liquid chromatography with a diode array detector (HPLC-DAD), HPLC-fluorescence (Agilent HPLC-type 1100 series), and headspace GC (Perkin Elmer Autosystem XL, HS40 headspace sampler, and flame ionization detector (FID). Solutions from column experiments were analyzed by GC–MS (Perkin Elmer Turbo Mass and Thermo Finnigan Trace DSQ) and samples from field measurements were analyzed following appropriate extraction processes using GC–MS and liquid chromatography–mass spectrometry (LC–MS/MS) (Agilent 6430 Triple Quadrupol with Agilent HPLC-type 1200 series). In addition to the analysis for organic compounds, on-site parameters (pH, conductivity, temperature, oxygen concentration, and redox potential; instrumentation from WTW, Weilheim, Germany) and the concentrations of anions (ion chromatography, Dionex DX 120) and cations (ICP, Perkin Elmer Optima 3300RL) were obtained for groundwater samples. Cyanide analysis was performed at the University of Cologne (Germany).

### 11.2.2 Analytics

Most samples were extracted using liquid–liquid extraction with pentanoic acid methyl ester [20] and analyzed by GC–MS [20–22] and a very brief description of this method is given only here. Extraction efficiencies, detection limits, and more details of this method besides a more sophisticated solid-phase extraction (SPE) [23] for low concentrations in combination with

LC–MS/MS used for some samples from reactive barriers can be found elsewhere [24–26].

### 11.2.2.1 GC–MS Analysis

Samples were dissolved in toluene or toluene/acetonitrile mixtures and analyzed using the following conditions: column OPTIMA 5 MS (Macherey Nagel), 30 m × 0.25 mm ID, 0.25 µm film; splitless injection with an injection volume of 0.5 µL; carrier gas helium with a flow of 1 mL/min; temperature program: 50°C for 3 min → 90°C, 20°C/min for 3 min → 150°C, 4°C/min → 300°C, 8°C/min for 10 min; and single ion monitoring (SIM)-mode detection.

### 11.2.2.2 LC–MS/MS Analysis

LC–MS/MS analysis was performed with the following conditions: column-Nucleodur C18 PAH, 125 × 3 mm ID, 3 µm particle size (Macherey Nagel); injection volume: 5 µL; flow 0.4 mL/min; binary eluent (A) acetonitrile (B) 1 mM $NH_4OOCCH_3$ in water (pH 7), starting with 90% B, decrease to 20% B within 12 min, hold for 2 min, increase to 90% B within 0.1 min and hold for 5.9 min; temperature 40°C; postcolumn derivatization with 10 µL/min formic acid in water (1%); and detection by MS/MS (electrospray ionization, multiple reaction monitoring (MRM) mode).

### 11.2.2.3 Liquid–Liquid Extraction for High Concentrations: GC–MS Only

A sample volume of 20 mL was fortified with a 50 µL internal standard in a crimp-neck vial. After shaking the sample to dissolve the internal standard, a triple-distilled pentanoic acid methyl ester (950 µL) was added. After shaking the airtight crimped vial for 60 min in a head-over-head shaker, the sample was centrifuged to facilitate phase separation. An aliquot of 600 µL of the organic phase was dried over sodium sulfate; 500 µL was used as a retained sample (stored at –20°C); and 100 µL was used for analysis (in 200 µL amber-glass micro vials).

### 11.2.3 Activated Carbons

Many different types of activated carbons are commercially available and activated carbons from three suppliers were used: "Epibon Y12×40," "Hydraffin 30N," "Hydraffin NA15" and "Hydraffin regenerated" from Donau Carbon, "ROW supra" and "GAC 1240" from Norit, and "F200" and "F400" from Chemviron. Properties of all activated carbons are available from the supplier. Comparable ACs were used in both of the reactive barriers being studied (see below). However, most experiments in the laboratory (isotherms for single compounds, column runs) were performed with the AC Epibon Y12×40 (Donau Carbon), whereas a comparison of different activated carbons was performed in column and batch experiments using mixtures. The choice of Epibon Y12×40 resulted from preliminary column tests conducted in an on-site 6.1-m research container using groundwater at a contaminated site in Germany (Zeche Viktoria, Lünen, well 12Q) early in the project and repeated for comparison purposes throughout the project. However, from subsequent comparisons in the laboratory, AC F400 (Chemviron) and not AC Epibon Y12×40 was found to be the most efficient activated carbon with respect to the adsorption parameters from isotherm experiments using artificial mixtures. For the activated carbons investigated, the adsorption characteristics were found to be comparable (at least within a factor of 2), and therefore Epibon Y12×40 was considered to be representative of the activated carbons used in the remediation of polycyclic aromatics and related compounds.

A detailed protocol was developed to investigate the adsorption isotherms of single compounds and mixtures in an aqueous solution (deionized water, buffered solutions, and real groundwater). A very brief overview is given here. Activated carbons were washed, dried, sieved (63–125 µm), and stored at 50% relative humidity. Batch experiments were performed in airtight crimped headspace vials (22.4 or 116 mL, tempered before use at 400°C). A mass of 50 µg–100 mg of AC was placed into a vial and filled up with the corresponding solution (head space <1%). Freshly prepared solutions with single compounds or mixtures of at least a factor of 2 below the solubility in water were prepared in either deionized water or in buffer solution. Solutions were checked before use with the suitable analytical method. In addition to 2–4 blank samples, 10–12 samples prepared for one-batch experiment were shaken in a head-over-head shaker (Heidolph Reax 2) for 72 h at room temperature ($T = 20 \pm 3°C$). After being centrifuged 2 times (Heraeus Megafuge 1, 3300×g), the samples were analyzed without further enrichment by either HPLC or headspace GC. Care had to be taken with respect to the storage of the activated carbon. Water is adsorbed on activated carbon, quantified by water isotherms using a differential scanning calorimetry with thermogravimetric analysis (DSC/TGA) with an instrument from Mettler/Toledo. In addition, volatile compounds such as benzene or toluene can be lost during sample preparation. Furthermore, the efficient removal of activated carbon from the aqueous solution is essential, especially for low-equilibrium concentrations.

Equilibrium concentrations $c$ (in mg/L) were always obtained by HPLC or headspace GC without any corrections. The loading $q$ (in mg/g) was calculated from Equation 11.2:

$$q = \frac{V}{m}(c_0 - c) \tag{11.2}$$

where
 $V$ = volume in L
 $m$ = mass-activated carbon in g
 $c_0$ = start concentration in mg/L

### 11.2.4  Column Experiments

A number of column experiments were performed using artificial solutions and real groundwater. Column experiments were performed with the activated carbon Epibon Y12×40 (Donau Carbon) in identical stainless-steel columns with tubings made of teflon (PTFE) for influent, effluent, and sampling ports (column diameter: 27.6 mm, length: 267 mm, see Figure 11.3).

Sieved (Fritsch Analysette 3 Pro with 1 mm sieve) activated carbon was washed in an acrylic glass tube (10 × 100 cm) with a suitable flow of water from the bottom to remove small particles. Columns were filled and packed with the prepared wet-activated carbon. In addition, one column was used with 6.5% (w/w) of activated carbon from the reactive barriers in Karlsruhe (Germany) and Brunn am Gebirge (Austria) to establish a biocenosis. After filling each column with 76 g of activated carbon, the columns were equilibrated in a water bath for approximately 10 h with a gentle flow of water through the columns. Numerous experiments were performed to characterize the activated-carbon packing with the following results: $d(corn) \pm \sigma = 1.0 \pm 0.2$ mm, $m(corn) \pm \sigma = 0.77 \pm 0.16$ mg, $\rho(corn) \pm \sigma = 1.01 \pm 0.15$ g/cm, $\rho(carbon\ particle): = 1.83$ g/cm, $\rho(bed) = 0.48$ g/cm, $\varepsilon(corn) = 0.44$, and $\varepsilon(bed) = 0.52$.

Artificial mixtures and contaminated water from different wells at the site Zeche Viktoria in Lünen (Germany) were used in column experiments. Groundwater was obtained using a pump (MP-1, Grundfos, Germany) and

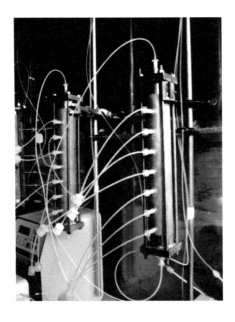

**FIGURE 11.3**
Two columns running in parallel to investigate breakthrough curves, adsorption ranking, and the influence of biotic degradation of adsorbed compounds on activated carbon.

transported and stored in intermediate-bulk containers (IBC), made of high-density polyethylene (HDPE) with a volume of 1000 L. Carbon dioxide gas was initially used to maintain an anoxic atmosphere during the transport, and was later replaced with nitrogen gas (the pH did not change using carbon dioxide). Columns, the sampling equipment, gases, and 2000 L of groundwater from the contaminated site in the Ruhr area were placed in a climatized research container on the campus and held at 12°C. An oxygen-free atmosphere (at excess pressure in the tank) was maintained using nitrogen gas. Hence, the headspace in the tank was always flushed with nitrogen gas to achieve an anaerobic atmosphere. With the aid of a peristaltic pump, a flow rate of 200 μL/min was maintained for both columns for a period of 98 days.

In addition to the contaminated groundwater, a nutrient mixture (ammonium and phosphate as N and P sources) was used in the first column (biotic column), and a sodium azide solution was used in the second column (abiotic column) to prevent microbial activity. The efficiency of this method was checked for incubated samples using effluent water from both columns and in contrast to the "biotic-column," no colonies were found in the effluent water from the "abiotic-column." Weekly samples were taken at seven different ports (port 7 represented 80% of the activated-carbon mass in the column) and from the influent flow. The influent concentrations were found to be stable, for instance a loss <10% within 100 days was found for carbazole, whereas for the highly adsorbing compound fluoranthene, a loss of about 40% was found over the same time period.

## 11.2.5 Contaminated Sites and Investigated Reactive Barriers

Two PRBs in Germany (Karlsruhe) and in Austria (Brunn am Gebirge, near Vienna) were investigated using detailed sampling campaigns. In addition to wells located in the inflow areas of both PRBs, influent and effluent areas of the reactors located in the gates of both PRBs were sampled. Samples analyzed for anions were transported without a stabilizer; samples analyzed for cations were stabilized with nitric acid; and samples analyzed for cyanide were stabilized with sodium hydroxide. In some cases, total organic carbon (TOC) was determined and such samples were stabilized with phosphoric acid. Field campaigns commenced in summer 2006 and individual samples analyzed for organic compounds were transported in 2.5 L amber-glass bottles. For GC analysis, samples were stabilized by sodium azide. All bottles containing the samples were transported in ice-cooled transport boxes.

Operating parameters of both PRBs are summarized in Table 11.1. More information on the PRB in Karlsruhe (Germany) can be found in References 27 and 28, and additional information on the PRB in Brunn am Gebirge (Austria) can be found in Reference 29. Four sampling campaigns were organized (the dates are presented in Table 11.2), and samples were taken as described above.

**TABLE 11.1**

Descriptions of the Two Reactive Barriers in Austria and Germany Investigated in This Study

| | Reactive Barrier | |
|---|---|---|
| Parameter | Karlsruhe (Germany) | Brunn am Gebirge (Austria) |
| Constructed | 2000/2001 | 1999 |
| barrier type | Full-scale, GAC (granulated activated carbon) | Full-scale, GAC |
| Concentration sum of the contaminants | About 2000 µg/L | About 8000 µg/L |
| Length of barrier | 240 m | 220 m |
| # of gates (reactors) | 8 | 4 |
| Dimension reactors | Diameter 1.8 m, height 15–18 m (volume for AC filling 21–27 m$^3$) | Diameter 2 m, height 6–8 m |
| AC used | GAC TL830 (Chemviron) for gates 1, 2, 7, and 8 and D 43/1 (CarboTech) for gates 3, 4, 5, and 6 | CC 15 (Donau Carbon) for all four gates |
| Sum mass AC | About 100 tonnes | About 20–24 tonnes |
| Water flow | 9–10 L/s (mean, total flow through the whole system) variable flow through individual gates 1–8:1.0–1.4 L/s | 0.5–2 L/s (total flow through the whole system) |

**TABLE 11.2**

Field Campaigns Performed at the Reactive Barriers in Austria (Brunn am Gebirge) and Germany (Karlsruhe)

| Reactive Barrier | Field Campaigns 1 + 2 | # Samples | Field Campaigns 3 + 4 | # Samples |
|---|---|---|---|---|
| Brunn am Gebirge | November 17–19, 2007 | 17 | July 10–14, 2008 | 13 |
| Karlsruhe | July 24–26, 2007 | 17 | September 14–16, 2009 | 21 |

## 11.3 Results

### 11.3.1 Extraction and Analysis of Organic Compounds

The accuracy of the analysis applied was demonstrated for a dense non-aqueous phase liquid (DNAPL) tar oil from the bottom of well 12Q at the contaminated site "Zeche Viktoria" in Lünen, Germany. A completely dissolved sample in dichloromethane was diluted with toluene and analyzed by GC–MS. The mass calculated from all of the concentrations analyzed in the sample yielded up to 98.7% of the mass of the tar oil (see Figure 11.4a). Hence, the spectrum of compounds analyzed covered almost all of the main

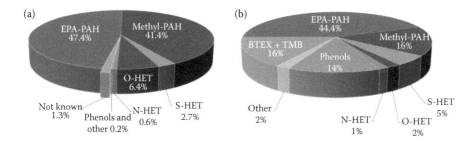

**FIGURE 11.4**
Analysis of a sample from the contaminated site Zeche Viktoria in Lünen (Germany, well 12Q). (a) Tar oil (weight %) from the bottom of the well. The mass can be explained by 99% using GC–MS for all compounds investigated (without BTEX). (b) Groundwater (% mass concentration in relation to the sum concentration of 16 mg/L).

components present in this specific tar oil that had been in equilibrium with the surrounding groundwater for decades.

Toluene as a solvent excludes benzene, toluene, ethylbenzenes and xylens (BTEX) from the analysis by GC–MS and therefore headspace analysis is not applicable. One can argue that, due to the high water solubility and vapor pressure, BTEX and phenols are removed by an efficient natural extraction of the tar oil by the surrounding water. This interpretation is in accordance with the very low concentrations of phenols found in such samples. In contradiction to the concentrations in tar oils, relatively high concentrations of BTEX and phenols were found in the corresponding water of well 12Q from "Zeche Viktoria" (Figure 11.4b).

## 11.3.2 Adsorption Isotherms on Activated Carbon

Using suitable models, a fit of $q = f(c)$ leads to parameters describing the adsorption process in a phenomenological approach. Different models are known [30,31] to describe the relationship between equilibrium concentration and adsorbed amount of the compound. The Freundlich model

$$q = K \cdot c^n \tag{11.3}$$

is often used in the logarithmic form:

$$\log(q) = \log(K) + n \cdot \log(c) \tag{11.4}$$

However, mixed models are sometimes necessary to describe the dependence of $q = f(c)$, for example, the Langmuir–Freundlich model:

$$q = \frac{K \cdot c^n}{1 + a \cdot c^n} \tag{11.5}$$

Nevertheless, Freundlich parameters are often used in models and many data obtained for activated carbons are available in the literature [32–37]. Only a few data for NSO-heterocycles are known, and isotherms were determined especially for NSO-heterocycles on the activated carbon Epibon Y12×40.

To assess the adsorption of N-HETs in PRBs, the influence of the pH value was investigated for 2-methylquinoline ($pK_a = 5.86$) as a model compound.

Knowledge of the surface properties of activated carbons is essential in the explanation and comparison of adsorption parameters. Isotherms differ according to the type of activated carbons. A comparison of adsorption parameters was performed for activated carbons usually employed for the remediation of organic compounds in PRBs. Furthermore, experiments were performed with artificial mixtures of six compounds, evaluated by the the ideal adsorbed solution (IAS)-model [38–40] with basic algorithms from Reference 31.

### 11.3.2.1 Adsorption Isotherms of Single Compounds in Water

Equilibrium concentrations ($c$), and adsorbed loadings ($q$), were obtained for a number of compounds using Epibon Y12×40 (Donau Carbon) as activated carbon. A plot of log ($q$) versus log ($c$) was performed and Freundlich parameters are presented in Table 11.3.

The dependence of $q = f(c)$ is always nonlinear, and Freundlich exponents ($n$), are in the range of 0.1–0.4. For N-HETs, $K$ is always lower compared to the parent PAH (e.g., quinoline and isoquinoline vs. naphthalene or phenanthridine vs. phenanthrene). A few examples available in the literature suggested an order $K$ (S-HET) > $K$ (O-HET) > $K$ (N-HET) for the same parent structure, verified in this study for all compounds shown in Table 11.3.

Freundlich parameters are valid only in a limited concentration range. Unlike benzene, the strong adsorbing compounds acenaphthene, 9-methylacridine, phenanthrene, and dibenzothiophene (Figure 11.5) can serve as examples with adsorption isotherms evaluated with, for example, the combined Langmuir–Freundlich model.

### 11.3.2.2 Adsorption Isotherm of 2-Methylquinoline: Dependence on pH

The pH-dependent adsorption of 2-methylquinoline was investigated in the pH range of 1.9–8. The load of 2-methylquinoline (shown in Figure 11.6), was calculated from 12 experimental adsorption isotherms on Epibon Y12×40 (see also Table 11.3) using the Freundlich model for $c_0 = 10$ mg/L and the fraction of the neutral molecule (calculated from $pK_a = 5.86$). The load of the neutral molecule (e.g., at pH $\gg pK_a$) is approximately a factor of 4 higher compared to the load of the cation (e.g., at pH $\ll pK_a$). Both curves for the fractions of the neutral compound in the solution, and the experimental loads proceed nearly parallel and are shifted to each other by approximately 1.5 pH-units. The two-component mixture consists of the pH-dependent neutral and cationic form of 2-methylquinoline. Using the data from Table 11.3,

**TABLE 11.3**

Freundlich Parameters ($\log (q) = \log (K) + n \log (c)$) Obtained for the Activated Carbon Epibon Y12×40 (Sieved 63–125 μm)

| Compound | AC Lot | Analytics | $K$ | $\pm\sigma$ | $N$ | $\pm\sigma$ | $r^2$ | $N$ | $c_0/\text{mg L}^{-1}$ |
|---|---|---|---|---|---|---|---|---|---|
| Phenol | New | HPLC | 42 | 1 | 0.34 | 0.01 | 0.9899 | 11 | 16.17 |
| Benzene | New | GC-HS | 21 | 1 | 0.42 | 0.01 | 0.9879 | 44 | 1.02–297 |
| Toluene | New | GC-HS | 51 | 1 | 0.38 | 0.02 | 0.9704 | 18 | 7.8/12.5 |
| p-Xylene | New | GC-HS | 122 | 3 | 0.41 | 0.01 | 0.9909 | 10 | 11.28 |
| Ethylbenzene | New | GC-HS | 95 | 3 | 0.37 | 0.02 | 0.9798 | 11 | 10.58 |
| Indane | New | GC-HS | 147 | 3 | 0.31 | 0.01 | 0.9837 | 12 | 17 |
| Indene | New | GC-HS | 294 | 32 | 0.27 | 0.03 | 0.8309 | 17 | 11.9/19.0 |
| Naphthalene | Old | GC-HS | 293 | 14 | 0.31 | 0.02 | 0.9739 | 11 | 20.59 |
| Naphthalene | New | GC-HS | 201 | 3 | 0.24 | 0.01 | 0.9906 | 9 | 14.81 |
| Acenaphthene | New | HPLC | 339 | 30 | 0.28 | 0.04 | 0.9249 | 9 | 1.25 |
| Phenanthrene | New | HPLC | 513 | 47 | 0.32 | 0.02 | 0.9621 | 21 | 0.6/0.78 |
| Benzofuran | Old | GC-HS | 139 | 2 | 0.35 | 0.01 | 0.9946 | 10 | 19.98 |
| Benzofuran | New | HPLC | 90 | 2 | 0.32 | 0.01 | 0.9960 | 12 | 18.96 |
| Benzofuran | New | GC-HS | 96 | 1 | 0.30 | 0.00 | 0.9989 | 12 | 18.96 |
| 2-Methylbenzofuran | New | GC-HS | 204 | 4 | 0.26 | 0.01 | 0.9891 | 12 | 20.20 |
| Dibenzofuran | New | HPLC | 396 | 25 | 0.22 | 0.02 | 0.9397 | 10 | 2.02 |
| Benzothiophene | Old | GC-HS | 226 | 10 | 0.35 | 0.02 | 0.9759 | 12 | 17.37 |
| Benzothiophene | New | GC-HS | 149 | 6 | 0.27 | 0.03 | 0.9410 | 9 | 16.72 |
| 3-Methylbenzothiophene | New | HPLC | 243 | 12 | 0.20 | 0.02 | 0.8963 | 12 | 19.32 |
| Dibenzothiophene | New | HPLC | 692 | 118 | 0.38 | 0.04 | 0.9639 | 11 | 0.42 |
| Indole | New | HPLC | 117 | 2 | 0.18 | 0.01 | 0.9799 | 13 | 42.59 |
| Quinoline | New | HPLC | 112 | 4 | 0.21 | 0.02 | 0.9476 | 11 | 19.88 |
| Isoquinoline | New | HPLC | 136 | 3 | 0.18 | 0.01 | 0.9876 | 12 | 20.66 |

*continued*

**TABLE 11.3 (continued)**

Freundlich Parameters ($\log (q) = \log (K) + n \log (c)$) Obtained for the Activated Carbon Epibon Y12×40 (Sieved 63–125 µm)

| Compound | AC Lot | Analytics | K | ±σ | N | ±σ | $r^2$ | N | $c_0$/mg L$^{-1}$ |
|---|---|---|---|---|---|---|---|---|---|
| 2-Methylquinoline (buffer pH = 2.9) | New | HPLC | 48 | 2 | 0.20 | 0.02 | 0.9333 | 13 | 16.35 |
| 2-Methylquinoline (buffer pH = 7) | New | HPLC | 145 | 6 | 0.23 | 0.02 | 0.9146 | 11 | 24.22 |
| 2-Methylquinoline (in water) | New | HPLC | 167 | 8 | 0.19 | 0.01 | 0.9562 | 11 | 13.02 |
| Phenanthridine (buffer pH = 7.0) | New | HPLC | 171 | 15 | 0.29 | 0.03 | 0.9145 | 11 | 2.00 |
| Phenanthridine (buffer pH = 2.5) | New | HPLC | 144 | 3 | 0.11 | 0.01 | 0.8829 | 10 | 9.37 |
| Acridine | New | HPLC | 298 | 42 | 0.26 | 0.04 | 0.8615 | 8 | 13.29 |
| 9-Methylacridine | New | HPLC | 427 | 20 | 0.12 | 0.01 | 0.9497 | 13 | 2.49 |
| Carbazole | New | HPLC | 319 | 19 | 0.22 | 0.01 | 0.9473 | 16 | 0.58/0.66 |

*Note:* Measurements were performed in deionized water and in buffer mixtures; temperature for all measurements $T \pm \sigma = 20 \pm 3°C$. Standard deviations are errors from linear regressions performed. AC lot = two lots obtained from the supplier Donau Carbon were used (old lot 1 kg; (April 2006), new lot 20 kg (July 2008); Analytics = analytical method used; $K$ (unit mg$^{1-n}$ g$^{-1}$ L$^n$) and $n$ (without a unit) are the Freundlich parameters with standard deviations; $r^2$ = goodness of fit of the linear regression from log ($q$) versus log ($c$); $N$ = number of data points; and $c_0$ = start concentration in mg/L.

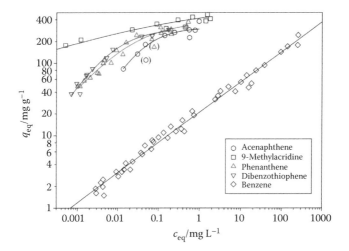

**FIGURE 11.5**

Adsorption isotherms for acenaphthene, 9-methylacridine, phenanthrene, and dibenzo-thiophene and for benzene (for comparison) on the activated carbon Epibon Y12×40 (sieved 63–125 μm) in deionized water at $T \pm \sigma = 20 \pm 3°C$. Curves are from the results of the Langmuir–Freundlich model. Data in parenthesis are defined as outliers and not used. Linear regression for benzene represents the Freundlich model.

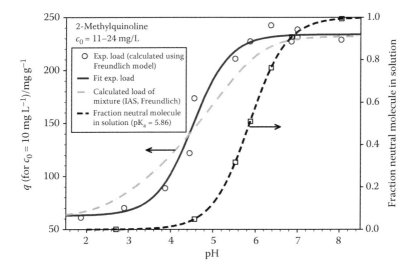

**FIGURE 11.6**

The adsorption loads of 2-methylquinoline on AC Epibon Y12×40 evaluated from pH-dependent Freundlich isotherms ($T = 20°C$). In addition, loads calculated with the IAS-model and the neutral fraction present in the solution are shown.

pH-dependent loads were calculated from the IAS-model for $c_0 = 10$ mg/L and concentrations were obtained from the dissociation curve, defined by the $pK_a$-value. The IAS-model [38] and algorithms are described in detail in Reference 31 and a Java program was developed to calculate the mixture loadings used in batch and column experiments. In addition, a fit routine with the Marquardt algorithm was incorporated and used to fit isotherms in mixtures (see below). The theoretical pH-dependent loads, included as a dotted line in Figure 11.6, were found to be in satisfactory agreement with the experimental results.

The shift of the pH-dependent load with respect to the corresponding fraction of the neutral molecule in the solution (shown in Figure 11.6) is important to assess the influence of pH on the adsorption for 2-methylquinoline (in particular), and of N-HETs (in general) on activated carbon in PRBs. Due to the shift in both curves, a practically maximum load (of the neutral molecule) is found at pH = 5. As a consequence, the maximum load of 2-methylquinoline should be found even for pH-values down to 5. This should be true for all N-HETs investigated due to the lower or comparable $pK_a$-values compared to the $pK_a$ of 2-methylquinoline. Furthermore, phenols with $pK_a \gg 7$ should be adsorbed in the neutral form without any influence of the much more polar anions.

### 11.3.2.3 Adsorption Isotherms of Mixtures

The knowledge of adsorption processes of mixtures on the activated carbon is important in the remediation of aromatic compounds in a PRB. The adsorption of a compound and therefore its load, is strongly influenced by other compounds that may be present in the mixture. Multiadsorption processes were investigated in batch experiments and are present in column experiments performed with artificial mixtures and real groundwater. A model was developed to fit the isotherms, and Freundlich parameters were obtained, which were compared with corresponding parameters from both the single-isotherm experiments and column runs.

#### 11.3.2.3.1 Model to Fit Multiadsorption Processes

A maximum number of six compounds can be used in commercial models (LDF [41] and AdDesignS [36]) to evaluate breakthrough curves in column runs. Hence, to compare batch experiments with column runs, a mixture consisting of phenol, benzene, toluene, 2-methylquinoline, benzofuran, and benzothiophene was used on six different activated carbons: Epibon Y12×40 and Hydraffin 30N (Donau Carbon), Norit ROW supra and Norit 1240 (Norit), and F200 and F400 (Chemviron). The composition was chosen in such a way that two aromatic hydrocarbons, phenol as the prototype of a phenolic compound, and three NSO-heterocycles were present in the mixture. As a representative example, adsorption isotherms of the mixture investigated using the activated carbon F200 (Chemviron) are shown in Figure 11.7. The extreme nonlinear behavior for weak adsorbing compounds

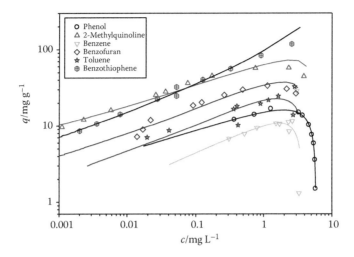

**FIGURE 11.7**
The experimentally determined mixture loadings (symbols) of a mixture consisting of six components on AC F200 (Chemviron) at $T = 20 \pm 3°C$ in water and a nonlinear global fit (lines) using the IAS-model.

(phenol or benzene) is typical for all activated carbons in the concentration range investigated.

The experimental data points of multiadsorption isotherms were evaluated by the IAS-model using the Freundlich isotherm as input parameters. If all parameters for all compounds ($K$ and $n$) are known, the IAS-model estimates the loads of the corresponding equilibrium concentrations.

For $n$ compounds, a fit with $2n$ parameters ($K$ and $n$) is necessary, for example, for six compounds, 12 parameters have to be fitted. However, two parameters are always connected with the experimental data (equilibrium concentrations and corresponding equilibrium loads) for each isotherm. In total, 60 data points are typically available. The criteria in the fit are defined in Equations 11.6 and 11.7:

$$\chi^2 = \sum_{i=1}^{60} (q_{IAS,i} - q_{eq,i})^2 \rightarrow \text{minimum} \tag{11.6}$$

$$q_{IAS,i} = f(K, n, c_{eq}) \quad \text{for all compounds} \tag{11.7}$$

where $c_{eq,i}$ and $q_{eq,i}$ denote the experimental data and $q_{IAS,i}$ denotes the fitted load from the IAS-model for compound i.

The load obtained by the IAS-model $q_{IAS,i}$ depends on Freundlich parameters $K$ and $n$ for all compounds, and parameters are varied until the minimum value for $\chi^2$ is found. The lines shown in Figure 11.7 are the result of such a fit obtained for the activated carbon F200.

When *K* and *n* were known, a reverse fit was used to simulate multiadsorption processes for mixtures in batch experiments. Although six compounds were used for comparison purposes, the number of compounds is only restricted by the computation time and not by the number of components.

### 11.3.2.3.2 Comparison of Activated Carbons

A number of experiments are necessary to compare different activated carbons for a number of selected compounds. The time-consuming procedure to measure adsorption isotherms was the main motivation for the development of an applicable computer program to evaluate Freundlich parameters from mixtures, discussed in the previous section. Hence, for a comparison of six compounds on six activated carbons, a potential total of 36 experiments was reduced to only six experiments using mixtures with six compounds. However, suitable analytical methods for the evaluation of mixtures were required for all of the components. Freundlich parameters were evaluated for six mixtures using six activated carbons and the fit procedure described in the previous section was also implemented. The mean Freundlich exponent $\bar{n}$ for all 36 adsorption isotherms was found to be $\bar{n} \pm \sigma = 0.34 \pm 0.07$ with $\bar{n}_{min} = 0.29$ (benzofuran) and $\bar{n}_{max} = 0.45$ (benzene) respectively, for the six activated carbons. The mean values $\bar{K}$ for the six compounds investigated are presented in Figure 11.8.

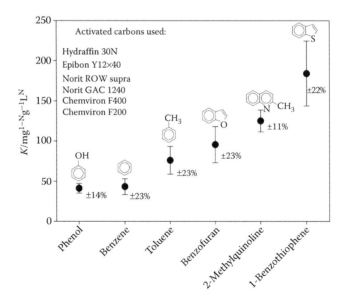

**FIGURE 11.8**
The mean Freundlich adsorption parameter *K* for six compounds obtained on six different activated carbons with relative standard deviations. Data were obtained from adsorption isotherms of the corresponding mixtures at $T = 20 \pm 3°C$.

Comparing the six activated carbons, an approximate order with respect to the adsorption parameter $K$ was as follows: AC(30N) ~ AC(Epibon) < AC (ROW) < AC(F200) < AC(1240) ~ AC(F400). However, relative standard deviations of $K$ vary in the range of 20%–25% only, surprisingly similar for all compounds.

### 11.3.2.4 Isotherm Parameters from Column Runs: Comparison with Batch Experiments

As already discussed, contaminated groundwater is a complex mixture of compounds with highly variable intrinsic properties and concentrations. To our knowledge, there is no available model that calculates breakthrough curves in column experiments on activated carbon for such a complex mixture. This is especially true for mixtures consisting of the number of compounds investigated in this study, although they are well characterized with respect to their concentration. Furthermore, numerous additional parameters (e.g., diffusion coefficients) have to be known to calculate the breakthrough curves. Another possibility is the approximation of breakthrough curves using equilibrium models, neglecting time-dependent processes and therefore kinetic parameters. Nevertheless, commercially available models (e.g., AdDesignS [36] or LDF [41]) take both breakthrough and equilibrium curves into account only for a limited number of compounds.

Artificial mixtures were used to verify the experimental methods and to compare batch and column experiments. Two column runs should serve as instructive examples.

In the first example, breakthrough curves of a simple artificial three-component mixture (benzene, toluene, and benzofuran), shown in Figure 11.9, were obtained from port 2 with the column described in the experimental section (AC Epibon Y12×40, $m$(AC at port 2) = 15.55 g, flow = 7.1 mL/min, $c_0 = 10$ mg/L for each component, $T = 11°C$, and run time 63 days). Breakthrough curves were calculated with the LDF-model [41], equilibrium curves using the IAS-approach described above. Freundlich parameters as the main input data for both models were taken from batch experiments, as already presented in Table 11.3. Breakthrough curves from LDF were adjusted slightly using the necessary (and estimated) diffusion coefficients. Concentrations above the inflow concentrations for benzene and toluene are typical for desorption processes due to stronger adsorbing compounds and were found in all column runs.

In extension to this three-component mixture, breakthrough curves were measured for an artificial six-component mixture (phenol, benzene, toluene, benzofuran, 2-methylquinoline, and benzothiophene, concentrations were sampled at eight ports, AC Epibon Y12×40, $m$(AC at port 8) = 70.9 g, flow = 6.9 mL/min, $c_0 = 3$–10 mg/L, $T = 12°C$, and run time = 80 days). Concentrations from port 3 were used to evaluate Freundlich parameters. From the LDF-model, the best consistency was found after manual adjustment

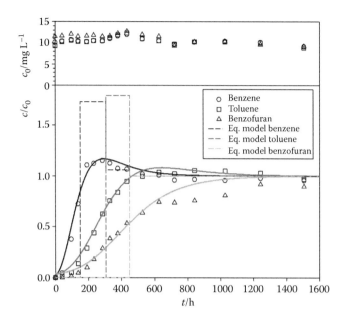

**FIGURE 11.9**
The results of a column run (concentrations obtained from port 2) of a mixture consisting of benzene, toluene, and benzofuran on AC Epibon Y12×40 at $T = 11°C$ (run time 63 days). Experimental data (upper part inflow concentrations) and breakthrough curves using the LDF-model and equilibrium curves (dashed lines) are shown.

(with respect to diffusion coefficients) of the breakthrough curves with the Freundlich parameters summarized in Table 11.4.

From three different experimental methods and associated evaluations presented in Table 11.4, the variability of the Freundlich parameters obtained in this study was estimated to be in the range of 20%–30%.

## 11.3.3 Compound Ranking from Column Runs

A number of column runs were performed for "biotic" and "abiotic" conditions as explained and described in the experimental section. Mean concentrations of all compounds at the beginning of the experiment, and the standard deviations within the 98-day run time for both columns are summarized in Table 11.5. Concentrations were stable within 10%–20%. However, due to assumed adsorption processes in the storage container, concentrations at the beginning varied by about a factor of 2 (see Table 11.5) for the strong adsorbing compounds (e.g., 2-methylnaphthalene or the high molecular PAH).

For comparison purposes, individual front velocities were defined for all compounds in the columns. Therefore, nine samples at intervals of 10 days from eight individual positions (seven ports and an influent), each

**TABLE 11.4**

A Comparison of Freundlich Adsorption Parameters $n$ and $K$ ($\mathrm{mg^{1-n}\,g^{-1}\,L^n}$) of Benzene, Toluene, Phenol, 2-Methylquinoline, Benzofuran, and Benzothiophene for AC Epibon Y12×40, Using Three Different Experimental Methods

| | Adsorption Isotherm—Freundlich Parameters | | | | | |
| | Batch Single Compound | | Batch Mixture | | Column Mixture | |
| Compound | $K$ | $N$ | $K$ | $N$ | $K$ | $N$ |
|---|---|---|---|---|---|---|
| Benzene | 21 | 0.42 | 30 | 0.52 | 33 | 0.42 |
| Toluene | 51 | 0.38 | 57 | 0.33 | 63 | 0.38 |
| Phenol | 42 | 0.34 | 43 | 0.15 | 42 | 0.33 |
| 2-Methylquinoline | 48 (pH ≪ pK$_a$) | 0.20 | 163 | 0.15 | 100 | 0.25 |
| | 156 (pH ≫ pK$_a$) | 0.21 | | | | |
| Benzofuran | 93 | 0.31 | 71 | 0.26 | 92 | 0.31 |
| Benzothiophene | 149 | 0.27 | 136 | 0.31 | 142 | 0.35 |

*Note:* Batch single compound: See results summarized in Table 11.3 (for benzofuran and 2-methylquinoline mean values were used); batch mixture: from a fit of a mixture of the six compounds specified using the IAS-model; column mixture: from a fit using the LDF-model [61] of a column experiment with the six compounds specified. Note the different temperatures used: T(batch) = 20 ± 3°C and T(column) = 12 ± 2°C.

representing a given mass of activated carbon ($m_{AC}$), were taken and subsequently analyzed. A total of 58 compounds were analyzed and more than 4000 concentrations were obtained. The fraction $c_i/c_0$ was calculated from the individual influent concentration ($c_0$) obtained at time t and port $i$ (index $i = 0$–7 represents the influent and port 1–7). The mass of activated carbon ($m_{AC}$) for a given port was calculated from the position of the port and the column dimensions. The boundary condition $c_i/c_0 = 0.5$ defines the amount of loaded carbon ($m_L$) for an individual compound. Using a Fortran-program $m_L$ at time $t$ was obtained from a plot of $c_i/c_0$ versus $m_{AC}$ by interpolation below and above $c_i/c_0 = 0.5$. The linear regression of $m_L$ versus $t$ leads to the mean adsorption front velocity $v = m_L/t$ (in mg of loaded activated carbon per hour). The number of data points in linear regressions performed as described above varied between 2 and 9. The breakthroughs of phenol and benzene were rapid and only the first two measurements could be used. However, almost all of the data points were applicable for strong adsorbing compounds such as anthracene or the corresponding N-HET acridine.

A strong linearity was found with a mean correlation coefficient for all linear regressions performed of R = 0.97. However, errors for strong adsorbing compounds were high, with the worst correlations found for pyrene (R = 0.58/0.74).

Front velocities of all compounds analyzed for both columns are presented in Figure 11.10. In addition, the errors of slopes from the plot of m$_L$ versus t are included as error bars in Figure 11.10. Front velocities for both columns

**TABLE 11.5**

Composition of the Groundwater Used in Both Column Experiments ("Biotic-Column" and "Abiotic-Column")

| Compound | $c \pm \sigma$ μg/L | σ | Compound | $c \pm \sigma$ μg/L | σ | Compound | $c \pm \sigma$ μg/L | σ |
|---|---|---|---|---|---|---|---|---|
| Phenol | 302 | 27 | 1-Aminonaphthalene | 4.0 | 0.8 | 2-Hydroxycarbazole | 3.1 | 0.9 |
| Benzene[a] | 708 | 165 | 2-Naphthol | 62 | 6 | 3-Methylbenzothiophene | 16 | 3 |
| 2-Methylphenol | 338 | 43 | 1-Indanone | 74 | 13 | 2-Hydroxydibenzofuran | 31 | 5 |
| 3/4-Methylphenol | 265 | 30 | Indane | 381 | 46 | 2-Methylnaphthalene | 50 | 22 |
| 3,5-Dimethylphenol | 220 | 40 | Indane[a] | 367 | 86 | 1-Methylnaphthalene | 262 | 74 |
| 2,3-Dimethylphenol | 48 | 5 | 1-Naphthol | 63 | 8 | Biphenyl | 58 | 9 |
| Toluene[a] | 336 | 32 | Benzothiophene | 664 | 67 | Acenaphthene | 72 | 11 |
| 2,4/2,5-Dimethylphenol | 203 | 30 | 2-Methylbenzofuran | 34 | 3 | Acenaphthylene | 4.8 | 0.9 |
| 2,6-Dimethylphenol | 136 | 14 | 1,2,3-Trimethylbenzene[a] | 43 | 5 | 1,3-Dimethylnaphthalene | 25 | 5 |
| 3,4-Dimethylphenol | 53 | 5 | Benzothiophene[a] | 886 | 82 | Anthracene | 1.7 | 0.8 |
| 2,4,6-Trimethylphenol | 57 | 5 | Isoquinoline | 15 | 4 | 1,6-Dimethylnaphthalene | 11 | 3 |
| 2,3,6-Trimethylphenol | 26 | 2 | Acridine | 1.9 | 0.5 | 2-Ethylnaphthalene | 5.7 | 1.4 |
| Benzofuran[a] | 216 | 16 | 1,3,5-Trimethylbenzene[a] | 53 | 7 | 1,4/2,3-Dimethylnaphthalene | 5.7 | 1.4 |
| Benzofuran | 182 | 19 | 2-Methylquinoline | 24 | 4 | Dibenzofuran | 53 | 15 |
| Ethylbenzene | 132 | 16 | 2-Phenylpyridine | 44 | 3 | Xanthone | 0.83 | 0.2 |
| o-Xylene[a] | 173 | 19 | 1,2,4-Trimethylbenzene[a] | 87 | 14 | 1,8/2,7-Dimethylnaphthalene | 17 | 4 |
| 2,3,5-Trimethylphenol | 41 | 5 | 1,2-Dimethylnaphthalene | 9.7 | 0.9 | Fluorene | 18 | 6 |
| 3,4,5-Trimethylphenol | 10 | 2 | 6/7-Methylquinoline | 4.2 | 0.5 | Phenanthrene | 5.3 | 2.5 |
| Indene[a] | 1149 | 198 | Dibenzothiophene | 1.0 | 0.3 | Pyrene | 0.19 | 0.1 |
| Indene | 1309 | 139 | Naphthalene[a] | 5864 | 652 | Fluoranthene | 0.44 | 0.3 |
| m,p-Xylene[a] | 494 | 51 | 1-Cyanonaphthalene | 72 | 6 | | | |

*Note:* Mean values with standard deviations are shown, obtained from 10 concentrations within the runtime of 3 months. Order of compounds as found in column experiments (e.g., phenol was the fastest compound, fluoranthene was the slowest in both columns).

[a] Compounds were analyzed by headspace GC, all other compounds were analyzed by GC–MS. Concentrations sum up to $\Sigma c = 13.3$ mg/L (for benzofuran, indene, indane, and benzothiophene mean values were used).

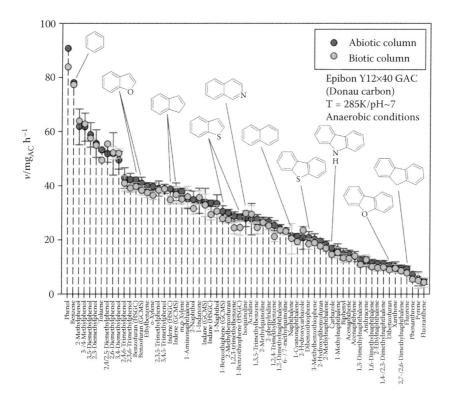

**FIGURE 11.10**

A comparison of front velocities in the "biotic" and "abiotic" columns. Front velocities are defined as the mass-loaded activated carbon per time (in mg/h). Front velocities obtained from nine concentrations in maximum were found to be linear within the run period for both columns (98 days). Error bars are uncertainties of the slope of the velocity functions. A fast breakthrough for phenol and benzene was found and the front velocities of both compounds were obtained only from two concentrations. See text for the definition of velocities used.

were comparable and it was not possible to detect differences for any compound within the statistical uncertainties. Nevertheless, for the "abiotic column" any potential biological degradation was inhibited as discussed in the experimental section. Hence, processes on this column, although using real groundwater, are not influenced by degradation reactions of microorganisms, and pure adsorption was observed. The order of compounds, sorted by front velocities and presented in Figure 11.10 was used to rank all compounds (see Table 11.5) with respect to their adsorption behavior.

Qualitatively and as assumed from their adsorption parameters, phenol and benzene were the worst-adsorbing compounds (see order in Table 11.5), and pyrene and fluoranthene were the best. The worst-adsorbing heterocycles were benzofuran (O-heterocycle), benzothiophene (S-heterocycle), and isoquinoline (N-heterocycle). Unfortunately for many of the compounds, Freundlich parameters $N$ and $K$ were not determined in this study and are not available in the literature. It is worth noting that the ranking shown in Figure 11.10 is not correlated with concentration (see also Table 11.5). As an example and in agreement with the Freundlich parameters summarized in Table 11.3, the front velocity of naphthalene ($c_0 = 5864$ µg/L) was lower compared to isoquinoline ($c_0 = 15$ µg/L).

### 11.3.4 Efficiencies of Permeable Reactive Barriers

The efficiencies of two PRBs in Karlsruhe (Germany) and Brunn am Gebirge (Austria) were investigated with respect to the adsorption of PAHs, phenols, NSO-heterocycles, BTEX, and related compounds in four field campaigns. In addition, concentrations at both sites were compared with data from the site Lünen (Germany), where a reactive barrier was originally planned but still not realized.

#### 11.3.4.1 Characterization of Sites

Samples from wells on the contaminated sites as well as from the influent and effluent waters of the PRB itself were analyzed. In addition, on-site parameters and concentrations of anions and selected cations (median values) are presented in Table 11.6. Laboratory experiments were performed with groundwater from another site in Lünen (Germany) and data from this site were also included in Table 11.6.

Although the composition of water samples from different wells on a given site can vary, the median values of all samples taken over a number of years (always within the period July to November) should serve as a general overview of the chemical characterization. Redox potentials are

**TABLE 11.6**

On-Site Parameters, Concentrations of Anions, and Selected Cations (Median Values, N = 30–40, Site Dependent) for the Contaminated Sites

| Site | T (°C) | $c(O_2)$ mg/L | κ mS/cm$^1$ | $V_{Red}$ (mV) | pH | $c(Cl^-)$ | $c(NO_3^-)$ mg/L | $c(SO_4^{2-})$ | $c(CN^-)$ |
|------|--------|---------------|-------------|----------------|-----|-----------|------------------|----------------|-----------|
| Lünen | 12.5 | 0.1 | 2.5 | −232.0 | 7.1 | 44.0 | (0.1) | 761.0 | 0.66 |
| Karlsruhe | 14.2 | 0.2 | 1.1 | −149.0 | 6.9 | 51.0 | (0.1) | 123.0 | 0.21 |
| Brunn | 15.1 | 0.2 | 1.9 | −78.5 | 6.9 | 370.5 | (0.1) | 122.5 | 0.07 |

| Site | $c(Ba)$ | $c(Ca)$ | $c(Fe)$ | $c(K)$ | $c(Mg)$ mg/L | $c(Mn)$ | $c(Na)$ | $c(Sr)$ |
|------|---------|---------|---------|--------|--------------|---------|---------|---------|
| Lünen | 0.1 | 245.5 | 0.3 | 11.8 | 69.1 | 1.1 | 96.4 | 3.1 |
| Karlsruhe | 0.3 | 172.5 | 3.8 | 12.2 | 23.6 | 0.6 | 26.2 | 0.5 |
| Brunn | 0.2 | 205.5 | 1.8 | 7.1 | 55.6 | 1.2 | 93.1 | 2.2 |

*Note:* Values in parentheses for nitrate are detection limits.

negative with very low oxygen concentrations ($c(O_2) < 0.5$ mg/L), neutral pH values in the range of pH = 7, conductivities between 1 and 3 mS/cm, and temperatures of 12–15°C are typical values for the on-site parameters of the three sites investigated. Nitrate concentrations are typically low and sulfate concentrations are high in Lünen. High concentrations of chloride are typically present in Brunn. Cyanide concentrations (analyzed as the sum of bound cyanide and free cyanide) were analyzed in Lünen and found without exception to be below detection limits. Concentrations of cyanide in Lünen with 660 µg/L are approximately a factor of 3 higher compared to the site in Karlsruhe and a factor of 10 higher compared to the site in Brunn.

Typical Fe and manganese concentrations were found to be in the range of 0.5–4 mg/L. However, severe problems with respect to clogging were not reported for both PRBs in Karlsruhe and Brunn.

The maximum concentrations found at the three sites investigated are presented in Table 11.7, based on more than 20 samples from different wells. Compounds summarized in Table 11.7 represent the spectrum of all compounds analyzed. The maximum concentrations were determined from different contaminated wells (wells 12Q, 22Q, and 23Q for site Lünen, well AB12 for site Karlsruhe and the influent at gate 2, and well B7 and B2006/31 for site Brunn am Gebirge). However, it has to be noted that at least in one sample from each site, concentrations for all compounds were found below detection limits. In addition to some potential degradation products such as hydroxylated hydrocarbons, nitroaromatics were analyzed in field campaigns in 2007. Concentrations of these compounds (often below detection limits) are also presented in Table 11.7.

### 11.3.4.2 Reactive Barrier in Karlsruhe (Germany)

Although the total concentrations in one of the most contaminated zones in Karlsruhe are in the range of 1800 µg/L, the influent concentrations at the gates of the barrier are low (maximum values of ~50 µg/L). Moreover, most of the mass (about 90%) at the gates results from the concentration of acenaphthene. Heterocycles were present in the main contaminated region (~16% of the total); however, they play no role at the gates, with concentrations below 1 µg/L. Nevertheless, for all compounds, effluent concentrations were found to be below detection limits and the efficiency of the PRB was >98% after more than 8 years of operation. Using a mean flow of approximately 1 L/s for one gate and a maximum total concentration of 50 µg/L, the mass inflow was in the range of 1–2 kg/year. This was found to be valid for one gate (and therefore one reactor) for approximately 12 tonnes of activated carbon. Hence, the estimated mean load of activated carbon was ≪1% even after a PRB operating time of 8 years. In contrast to organic compounds, the retardation of cyanide was low and concentrations of cyanide in the influent and effluent were found to be comparable.

**TABLE 11.7**

Maximum Concentrations of Organic Compounds at the Three Contaminated Sites in Lünen, Karlsruhe (Karl), and Brunn am Gebirge (Brunn) Obtained from 20–32 Samples from Different Wells, Respectively

| Compound | Lünen | Karlsruhe | Brunn | Compound | Lünen | Karlsruhe | Brunn |
|---|---|---|---|---|---|---|---|
| | $c(max)/\mu g\ L^{-1}$ | | | | $c(max)/\mu g\ L^{-1}$ | | |
| Benzene | 633 | 46 | 12 | 2-Ethylnaphthalene | <LOD | 1 | 4 |
| Toluol | 379 | 37 | 5 | 1-Ethylnaphthalene | 10 | 1 | 13 |
| Ethylbenzene | 1571 | 29 | 38 | 2,6-/2,7-Dimethylnaphthalene | 68 | 11 | 12 |
| m,p-Xylene | 446 | 53 | 17 | 1,3-Dimethylnaphthalene | 129 | 7 | 28 |
| O-xylene | 257 | 19 | 24 | 1,6-Dimethylnaphthalene | 87 | 2 | 4 |
| 1,3,5-Trimethylbenzene | 421 | 10 | 11 | 1,4-/2,3-Dimethylnaphthalene | 74 | 10 | 10 |
| 1,2,4-Trimethylbenzene | 113 | 16 | 34 | 1,5-Dimethylnaphthalene | 29 | 4 | 27 |
| 1,2,3-Trimethylbenzene | 633 | 6 | 34 | 1,2-Dimethylnaphthalene | 31 | 2 | 35 |
| Phenol | <LOD | <LOD | <LOD | 1,8-Dimethylnaphthalene | 1 | <LOD | 1 |
| 2-Methylphenol | 1 | 15 | <LOD | 1-Benzothiophene | 1566 | 35 | 717 |
| 3-/4-Methylphenol | <LOD | 14 | <LOD | 2-Methylbenzothiophene | 2 | 0.3 | NA |
| 2,6-Dimethylphenol | 45 | 4 | 6 | 3-Methylbenzothiophene | 65 | 2 | 13 |
| 2,4-/2,5-Dimethylphenol | 4 | 22 | <LOD | Dibenzothiophene | 6 | 8 | 19 |
| 3,5-Dimethylphenol | 13 | 23 | 6 | 4,6-Dimethyldibenzothiophene | <LOD | 0.1 | NA |
| 2,3-Dimethylphenol | 7 | 2 | 1 | Benzo(b)furan | 297 | 31 | 3 |
| 3,4-Dimethylphenol | 11 | 5 | 0.4 | 2-Methylbenzofuran | 63 | 3 | 22 |
| 2,4,6-Trimethylphenol | 104 | 3 | 3 | 2,3-Dimethylbenzofuran | 14 | 0.4 | NA |
| 2,3,6-Trimethylphenol | 45 | 1 | 34 | Dibenzofuran | 651 | 125 | 488 |
| 2,3,5-Trimethylphenol | 19 | 2 | 3 | Xanthene | 1 | 1 | 2 |
| 3,4,5-Trimethylphenol | 32 | 1 | 1 | 2-Hydroxydibenzofuran | 71 | 6 | NA |
| Naphthalene | 15,586 | 508 | 5501 | Xanthone | 2 | <LOD | 4 |
| A:enaphthylene | 26 | 7 | 12 | Quinoline | 2 | <LOD | 1 |

| Compound | | | | Compound | | | |
|---|---|---|---|---|---|---|---|
| Acenaphthene | 2909 | 429 | 1196 | Isoquinoline | 37 | 6 | 1 |
| Fluorene | 1062 | 284 | 666 | Indole | 2 | 1 | NA |
| Phenanthrene | 180 | 124 | 182 | 2-Methylquinoline | 143 | 10 | 39 |
| Anthracene | 3 | 16 | 22 | 8-Methylquinoline | 1 | <LOD | <LOD |
| Fluoranthene | 3 | 14 | 43 | 3-Methylisoquinoline | <LOD | <LOD | NA |
| Pyrene | 1 | 8 | 17 | 1-Methylisoquinoline | <LOD | 2 | NA |
| Benzo(a)anthracene | <LOD | 1 | 1 | 6-/7-Methylquinoline | 2 | <LOD | <LOD |
| Chrysene | <LOD | 0.7 | 1 | 3-Methylquinoline | 1 | <LOD | <LOD |
| Benzo(b)fluoranthene | <LOD | 0.4 | <LOD | 4-Methylquinoline | 4 | 1 | 8 |
| Benzo(k)fluoranthene | <LOD | 0.4 | <LOD | 2,6-Dimethylquinoline | <LOD | 3 | NA |
| Benzo(a)pyrene | <LOD | 0.4 | <LOD | 2-Phenylpyridine | 9 | 2 | 5 |
| Indeno(1,2,3-cd)pyrene | <LOD | 0.1 | <LOD | 3-Phenylpyridine | <LOD | 0.4 | <LOD |
| Dibenzo(a,h)anthracene | <LOD | <LOD | <LOD | 4-Phenylpyridine | <LOD | <LOD | <LOD |
| Benzo(g,h,i)perylene | <LOD | 0.1 | <LOD | Acridine | 8 | 11 | 3 |
| Indane | 3575 | 66 | 352 | 9-Methylcarbazole | <LOD | <LOD | NA |
| Indene | 1751 | 60 | 97 | Phenanthridine | <LOD | 0.2 | 2 |
| 2-Methylindene | 218 | 9 | NA | Carbazole | 83 | 63 | 159 |
| 2-Methylnaphthalene | 1 | 24 | 9 | 9-Methylacridine | <LOD | 0.3 | NA |
| 1-Methylnaphthalene | 1444 | 63 | 324 | 2-Hydroxycarbazole | 2 | <LOD | NA |
| Biphenyl | 2 | 22 | 56 | Phenanthridinone | 201 | 12 | NA |
| 1-Indanone | 104 | 8 | 4 | 2-Naphthol | 30 | 8 | 1 |
| 1-Cyanonaphthalene | 101 | 6 | 12 | 1-Aminonaphthalene | 9 | 3 | 1 |
| 1-Naphthol | 64 | 25 | 14 | 9-Nitroanthracene | <LOD | <LOD | <LOD |
| Dibenzothiophensulfon | <LOD | <LOD | <LOD | 3-Nitrofluoranthene | <LOD | <LOD | <LOD |
| Coumarine | 33 | <LOD | 2 | 1-Nitropyrene | <LOD | <LOD | <LOD |
| 1,4-Naphthochinone | 4 | <LOD | <LOD | 1,5-Naphthalene diol | <LOD | 3 | 5 |
| 1,2-Acenaphthenchinone | 55 | <LOD | 1074 | | | | |

*continued*

**TABLE 11.7 (continued)**

Maximum Concentrations of Organic Compounds at the Three Contaminated Sites in Lünen, Karlsruhe (Karl.), and Brunn am Gebirge (Brunn) Obtained from 20–32 Samples from Different Wells, Respectively

| Compound | Lünen | Karlsruhe | Brunn |
|---|---|---|---|
| | $c(\max)/\mu g\ L^{-1}$ | | |
| 9-Fluorenone | 2 | <LOD | 7 |
| Anthrachinone | 1 | <LOD | 8 |
| 9(10 h)-Acridinone | 24 | <LOD | 381 |
| 1,8-Naphthalene anhydride | 65 | <LOD | 6 |
| Hydroquinone | <LOD | <LOD | <LOD |
| Benzanthrone | <LOD | <LOD | <LOD |
| Nitrobenzene | 5 | 2 | <LOD |
| 1-Nitronaphthalene | <LOD | <LOD | <LOD |
| 5-Nitroacenaphthene | <LOD | <LOD | <LOD |
| 2-Nitrofluorene | <LOD | <LOD | <LOD |

| Compound | Lünen | Karlsruhe | Brunn |
|---|---|---|---|
| | $c(\max)/\mu g\ L^{-1}$ | | |
| 1-Acenaphthenol | 8 | <LOD | 44 |
| 9-Fluorenol | 5 | <LOD | 6 |
| 2-Hydroxy-9-fluorenone | 1 | <LOD | 21 |
| 9-Phenanthrole | <LOD | <LOD | <LOD |
| 1-Pyrenol | 7 | <LOD | <LOD |
| 2-Hydroxyquinoline | 300 | <LOD | <LOD |
| 2-Hydroxybiphenyl | <LOD | <LOD | <LOD |
| 9-Cyanoanthracene | <LOD | <LOD | 12 |
| 9-Cyanophenanthrene | <LOD | <LOD | 7 |

*Note:* <LOD = below detection limit, NA = not analyzed. It has to be noted that for all compounds, at least in one, sample concentrations were found below detection limit (<LOD). For Karlsruhe, the concentrations for well AB12 in the 2007 field campaign were found to be in part a factor of 10 higher for comparable samples and were not used. High blank concentrations of phenol and isomers of cresol prevent a sensitive analysis, often leading to concentrations <LOD.

### 11.3.4.3 Reactive Barrier in Brunn am Gebirge (Austria)

The plume direction of the PRB in Brunn am Gebirge (Austria) was focused on gate 2 with a very high total concentration of approximately 5600 µg/L. Total concentrations at the neighboring gates 1 and 3 were a factor 100 (gate 3) and factor 1000 (gate 1) lower, respectively. The fraction of NSO-heterocycles at gate 2 was found to be approximately 7%, predominantly determined by benzothiophene (180 µg/L), dibenzothiophene (120 µg/L), and carbazole (50 µg/L).

Effluent concentrations were found to be below detection limits for all compounds, and the efficiency of the PRB was >99% after more than 9 years of operation. Using the minimum flow for the whole barrier (0.5 L/s) as an assumed value for gate 2 and a maximum total concentration of 5600 µg/L, the mass inflow at gate 2 was in the range of 50–100 kg/year. This mass is valid for gate 2 with approximately 6 tonnes of activated carbon. Hence, the estimated load for all compounds on activated carbon based on an operating time of 9 years of the PRB was approximately 7%–15%. The loads for gates 1 and 3 were much lower and assumed to be below 1%.

In agreement with the measurements at the PRB in Karlsruhe (Germany) and in contrast to organic compounds, the retardation for cyanide was low and the cyanide concentrations in the influent and effluent were comparable.

## 11.4 Conclusions

Analytical methods and extraction procedures were developed to investigate a complex spectrum of aromatic compounds that are typically found in groundwater samples. In addition to BTEX and the 16-EPA PAH, these include phenols, NSO-heterocycles, and further-related aromatic compounds, such as alkylated PAHs. This spectrum of compounds was detected in the analysis of tar oils and groundwater from different contaminated sites. However, in addition to a basic set (including 16-EPA PAH, NSO-heterocycles, BTEX, and phenols), the spectrum of compounds analyzed varied during the investigation.

In the field campaigns, BTEX, indane and indene, and several PAHs (e.g., naphthalene, acenaphthene, fluorene, and phenanthrene) were present. Methylnaphthalenes and dimethylnaphthalenes were found in high concentrations and should be included in sampling programs. However, concentrations of 1,8-dimethylnaphthalene were usually below the detection limit. The most commonly found heterocycles were benzothiophene, benzofuran, dibenzofuran, 2-methylquinoline, carbazole, acridine, and phenanthridinone. High molecular PAHs were generally found in low concentrations (for instance concentrations of dibenzo(a,h)anthracene were below the detection limit in all samples). Remarkable concentration differences were detected for isomers. Hence, naphthalenes substituted in 1-position (e.g., methylated, ethylated, or hydroxylated derivatives) were found in higher concentrations compared to the corresponding 2-substituted naphthalenes. Furthermore and in contrast to concentrations of 2-phenylpyridine, 3- and 4-phenylpyridine were usually found below detection limits.

Adsorption isotherms on activated carbons (using especially Epibon Y12×40 from Donau Carbon) were obtained from batch experiments for both single compounds and mixtures. In addition, Freundlich parameters for both experimental methods were compared with column runs. Freundlich parameters were in satisfactory agreement using different experimental methods with a 20%–30% variation. However, adsorption parameters for many compounds found in contaminated groundwater (e.g., methylated naphthalenes and methylated NSO-heterocycles) have not been published in the literature. From the few values available, qualitatively, the adsorption of S-heterocycles seems to be comparable with the corresponding parent hydrocarbons, whereas the adsorption of N-HETs was generally found to be lower. Adsorption parameters of O-heterocycles could be qualitatively classified between S- and N-HETs (for the same parent structures). These rules of thumb are in agreement with analytical results. O- and S-heterocycles were analyzed with comparable detection limits for GC–MS, whereas LC–MS/MS was found to be a method with the lowest detection limit for the much more polar N-HETs.

The pH-dependent adsorption of 2-methylquinoline as an example of an N-HET was determined. As assumed and in agreement with results from comparable amines and carboxylic acids, the adsorption of the neutral compound (e.g., at $pH \gg pK_a$) was found to be higher compared to the corresponding cation at $pH \ll pK_a$. However, even for 2-methylquinoline with $pK_a = 5.86$, a maximum adsorption was found for $pH > 6$. Hence, a pH-dependence is not considered to be important for the N-HETs investigated in a theoretical discussion of the efficiency of PRBs.

In addition to the efficiency of PRBs for organic compounds using adsorption parameters for activated carbon, NSO-heterocycles were also investigated in the study. Only a few adsorption isotherms are available. However, an overall assessment and a comparison of adsorption parameters of NSO-heterocycles with their parent hydrocarbons is possible. Benzene and phenol were the compounds with the lowest adsorption capacities and with the fastest breakthrough in column runs. However, although not investigated, pyridine and methylpyridines were assumed to be N-HETs with even lower Freundlich parameters and lower adsorption capacities. Hence, pyridine and derivatives should be included in future investigations. A ranking was conducted based on column runs and benzofuran was found to be the heterocycle with the lowest adsorption capacity. The adsorption capacity of benzothiophene was higher (compared to benzofuran) and comparable to indane.

From the results obtained, general conclusions can be drawn with respect to the design and operation of reactive barriers using activated carbon. In the two PRBs investigated and found in all measurement campaigns, concentrations were reduced by more than 98%, a value even true for the polar phenols and heterocyclic compounds. The successful remediation was obtained after more than 9 years of operation. The concept of using activated carbon as an adsorbent in PRBs appears suitable for the removal of nonpolar aromatics and corresponding polar polyaromatic hydrocarbons and related aromatics and phenols.

Field measurements were qualitatively confirmed in column experiments and by adsorption measurements conducted in the laboratory. As already discussed above, lead substances such as phenol, benzene, and benzofuran were identified. In addition, pyridines should be included in future monitoring campaigns.

Measurements of the PRB influents showed a strong spatial and temporal variability in the concentration and associated loads of the investigated compounds. Strong differences at different positions of the activated-carbon reactors have to be considered when designing a reactive barrier. To assess the capacity of a PRB, knowledge of the concentrations and loads at different positions of the planned site of the reactive barrier is essential. However, extrapolating concentration measurements from localized hot spots on the contaminated site is not recommended.

## Acknowledgments

Funding from the Federal Ministry of Education and Research in Germany (BMBF) within the projects KORA (Funding No. 02WN0366) and RUBIN (Funding No. 02WR0763) is gratefully acknowledged. Within the RUBIN project, we thank for the contributions from the Technical University of Dresden (Germany, Professor P. Werner, H. Lorbeer, and S. Schönekerl) and the University of Cologne (Germany, Professor T. Mansfeldt). Some experiments in the analysis of N-HETs were funded by the Lower Saxony Water Management, Coastal Defence and Nature Conservation Agency (NLWKN); many thanks to Dieter Steffen. Furthermore, we offer our special thanks to Volker Birke, one of the coordinators of the RUBIN project. Last but not least, guests within the exchange program "International Association for the Exchange of Students for Technical Experience" (IAESTE), PhD students, and Bachelor Masters and Diploma students from Leuphana University contributed to the results presented, namely Sven Jerofke, Arne Kappenberg, Meno-Alexander Kersbaum, Mustapha Koroma, Adrian David Kyburz, Dietmar Meyer, Eric Alexander Naumann, Nadiia Nikulina, Debora Reis Riberiro, Ina Schlanges, Anja Talke, and Lars Tangermann.

# References

1. Powell, R. M., D. W. Blowes, R. W. Gillham, D. Schultz, T. Sivavec, R. W. Puls, J. L. Vogan, P. D. Powell, and R. Landis. 1998. *Permeable Reactive Barrier Technologies for Contaminant Remediation.* Washington: United States Environmental Protection Agency (EPA/600/R-98/125).
2. Puls, R. W., C. J. Paula, and R. M. Powell. 1999. The application of in situ permeable reactive (zero-valent iron) barrier technology for the remediation of chromate-contaminated groundwater: A field test. *App. Geochem.* 14: 989–1000.
3. Gavaskar, A. R. 1999. Design and construction techniques for permeable reactive barriers. *J. Hazard. Mat.* 68: 41–71.
4. Scherer, M. M., S. Richter, R. L. Valentine, and P. J. J. Alvarez. 2000. Chemistry and microbiology of permeable reactive barriers for in situ groundwater clean up. *Crit. Rev. Environ. Sci. Technol.* 30: 363–411.
5. Simon, F.-G., T. Meggyes, and C. McDonald. 2002. *Advanced Groundwater Remediation: Active and Passive Technologies.* London: European Science Foundation, Thomas Telford Publishing.
6. U.S. Environmental Protection Agency. 2014. Contaminated site clean-up information. http://www.clu-in.org (accessed March 20, 2014).
7. Thiruvenkatachari, R., S. Vigneswaran, and R. Naidu. 2008. Permeable reactive barrier for groundwater remediation. *J. Ind. Eng. Chem.* 14: 145–156.
8. Meggyes, T., M. Csövári, K. E. Roehl, and F.-G. Simon. 2009. Enhancing the efficacy of permeable reactive barriers. *Land Contam. Reclam.* 17: 635–650.
9. RUBIN. 2007. German Permeable Reactive Barrier Network. http://www.rubin-online.de (accessed March 20, 2014).
10. KORA. 2008. Retention and degradation processes to reduce contaminants in groundwater and soil. http://www.natural-attenuation.de (accessed March 20, 2014).
11. Ward, T. M. and F. W. Getzen. 1970. Influence of pH on the adsorption of aromatic acids on activated carbon. *Environ. Sci. Technol.* 4: 64–67.
12. Laszlo, K., E. Tombacz, and C. Novak. 2007. pH-dependent adsorption and desorption of phenol and aniline on basic activated carbon. *Colloids Surf. A.: Physicochem. Eng. Asp.* 306: 95–101.
13. Worch, E. 1986. Untersuchungen zur Einzel- und Gemischadsorption von Phenolen an Aktivkohlen. Teil 3: Zur pH-Abhängigkeit der Adsorptionsgleichgewichte (Studies on single and mixed adsorption of phenols on activated carbons. Part 3: The pH dependence of adsorption equilibria). *Acta Hydrochim. Hydrobiol.* 14: 407–413
14. Muller, G., C. J. Radke, and J. M. Prausnitz. 1980. Adsorption of weak organic electrolytes from aqueous solution on activated carbon. Effect of pH. *J. Phys. Chem.* 84: 369–376.
15. Seidel, A. and K.-H. Radeke. 1990. Effect of pH on adsorption equilibria for dissolved weak organic electrolytes on activated carbon. *Acta Hydrochim. Hydrobiol.* 18: 691–699.
16. U.S. Environmental Protection Agency. 2012. Estimation Programs Interface Suite™ for Microsoft® Windows, v 4.11.

17. Völgyi, G., E. Baka, K. J. Box, J. E. Comer, and K. Takács-Novák. 2010. Study of pH-dependent solubility of organic bases. Revisit of Henderson–Hasselbalch relationship. *Anal. Chim. Acta.* 673: 40–46.
18. Lide, D. R. 2007. *Handbook of Chemistry and Physics on CD-ROM*. Boca Raton, FL: CRC Press.
19. Iliescu, T., M. Vlassa, M. Caragiu, I. Marian, and S. Astilean. 1995. Raman study of 9-methylacridine adsorbed on silver. *Vibr. Spectros.* 8: 451–456.
20. Meyer, D. 2005. Untersuchung zum Ausbreitungsverhalten polyzyklischer aromatischer Verbindungen an teerkontaminerten Standorten (Investigations of the distribution of polycyclic aromatic compounds at tar contaminated-sites). PhD dissertation, Leuphana University, Lüneburg.
21. Schlanges, I. 2011. Polyzyklische aromatische Verbindungen (PAV) im Grundwasser teerölkontaminierter Altlaststandorte—reaktive Wandsysteme zur Grundwasserreinigung (Polycyclic aromatic compounds in groundwaters of tar oil contaminated sites—Reactive barriers in the remediation of groundwater). PhD dissertation, Leuphana University, Lüneburg.
22. Schlanges, I., D. Meyer, W.-U. Palm, and W. Ruck. 2008. Identification, quantification and distribution of PAH-metabolites, NSO-(hetero)PAH and substituted PAH in groundwater samples of tar-contaminated sites from Germany. *Polycyclic Arom. Comp.* 28: 320–338.
23. Rivera, L., M. J. C. Curto, P. Pais, M. T. Galceran, and L. Puignou. 1996. Solid-phase extraction for the selective isolation of polycyclic aromatic hydrocarbons, azaarenes and heterocyclic aromatic amines in charcoal-grilled meat. *J. Chrom. A.* 731: 85–94.
24. Mänz, J. S. 2012. NSO-Heterocyclen und verwandte Verbindungen im Grundwasser von Altlaststandorten und angrenzenden Fließgewässern—Analytik, Vorkommen und Adsorption auf Aktivkohle (NSO-heterocycles and related compounds in groundwater of contaminated sites and adjacent rivers—Analysis, sources and adsorption on activated carbon). PhD dissertation, Leuphana University, Lüneburg.
25. Mänz, J. S., E. Naumann, W.-U. Palm, V. Birke, and W. Ruck. 2009. Impact and importance of heterocyclic PAH in remediation: Adsorption of activated carbon. *Paper Presented at the 3rd International Contaminated Site Remediation Conference,* CRC for Contamination Assessment and Remediation of the Environment, September 11–15, Adelaide (Australia).
26. Mänz, J. S., A.-K. Siemers, L. Tangermann, W.-U. Palm, and W. Ruck. 2011. Analytical method for the combined determination of NSO-heterocycles, phenols and PAH from solid and aqueous samples using GC–MS and LC–MS/MS. *Paper Presented at the ANAKON Conference,* March 22–25, Zürich (Switzerland).
27. Schad, H., B. Haist-Gulde, R. Klein et al. 2000. Funnel and gate at the former manufactured gas plant site in Karlsruhe: Sorption test results, hydraulic and technical design, construction. *Paper Presented at the 7th International FZK/TNO Conference on Contaminated Soil,* September 18–22, Leipzig (Germany).
28. Birke, V., V. H. Burmeier, and D. Rosenau. 2002. PRB technologies in Germany: Recent progress and new developments. In: *Proceedings of the 3rd International Conference on Remediation of Chlorinated and Recalcitrant Compounds,* eds., A. R. Gavaskar and A. S. C. Chen, 452–459. Columbus: Battelle Press.

29. Niederbacher, P. and M. Nahold. 2005. Installation and operation of an adsorptive reactor and barrier (AR&B) system in Brunn am Gebirge, Austria. In: *Long-Term Performance of Permeable Reactive Barriers*, eds., K. E. Roehl, T. Meggyes, F.-G. Simon, and D. J. Stewart, 283–310. Amsterdam: Elsevier.

30. Yang, R. T. 2003. *Adsorbents: Fundamentals and Applications*. Hoboken: John Wiley & Sons.

31. Kümmel, R. and E. Worch. 1990. Adsorption aus wässriger Lösung (Adsorption from aqueous solution). Leipzig: VEB Deutscher Verlag für Grundstoffindustrie.

32. Dobbs, R. A. and J. M. Cohen. 1980. *Carbon Adsorption Isotherms for Toxic Organics*, EPA-600/8-80-023. Cincinnati: U.S. Environmental Protection Agency.

33. Abe, I., K. Hayashi, T. Hirashima and M. Kitagawa. 1982. Relationship between the Freundlich adsorption constants K and 1/N for hydrophobic adsorption. *J. Am. Chem. Soc.* 104: 6452–6453.

34. Walters, R. W. and R. G. Luthy. 1984. Equilibrium adsorption of polycyclic aromatic hydrocarbons from water onto activated carbon. *Environ. Sci. Technol.* 18: 395–403.

35. Crittenden, J. C., S. Sanongraj, J. L. Bulloch et al. 1999. Correlation of aqueous-phase adsorption isotherms. *Environ. Sci. Technol.* 33: 2926–2933.

36. Hokanson, D. R., D. W. Hand, J. C. Crittenden, T. N. Rogers, and E. J. Oman. 2005. *Adsorption Design Software for Windows (AdDesignS). Version 1.0.45.* Houghton: Michigan Technological University.

37. Mezzari, I. A. 2006. Predicting the adsorption capacity of activated carbon for organic contaminants from fundamental adsorbent and adsorbate properties. MSc thesis, North Carolina State University, Raleigh.

38. Radke, C. J. and J. M. Prausnitz. 1972. Thermodynamics of multi-solute adsorption from dilute liquid solutions. *AIChE J.* 18: 761–768.

39. Crlttenden, J. C., P. Luft, D. W. Hand et al. 1985. Prediction of multicomponent adsorption equilibria using ideal adsorbed solution theory. *Environ. Sci. Technol.* 19: 1037–1043.

40. Benjamin, M. M. 2009. New conceptualization and solution approach for the ideal adsorbed solution theory (IAST). *Environ. Sci. Technol.* 43: 2530–2536.

41. Worch, E. 2008. *Linear Driving Force (LDF)-Model.* Version 2.4. Dresden: Technische Universität.

# 12

## Case Study of PRB Application for the Remediation of Groundwater

**James Stening**

### CONTENTS

## 12.1 Introduction

In February 1999, Orica Australia Pty Ltd (Orica) constructed the first reactive iron barrier in Australia, and indeed in the Southern Hemisphere. Although at that time it was the 36th such barrier that the technology licensors, EnviroMetal Technologies, Inc. (ETI), had designed, it was the first time they had designed one for such high volatile chlorinated hydrocarbon (CHC) concentrations in a high-sulfate low-pH aquifer.

It was also one of the first times Orica had undertaken a groundwater remediation project. In October 1996, Orica's predecessor, ICI Australia Pty Ltd*

---

* When the parent company ICI plc in the United Kingdom divested its major shareholding in ICI Australia in 1997, a new independent Australian company was formed, which became known as Orica in February 1998.

had issued a report on site investigation, the ICI Botany Groundwater Stage 2 Survey (Woodward-Clyde, 1996), which was a thorough investigation of contemporaneous and historical operations, hydrogeology, impacted soil, groundwater, surface water, sediment, marine biota, soil vapor emissions, and related human health risks in and around ICI Australia's Botany site. The Stage 2 Survey also included a section on remediation technology options, including zero-valent iron (ZVI) permeable reactive barriers (PRBs). ZVI PRB technology was identified as the preferred technology for remediating shallow groundwater discharging into a surface drain known as Springvale Drain. To evaluate this and a number of other cleanup options, ICI Australia (and subsequently Orica) embarked on a program of remediation technology review, research, and development, which continues today.

As one of the first steps in this remediation technology research program, ICI Australia convened a workshop in April 1997, inviting a number of leading Australian and international experts on contaminated sites. One of the workshop participants was John Vogan, president of ETI, who discussed potential application of the reactive iron barrier. ETI presented ICI Australia a proposal for a staged process to evaluate reactive iron barrier technology in the context of the Botany groundwater.

The purpose of the barrier was to evaluate the efficacy of the technology for destroying a number of specific aqueous-phase CHCs. Therefore, a pilot-scale barrier was designed conservatively—based on results of extensive laboratory column trials—and was constructed as precisely as possible, incorporating a large array of monitoring locations. An initial monitoring period of 9 months was extended to 19 months, although subsequent infrequent monitoring has revealed some significant developments in the barrier's cleanup capabilities.

Being the first of many kinds, this technology evaluation project presented a number of significant design, procurement, and construction challenges.

---

## 12.2 Background

Orica (and its predecessor ICI Australia) has manufactured a wide range of organic and inorganic chemicals at its Botany site—now known as the Botany Industrial Park (BIP)—since the early 1940s. The first stage of a ChlorAlkali Plant was commissioned in 1944, producing chlorine, caustic soda (sodium hydroxide), and hydrogen. To facilitate expansion of caustic soda production, downstream CHC plants were progressively added, including

- Carbon tetrachloride (CTC), 1945–1952
- Trichloroethylene (TCE), 1948–1977
- CTC and tetrachloroethene (perchloroethylene, PCE), 1963–1991

- Ethylene dichloride (EDC, also known as 1,2-dichloroethane or 1,2-DCA), 1966–2001, which included production of vinyl chloride monomer (VCM) and polyvinyl chloride (PVC) up to 1997

In the early years of operations at BIP, statutory controls for effluent management, treatment, and disposal were minimal. Indeed, up until 1958, no trade wastewater discharges were connected to sewers, and trade waste was discharged directly to a man-made unlined stormwater drain known as Springvale Drain, which flows to Botany Bay, or to on-site soak-away ponds. The equipment was likely to have been drained and decontaminated onto the plant floor, with the waste draining into effluent pits and drains that had some degree of leakage. As the site is underlain by sandy soil, liquids quickly drain away. Waste products from the CHC plants were typically stored in steel drums pending treatment or disposal—often for long periods in open unpaved areas. In some areas, the drums corroded and leaked their contents into the ground. As a result, extensive soil contamination occurred, which eventually led to contamination of the groundwater beneath the site.

The BIP is located approximately 10 km south of downtown Sydney and about 1.5 km hydraulically upgradient (northeast) of Botany Bay, at the southern end of the Botany Sands Aquifer—a high-yielding groundwater resource. (An abundant source of high-quality groundwater was one of the reasons the location was chosen to establish the site in the 1940s.) The Botany Sands Aquifer stretches in the form of an arc from Bondi Beach in the east, through Centennial Park and Moore Park in the north, St Peters and Tempe in the northwest, around the western side of Botany Bay through Brighton-Le-Sands and Ramsgate, and encompassing the Kurnell peninsula in the south.

Groundwater quality in the vicinity of BIP is affected by natural organic compounds and sulfides, and low pH. The natural contaminants are largely artifacts of the coastal swamps and dunes that characterized the area for the last 10–15,000 years. The aquifer lithology consists of fill, Aeolian sands intercalated with discontiguous peat layers, overlying a generally competent clayey sand layer immediately above the sandstone bedrock (Woodward-Clyde, 1996). Around the BIP, the aquifer thickness is typically 20–40 m and the depth to the groundwater is generally 4–6 m below ground surface (bgs), progressively reducing on Southlands and toward Botany Bay (Figure 12.1). Hydraulic conductivity of the aquifer is high, generally 10–40 m/day. Regional groundwater flow is southwesterly toward Botany Bay.

A number of environmental investigations and regular monitoring events have been conducted on and around the BIP since 1989. Several groundwater plumes have been identified originating from up to nine inferred source areas. These are shown in Figure 12.1, together with some key infrastructure, groundwater flow direction, and an indication of the overall 1 mg/L CHC plume envelope.

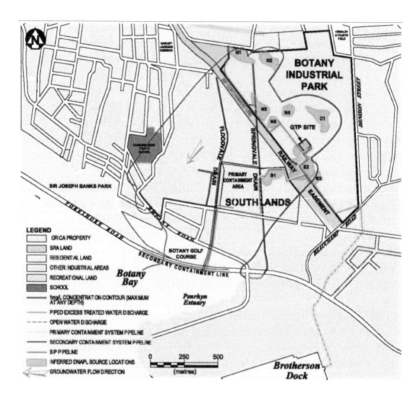

**FIGURE 12.1**
Botany groundwater location plan. (From Orica Ltd.)

All plumes originate from multicomponent source areas denoted with the letters S, C, and N (standing for Southern, Central, and Northern, respectively), corresponding with the grouping as follows:

- Southern Plumes—Comprising primarily CTC, PCE, TCE, and EDC (and their daughter products including VCM and chloroform) from former CTC, PCE, and TCE production.

- Central Plume—Comprising primarily EDC with lower concentrations of TCE, VCM, and other daughter products from the former EDC to PVC manufacturing processes.

- Northern Plumes—Comprising mainly EDC. It is believed that these plumes were mainly formed from the historical movement of the Central Plume toward the former large-scale industrial groundwater abstraction well fields north/northwest of the site in the 1960s to the 1980s. It is also likely that other CHCs were derived from corroding drums of CHC waste in the former open-air waste-storage areas. There is also a CTC plume associated with a former tank farm and loading bay (N4).

The Stage 2 Botany Groundwater Survey (Woodward-Clyde, 1996) included an assessment of soil and groundwater remediation technologies available at the time. A number of technologies were identified that could be applied to the range of CHCs associated with ICI Australia's historical operations. In an effort to develop a better understanding of some of the most promising remediation options, in August 1996, ICI Australia attended a Remediation Technologies Development Forum (RTDF) conference in the United States on PRB technologies, which introduced ICI Australia to some of the world's leading proponents and practitioners of those technologies.

Following the conference, ICI Australia invited a number of Australian and international experts in site investigation and cleanup technologies—including both biological and abiotic PRBs—to participate in a project-specific workshop in Sydney in April 1997. One of the participants of the workshop was John Vogan, president of ETI. ETI was the commercial arm of the University of Waterloo established to commercialize the ZVI technology developed by Professor Bob Gillham and the late Stephanie O'Hannesin at the University of Waterloo. At that stage, ZVI technology was in its infancy—the first commercial reactive iron barrier had been installed in Sunnyvale, California in 1995 (USEPA, 1998). Yet, it appeared that ZVI could be used in the Southern Plumes where CTC, PCE, and TCE were the dominant contaminants, and EDC was a relatively minor component. This was an important consideration because it was known that ZVI could not degrade EDC or dichloromethane (DCM, methylene chloride). At that time, discharge of the Southern Plumes in shallow groundwater to Springvale Drain represented the largest mass flux of contaminants from the site to the environment. However, the Botany site presented some significant hydrogeochemical challenges that had not been previously encountered with reactive iron barriers:

- Dissolved-phase CHC concentrations up to 220 mg/L
- Low pH (pH < 5)
- Elevated sulfide concentrations (>30 mg/L)
- High dissolved organic carbon (DOC) (>500 mg/L)
- A range of volatile fatty acids

To address these challenges and to enable a thorough evaluation of this emerging technology, it was agreed that a conservative and rigorous approach must be adopted. During the workshop, a conceptual design for a pilot-scale reactive iron barrier was developed. ICI Australia's environmental consultants Woodward-Clyde Pty Ltd, in consultation with ETI, subsequently prepared a proposal for ICI Australia setting out a staged process comprising

- Site selection
- Site, hydraulic, and geochemical characterization (to validate the selected site and to provide baseline data)

- Column trials in a University of Waterloo laboratory using ground-water from the Botany site
- Design and construction of a reactive iron barrier
- A 9-month monitoring program
- Detailed reporting and evaluation

While still in Australia for the workshop, John Vogan also met the New South Wales Environment Protection Authority (NSW EPA) and the Community Liaison Committee (CLC, a consultative committee comprising local residents, councils, industry, regulators, ICI Australia, and its consultants) to discuss the ZVI technology and how it was proposed to be evaluated in the context of the Botany Groundwater Project. Such consultative measures were an important element in gaining regulatory approval and community stakeholder acceptance for implementing the pilot-scale trials of this relatively novel technology.

## 12.3 Selected Site

The selected location for the pilot-scale reactive iron barrier was on a vacant block of land, known as Southlands, which was owned by ICI Australia and located adjacent to the main manufacturing site on the hydraulically down-gradient side. Figures 12.1 and 12.2 illustrate the location of Southlands relative to the manufacturing site. It was proposed to install the barrier near the northwestern corner of the eastern area (known as Block 1) of Southlands, hydraulically upgradient of Springvale Drain.

Springvale Drain was an unlined drain believed to have been excavated during the late nineteenth century to aid the dewatering of the swampy land in the area. Up until the commissioning of Orica's Groundwater Treatment Plant and hydraulic containment (groundwater extraction) network (see below) in 2006, Springvale Drain continued to intercept shallow ground-water in the vicinity of Southlands, and consequently provided an expedited pathway for contaminants in the shallow groundwater to flow into Penrhyn Estuary and subsequently into the Botany Bay. Today, as a consequence of the lower water table due to Orica's groundwater extraction, shallow groundwater does not typically discharge into the Southlands section of Springvale Drain. However, at the time of the pilot-scale barrier design and construction, it was anticipated that, if a full-scale reactive iron barrier were to be installed, it would be installed on Southlands upgradient of Springvale Drain so as to prevent the ongoing discharge of contaminants into the drain.

The position of the pilot-scale reactive iron barrier was selected to intercept part of the Southern Plumes contaminated predominantly with PCE,

**FIGURE 12.2**
BIP and surrounds. (From Orica Ltd.)

CTC, and TCE. The hydraulic and chemical properties of the groundwater at that location were confirmed by installing a large array of monitoring wells upgradient, cross-gradient, and downgradient of the proposed pilot-scale reactive iron barrier location. Several sacrificial wells were also installed in the proposed location of the barrier, and were subsequently destroyed when the barrier was constructed.

Predesign hydraulic characterization of the proposed pilot-scale site comprised:

- Installation of nine wells up to a depth of approximately 4 m (fully screened across the water table)
- Collection of water-level data to provide shallow groundwater flow direction (to align the pilot-scale barrier perpendicular to the flow regime)
- Installation of two multilevel wells to assess vertical gradients
- Detailed hydraulic testing of variation in aquifer permeability
- In situ groundwater velocity measurement using a velocity probe
- Electrical conductivity and $\gamma$-logs of the aquifer

Predesign geochemical characterization indicated that the upper 4 m of the shallow aquifer was relatively clean, and that high concentrations of the target CHCs were relatively consistent below 5 m.

## 12.4 Laboratory Column Trials

To evaluate the feasibility of employing a reactive iron barrier to degrade the target contaminants with negligible hydraulic and hydrogeochemical impacts on the aquifer, a series of column trials were conducted at the University of Waterloo under the supervision of ETI. The column trials were conducted using a 50 cm Plexiglass column constructed with seven sampling ports and filled with iron granules sourced from two different Australian suppliers (see Figure 12.3). Groundwater sourced from the ICI Australia Southlands site was pumped through the column at a flow rate similar to the groundwater velocity. Water samples were collected from the column influent and effluent, and the seven sampling ports.

The objectives of the column trial were

- Assess the feasibility of employing ZVI using the site's groundwater
- Determine the degradation rates (an important parameter for determining the appropriate flow-through thickness of a reactive iron barrier)

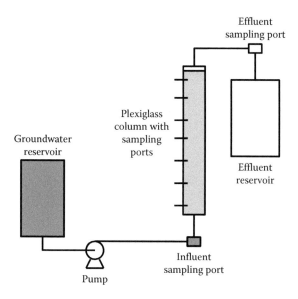

**FIGURE 12.3**
Schematic diagram of ZVI column trial arrangement. (Adapted from EnviroMetal Technologies (ETI). 1998. *Bench Scale Treatability Report of the EnviroMetal Process Using Groundwater from the Orica Botany Facility.* Prepared for SHE Pacific Pty Ltd. EnviroMetal Technologies, Inc., Waterloo, ON (unpublished).)

- Characterize the degradation products (to confirm contaminant destruction and to determine whether any more harmful substances might be formed)
- Quantify Eh and pH changes (although the redox potential of the groundwater samples changed following sampling and transport to Canada, the step change in Eh as the groundwater passed through the ZVI was a relevant parameter to monitor to evaluate the PRB performance)
- Monitor changes in inorganic geochemistry (in part to indicate whether mineral fouling of the reactive iron barrier might occur)
- Characterize microbial activity (which could result in biological fouling of the reactive iron barrier might occur)
- Compare Australian-sourced granular iron reactivity and performance (including potential longevity)
- Provide predesign data for a potential pilot-scale reactive iron barrier

ETI issued results of the completed column trials to Orica in August 1998. The trials concluded that the principal target contaminants CTC and PCE were degraded, but the reaction half-lives higher than ETI had previously experienced with other sites' groundwater and other iron sources. ETI attributed the slower degradation rates to the presence of very high concentrations of DOC (>500 mg/L) in the Botany groundwater. ETI postulated that the DOC could be coating the iron surface or entraining PCE in solution, preventing reaction. The column trial results also indicated typical Eh values, but the groundwater pH—in both the influent and effluent—was lower than at other clients' sites. There were no signs of biological fouling or mineral precipitation, which were important considerations for barrier longevity. Importantly, an Australian iron source—Master Builders Australia Pty Ltd—was also validated.

## 12.5 Pilot-Scale Barrier Design

As indicated above, the column trials were also intended to provide predesign data for a potential pilot-scale reactive iron barrier. On the basis of ETI's calculations of contaminant half-lives (45 h for PCE and 1–2 h for CTC) and the site conditions (such as groundwater velocity), the following design parameters were recommended:

- Barrier width (i.e., perpendicular to the groundwater flow): 5 m
- Barrier height: 3.5 m (installed 4 m bgs)

- Barrier flow-through thickness: 1.5 m (equivalent to about 10 days' residence time in the barrier based on a conservative 0.15 m/day estimate for groundwater velocity)

The large flow-through thickness was recommended to ensure complete degradation of the PCE and evaluation of the formation and destruction of degradation products through the barrier.

Groundwater modeling demonstrated that the intended iron had a suitable permeability relative to the aquifer material. Pea gravel was therefore not required to be installed upgradient or downgradient of the barrier.

The design of the barrier and the construction method were collaboratively developed by ETI, Woodward-Clyde, and Orica. Two key considerations were given to the design and construction method:

1. The precision of the barrier dimensions
2. The quality of the data that could be obtained from monitoring the barrier

To achieve these, it was decided to construct the barrier using sheet piling and to install the monitoring wells and bundle piezometers in the barrier attached to a frame (which ensured that the precise location—laterally and vertically—of each monitoring point was known). Installing the monitoring wells and piezometers attached to a frame also made construction significantly easier—the complete monitoring network could be inserted into the excavation before the iron and overlying sand were backfilled into the excavation. The construction sequence is shown in Figures 12.4 through 12.9.

**FIGURE 12.4**
Installing 12-m-long sheet piles (note the existing monitoring-well network to characterize the trial area). (From Orica Ltd.)

The pilot-scale barrier was constructed in February 1999. John Vogan visited the site to supervise the construction. He also updated the CLC during a construction site visit on the status of PRBs for groundwater treatment. By February 1999, there were 22 full-scale and 14 pilot-scale installations around the world (all in North America and Europe).

Construction of the pilot-scale barrier at Botany occurred in a number of steps:

1. Sheet piles of 12 m length were installed 11 m bgs (see Figure 12.4). Although this was the design depth, piling to this depth was difficult due to the dense-flowing sand. The water table in the construction area was approximately 1 m bgs or less.

2. The soil within the nominally $5 \times 1.5$ m sheet-piled "rectangle" (see Figure 12.10) was excavated to 7.5 m bgs.

3. Prior to backfilling the excavation, it was filled with water. (The excavation had only been partially dewatered to prevent sheet pile collapse.) The water served two purposes:
   - It prevented upwelling of sand and groundwater from the base of the excavation.
   - It enabled the backfilled iron and sand to settle more uniformly across the entire length and breadth of the excavation.

4. The three monitoring wells and 27 bundle piezometers were inserted into the excavation attached to a steel frame (see Figure 12.5).

5. Seventy-two 1-ton bags of granular iron were poured into the excavation (see Figure 12.6). It was backfilled to a depth of approximately

**FIGURE 12.5**
Inserting monitoring wells and bundle piezometers on a frame into the excavation. (From Orica Ltd.)

**FIGURE 12.6**
Installing granular iron from 1 tonne bulk bags (note dust masks and the excavated sand in the background). (From Orica Ltd.)

4 m bgs. Some of the bulk bags had an inner plastic liner. After two of these liners inadvertently slipped into the excavation while discharging the iron (and could not be recovered), reinforcement mesh was installed on top of the excavation to capture other liners (see Figure 12.7).

6. The excavation was filled to the surface with clean sand (see Figure 12.8). The excavated sand was not reused to avoid complicating results from the trial with possible contamination in the excavated material.

**FIGURE 12.7**
Installing granular iron into the excavation. (From Orica Ltd.)

**FIGURE 12.8**
Backfilling the top 4 m with fresh sand. (From Orica Ltd.)

7. The finished reactive iron barrier was then capped with concrete and well monuments were installed for each of the groundwater-monitoring locations (see Figure 12.9).

Further construction details are illustrated in Figures 12.10 and 12.11. Figure 12.10 shows the irregular shape of the sheet piling. As the critical dimension in terms of barrier performance was the flow-through thickness, the opposing sheet piles were aligned and separated by a minimum of 1535 mm—slightly more than the design thickness, but dictated by the piles' configuration and dimensions.

**FIGURE 12.9**
The completed pilot-scale reactive iron barrier with a protective concrete cap. (From Orica Ltd.)

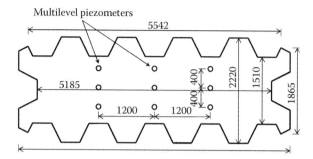

**FIGURE 12.10**
Plan view sketch of the sheet-piled excavation (nominal dimensions in millimeters). (From Orica Ltd.)

Figure 12.11 illustrates the array of bundle piezometers and monitoring wells within and immediately adjacent to the reactive iron barrier. Three longitudinal transects (i.e., along the groundwater flow path) comprising piezometers and nested wells at three depths were used to evaluate the chemical composition of the groundwater. Three transects were installed to evaluate whether "edge" effects were caused by differential permeability

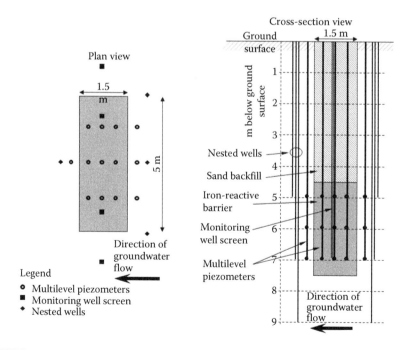

**FIGURE 12.11**
Plan and cross-section views of monitoring locations in and adjacent to the pilot-scale reactive iron barrier. (From Orica Ltd.) The groundwater table was approximately 1 m bgs.

between the iron and surrounding sandy soil matrix. The long-screened wells were installed with the intention of monitoring groundwater velocity and direction with in situ velocity probes (gauging groundwater depths ultimately became a more reliable method of doing this).

In October 2000, two bundle piezometers were added, installed in the iron bed 0.2 m from the front face (to measure CTC—see Section 12.6).

## 12.6 Groundwater Sampling and Analysis

The groundwater monitoring program was developed by Woodward-Clyde. Groundwater samples were typically collected and analyzed for

- Volatile CHCs
- Organics—Dissolved and total organic carbon (DOC and TOC), and chemical and biological oxygen demand (COD and BOD)
- Inorganics—Ferrous and total dissolved iron, dissolved and total manganese, sulfide, sulfate, alkalinity, total dissolved solids (TDS)
- Field parameters such as electrical conductivity, dissolved oxygen, redox potential (Eh), and pH

Several key performance parameters were closely monitored:

- Percent mass reduction (MR)—The percentage reduction of total CHCs across the barrier thickness (in a given vertical and horizontal transect) relative to the influent total CHC concentration
- Degradation of half-lives of key contaminants—particularly CTC and PCE—calculated assuming first-order reaction kinetics (exponential decay)
- TOC (elevated TOC concentrations could hinder degradation of target CHCs, especially PCE)
- Hydraulic gradient into, through, and out of the barrier (which could be indicative of fouling or preferential flow paths and also residence (flow-through) time in the barrier)
- Sulfide, sulfate, Eh, and pH (changes could be indicative of mineral formation, precipitation, and consequent fouling)
- Ferrous (dissolved) iron downgradient of the barrier (as dissolved iron could adversely impact surface water quality in Springvale Drain and then Botany Bay)

The initial sampling program comprised groundwater samples being collected at 1, 3, and 6 months (late March, early June, and early September

1999, respectively). At the completion of the first 6 months' monitoring, it was concluded by ETI that results were broadly consistent with the laboratory column trial results and also the geochemical modeling. However, given the variability of the results, it was recommended that two further rounds of monitoring be conducted, and, as noted above, that two additional multilevel monitoring points be installed at a distance of 20 cm into the iron zone alongside the center monitoring transect to better evaluate the CTC degradation.

Subsequent sampling rounds were added for 9 (early December 1999), 13 (late March 2000), and 19 months (early October 2000) to further evaluate the pilot-scale reactive iron barrier's performance. Two iron cores were also collected at month 17 (see Section 12.8) to examine the condition of the iron, particularly with regard to mineral precipitation and biological fouling. These works concluded the formal evaluation of the barrier. A number of ad hoc supplementary monitoring events were arranged to periodically check whether conditions or performance might have changed.

## 12.7 Groundwater Monitoring Results

Data from the first 13 months of performance monitoring have been reported previously (Duran et al., 2000; Stening et al., 2000).

### 12.7.1 Hydraulic Analyses

Groundwater velocities through the barrier were found to range from 0.1 to 0.8 m/day. The average velocity was determined to be around 0.6 m/day, which was greater than the velocity assumed for the design of the pilot-scale barrier, 0.15 m/day. The results confirmed that the iron was more permeable than the surrounding sand.

The higher relative permeability of the iron was believed to be the cause of apparent upwelling at the 1.5-m ports at 7 m bgs (i.e., at the back end of the iron barrier only 0.5 m above the bottom), evidenced by CHC concentrations increasing at this point to values similar to influent concentrations. (It has also been postulated that the irregular concentration profile might have been at least in part due to two bulk bag liners that had slipped into the bottom of the excavation during installation of the iron.)

### 12.7.2 Organic Analyses

The first 13 months of performance monitoring showed significant reduction of most CHCs compared with total influent concentrations, but, as expected, little reduction of EDC. As mentioned above, the inability of ZVI to degrade EDC has been historically recognized as a limitation of reactive iron barriers.

Nevertheless, results were very encouraging. CTC was found to degrade very rapidly—a half-life of <2 h—in the first 0.4 m of the barrier, and did not stall with formation of DCM, another aliphatic CHC that is not readily degraded by ZVI. The PCE was found to have a higher half-life (12 h to 2 days); so, a higher relative concentration of PCE would cause percent MR results to be lower and some degradation products (such as TCE, *cis*-1,2-dichloroethene, and vinyl chloride [VC]) to emerge from the barrier.

From Table 12.1 (which includes 19-month data), it can be noted that influent concentrations of volatile CHCs and TOC varied markedly—with time and depth. ETI observed that the PCE half-life decreased (i.e., degradation was faster) with depth, and that influent TOC concentrations were lowest in the 7 m bgs monitoring port. However, a relationship between PCE half-life and influent TOC concentration could not be clearly concluded over time at the 5 and 6 m bgs ports.

There was no notable correlation between the influent organic concentrations and percent MR; percent MR was more strongly influenced by the relative concentration of PCE in the influent and, subsequently, its half-life and the half-life of its degradation products. (Note that, due to the apparent effects of upwelling in the 1.5-m ports at 7 m bgs as mentioned above, data from the 1.5-m ports were disregarded when calculating MR; data from the 1.2-m ports were used instead.)

It is worth noting that the percent MR and PCE half-life values for month 19 were some of the best results achieved up to that time at 5 and 6 m bgs. This was not the case at 7 m bgs. Although TOC concentrations were not measured in the month 19 sampling event, it can be seen that at 7 m bgs, the influent concentration of CHCs was the lowest to date. The concentration data for the individual CHCs entering the barrier are not presented, but at that time, it was noted that the relative concentrations of CHCs poorly degraded by ZVI were higher than normal. Notwithstanding the inconsistencies in temporal trends for mass removal and half-lives at the three monitoring depths, the fact that the results showed no significant signs of deterioration over time was encouraging, suggesting that long-term CHC degradation could be achieved with the reactive iron barrier.

Consistently during 19 months of sampling, most of the MR of CHCs was found to occur within the first 0.4 m of the iron barrier. In particular, CTC was found to be completely degraded within the first 0.4 m. Sampling of the multilevel piezometers that installed 0.2 m into the iron barrier prior to the month 19 sampling event confirmed this: at 5 and 6 m bgs >90% of the CTC and >50% of the PCE were degraded within 0.2 m. However, at 7 m bgs, apparently anomalous data prompted ETI to note (in unpublished correspondence to Orica) "these new data from the 7 m interval reinforce our suspicion that system hydraulics (e.g. vertical gradients) may be influencing the results at this depth."

**TABLE 12.1**

Performance of the Pilot-Scale Reactive Iron Barrier to Month 19

| Month | 5 m bgs | | | | 6 m bgs | | | | 7 m bgs | | | |
|---|---|---|---|---|---|---|---|---|---|---|---|---|
| | Influent Concentration (mg/L) | | MR (%) | PCE Half-Life (h) | Influent Concentration (mg/L) | | MR (%) | PCE Half-Life (h) | Influent Concentration (mg/L) | | MR (%) | PCE Half-Life (h) |
| | CHCs | TOC | | | CHCs | TOC | | | CHCs | TOC | | |
| 1 | 150 | 551 | 96 | 6.3 | 81 | 495 | 81 | 10 | 69 | 249 | 86 | 1.7 |
| 3 | 193 | 454 | 89 | 23 | 111 | 468 | 88 | 27 | 87 | 248 | 88 | 3.3 |
| 6 | 202 | 407 | 85 | 20 | 99 | 527 | 81 | 21 | 72 | 205 | 81 | 6.7 |
| 9 | 140 | 367 | 74 | 23 | 140 | 504 | 83 | 12 | 89 | 154 | 79 | 15 |
| 13 | 114 | 281 | 79 | 45 | 149 | 454 | 86 | 11 | 60 | 115 | 69 | 21 |
| 19 | 167 | | 91 | 13 | 143 | | 88 | 10 | 51 | | 64 | 12 |

### 12.7.3 Inorganic Analyses

Field parameters Eh and pH were measured at the time of sampling. As expected in reactive iron barriers, Eh decreased (typically from around −100 to −200 mV), indicating reducing conditions. The pH increased (about two units), consistently over time at all depths. Monitoring results from 5 to 6 m bgs levels showed an increase in dissolved iron (mainly ferrous iron) concentrations hydraulically downgradient of the pilot-scale barrier. Owing to the lower pH in the influent groundwater, the pH within the barrier did not increase to the levels that would be expected to result in significant iron precipitation. Thus, as a result of the low pH at the Botany site, less iron was precipitating in the pilot-scale barrier compared to other sites where reactive iron barriers had been installed. This might be advantageous insofar as less precipitation should result in less porosity loss in the barrier, but it could also be disadvantageous if elevated iron concentrations could have esthetic effects on surface water quality at the groundwater discharge points in the downgradient Springvale Drain or if aquifer porosity was reduced by precipitation of iron minerals (likely iron oxyhydroxides) in the downgradient aquifer.

Sulfate and sulfide concentrations decreased significantly and rapidly (by one to three orders of magnitude) in the reactive barrier. Sulfide was likely reacting with iron to form iron sulfide (FeS). Sulfate was being reduced in the barrier, possibly forming hydrogen sulfide ($H_2S$) or FeS. Samples of the groundwater were analyzed for sulfate-reducing bacteria (SRB) to assess for possible biological reduction of the sulfate. Very minor populations were detected in only two samples, suggesting that the sulfate reduction is not being biologically mediated (Duran et al., 2000).

## 12.8 Analysis of Iron Cores

The possible formation of FeS suggested that mineral precipitation could be occurring in the pilot-scale barrier, although, the generally consistent CHC degradation and hydraulic gradient through the barrier over the 19-month monitoring period suggested that blinding of reactive sites on the iron particles and mineral fouling were not occurring to any significant extent. To evaluate this further, two core samples were extracted from the iron barrier in August 2000 (month 17) and sent to ETI for analysis.

The core samples were collected by driving a push probe at an angle into the ground a short distance before the front face of the iron zone and continuing a short distance into the iron zone. The cores were divided into various subsamples to be analyzed for carbonates, sulfur, and other selected inorganic constituents. Select subsamples were also subjected to RAMAN spectroscopy and analyzed for microbial activity.

One of the subsamples contained iron that was clumped together, which could be indicative of mineral precipitation. This subsample returned the highest percentage of sulfur—0.21%. Assuming that all of the sulfur was present as FeS, the resultant loss in porosity in the first 0.1 m of the barrier was estimated to be <0.5%. Carbonate analyses suggested that carbonate precipitates had caused <5% loss in porosity in the first 0.1 m of the barrier, with lower losses further into the iron. The RAMAN spectroscopy qualitatively confirmed the presence of carbonate, sulfur, and iron mineral species on the iron grains. Although, microbial populations—evaluated using standard microbial enumerations (plate counts) and phospholipid fatty acid (PLFA) analysis—were found to be larger than those measured in similar studies, there was no evidence of significant biofouling (EnviroMetal Technologies [ETI], 2001).

## 12.9 Full-Scale Reactive Iron Barrier Assessment

The first 19 months of sampling and analysis represented the formal assessment period for the pilot-scale reactive iron barrier. The data were assessed to evaluate the effect of the site's high influent concentrations of chlorinated solvents and TOC on the reductive dechlorination reaction kinetics. Reaction half-lives generally appeared to be higher than those published for other sites. This was thought to be a result of the relatively high TOC arising from the site's historical swampy conditions. However, the overall average chlorinated solvent MR was 80%–90%. Temporal trends were evident, but these were not consistent with increasing depth or TOC. Another factor that reduced apparent MR was the presence of EDC and (to a far lesser extent) DCM in the inflowing groundwater, which are not known to be degraded by the ZVI.

This pilot-scale evaluation had demonstrated that ZVI could be used successfully to degrade a broad range of dissolved CHCs in a geochemically complex aquifer. As a result, Orica began planning the installation of a full-scale reactive iron barrier, up to 300 m long, 0.4 m thick, and 7 m deep, to protect Springvale Drain. John Vogan of ETI had noted in a CLC meeting in late 1999 that at that time, a reactive iron barrier of that size would have been the biggest in the world. It also became apparent that no domestic supplier of ZVI could manufacture the required quantity within a suitable time frame; so, investigations into sourcing the material from overseas began in late 2000.

A number of design configurations were evaluated to maximize the interception of the non-EDC plumes, and also minimize the potential impediments to the future development of the land on which the barrier would be installed. Additional monitoring wells and multilevel piezometers were installed in 2001 and sampled in a number of monitoring events along possible alignments in Southlands to assess the types and concentrations of CHCs in those locations.

Concurrent with the assessment of potential impacts from the operation of a large-scale reactive iron barrier, in late 2001, Orica also sought expressions of interest from 18 local and overseas contractors for the construction of such a barrier. No single contractor had all the capabilities required; so, it was necessary for the contractors and consultants to form consortia before formal tendering could begin. Options for construction methods included trenching (with biodegradable guar polymer to hold the trench open prior to iron emplacement) and hydrofracking (azimuth-controlled hydraulic fracturing) with the injection of a mixture of biodegradable polymer and micron-sized iron particles. Unfortunately, no suitable technology could be found to reliably and predictably emplace such a large reactive iron barrier in the flowing sands characteristic of the Botany Sands Aquifer (i.e., the sands have poor structural stability; so, there was a high risk of sand slumping into the trench prior to iron emplacement).

Discussions continued through 2002 with a short list of contractors to find ways to overcome the technical challenges presented by the Botany site. In December 2002, Orica also attended the RTDF meeting on PRBs, which confirmed that no reactive iron barrier had been installed in an aquifer with the low pH, high sulfate, and high dissolved organic compounds that characterized the proposed location of the full-scale barrier.

Ongoing concerns about the potential for elevated dissolved iron concentrations to persist downgradient of the proposed full-scale barrier resulted in additional groundwater monitoring in the vicinity of the pilot-scale barrier being undertaken in mid-2003. The results confirmed that increased dissolved iron did persist in groundwater downgradient of the pilot-scale barrier. Surface water monitoring in March 2003 in Springvale Drain hydraulically downgradient of the pilot-scale barrier also identified elevated dissolved iron concentrations downstream of the inferred seeps from the barrier.

As a consequence of these factors, implementation of the proposed full-scale barrier was put on hold pending the outcome of a groundwater remediation strategy review. A number of investigation tasks were identified in the review to address concerns around dissolved iron impact, depth (e.g., shallow vs. deep installation to bedrock) and alignment of the proposed full-scale reactive iron barrier, alternative installation techniques, and alternative reactive barrier technology including in situ enhanced bioremediation.

In October 2003, as a result of a number of factors, the EPA issued a Notice of Clean Up Action to Orica. Although, the EPA considered the full-scale reactive iron barrier and bioremediation to be still important in the mid- to long term, they believed that Orica's stated cleanup goals were unlikely to be achieved within the stated time frames and that more certainty around the application of remediation technology was required. Among a range of very challenging cleanup targets and deadlines, the Notice required Orica to develop a Groundwater Cleanup Plan and to implement contaminant containment using hydraulic containment in two particular areas—Southlands and an area hydraulically downgradient of Southlands—and ex situ treatment to clean up the extracted groundwater.

The resultant "pump and treat" system started up in stages, commencing in October 2004 and culminating in January 2006 with the introduction of contaminated groundwater feed from three lines of extraction wells (114 wells in total) into a newly constructed Groundwater Treatment Plant on the BIP designed to treat up to 14.5 ML/day of groundwater to effectively drinking water quality. (The pump-and-treat system is described in some detail in the Orica website www.orica.com.) Operation of the Groundwater Treatment Plant and the associated hydraulic containment well network effectively made the reactive iron barrier a redundant remediation measure, as the reduction in groundwater levels caused by the extraction systems reduced/eliminated groundwater discharge to Springvale Drain. The EPA eventually permitted Orica to remove plans for a full-scale reactive iron barrier from the Groundwater Cleanup Plan, which was replaced by a Groundwater Remediation and Management Plan in October 2009.

## 12.10  Additional Sampling and Analysis

Follow-up investigations have been conducted on an ad hoc basis, and have shown significant changes in the influent and effluent groundwater quality. The aforementioned regulatory requirement to install and operate a large pump-and-treat system (which was commissioned in January 2006 and today typically extracts and treats approximately 6–6.5 ML/day of groundwater[*]) resulted in significant changes in the direction and speed of the CHC plumes.

These changes are illustrated in Table 12.2. First, the data for month 91—late November 2006, 10 months after the commissioning of the Groundwater Treatment Plant—show that concentrations of most of the CHCs into the reactive iron barrier have decreased to a total of <60 mg/L, but EDC concentrations have risen up to 127 mg/L. Second, the reactivity and permeability of the reactive iron barrier do not appear to have diminished significantly with time. Most notably, though, EDC is now being removed in the reactive iron barrier—by up to one or two orders of magnitude. Results in May 2002

---

[*] As mentioned above, the design capacity of the Groundwater Treatment Plant was 14.5 ML/day, which turned out to be much more than required. The excess design capacity resulted from incorrect groundwater level data provided by a third party and subsequently used in the hydraulic modeling during the design period. By the time the anomalous data were discovered, it was too late to alter the plant design. However, the additional capacity has proved useful during plant commissioning and maintenance.

**TABLE 12.2**

Pilot-Scale Reactive Iron Barrier Data for Months 39 and 91

| | 5 m bgs | | | | 6 m bgs | | | | 7 m bgs | | | |
|---|---|---|---|---|---|---|---|---|---|---|---|---|
| | Influent Concentration (mg/L) | | Mass Reduction (%) | | Influent Concentration (mg/L) | | Mass Reduction (%) | | Influent Concentration (mg/L) | | Mass Reduction (%) | |
| Month | EDC | Other CHCs | EDC | Other CHCs | EDC | Other CHCs | EDC | Other CHCs | EDC | Other CHCs | EDC | Other CHCs |
| 39 | – | – | – | – | 12 | 105 | 88 | 85 | – | – | – | – |
| 91 | 7.7 | 7.7 | 99.5 | 89 | 127 | 56 | 82 | 51 | 15 | 44 | 90 | 99.8 |

*Note:* MR is based on iron thickness of 1.2 m for 7 m bgs data.

(month 39) showed signs of EDC dechlorination (only the 6 m bgs transect was sampled); sampling in November 2006 (month 91) confirmed this.

It was postulated that the cause of the EDC mass removal was microbial colonization of the barrier by iron-reducing and/or hydrogen-utilizing bacteria (i.e., the hydrogen formed by the corrosion of the ZVI was acting as an electron donor for reductive dechlorination of the EDC and other CHCs). To evaluate this, a single core was retrieved in late February 2008 from the front face of the reactive iron barrier, collecting upgradient indigenous aquifer material and ZVI from the barrier. The core was taken at 20° from vertical using a Geoprobe push tube, and split into two samples for analysis. One sample (Sample A, representing indigenous aquifer material) was taken at a depth of 3 m bgs approximately 1 m upgradient of the reactive iron barrier. Another sample (Sample B) was taken approximately 0.15 m into the reactive iron barrier at a depth of approximately 6 m bgs.

The samples were analyzed by the Centre for Marine Bio Innovation at the University of New South Wales. A clone library of small subunit ribosomal ribonucleic acid (RNA) genes for each of the samples was constructed. The deoxyribonucleic acid (DNA) was extracted from both samples using the Fast DNA Spin Kit for Soil (Qbiogene). The small subunit ribosomal RNA genes present in the DNA extracts were amplified using the polymerase chain reaction (PCR) with universal bacterial primers and cloned for sequencing. The libraries consisted of 56 and 44 clones for Samples A and B , returning 49 and 44 sequences, respectively. Table 12.3 (Stening et al., 2008) summarizes the most abundant sequences in the clone libraries.

Despite the fact that the clone library generated from Sample B had slightly fewer clones than the library generated from Sample A, the former contained two sequences belonging to known dehalorespiring microbes (*Dehalococcoides* and *Dehalobacter*) while the latter contained none. This indicates that the reactive iron barrier harbors higher concentrations of dehalorespiring microbes than the aquifer upgradient. The fact that dehalorespiring microbes were detected in a clone library constructed using universal bacterial primers suggests that they constitute a larger percentage of

**TABLE 12.3**

Composition of Clone Libraries from Samples A and B

|  | Closest Relative | % of Library | Comments |
|---|---|---|---|
| Clone Library A | *Dechlorosoma suillum* | 48 | Chlorate respirer; Fe(II) oxidizer |
| (sand) | *Azospira* | 25 | Nitrate respirer |
|  | *Beta Proteobacteria* | 9 | Anaerobically cycles iron redox |
| Clone Library B | *D. suillum* | 16 | Chlorate respirer; Fe(II) oxidizer |
| (iron) | *Desulfovibrio* | 10 | Sulfate-reducing bacterium |
|  | *Desulfosporosinus* | 5 | Sulfate-reducing bacterium |
|  | *Dehalococcoides* | 2 | Dehalorespirer |
|  | *Dehalobacter* | 2 | Dehalorespirer |

the indigenous microflora than commonly observed. The most abundant sequence detected in the libraries belonged to a member of the *Dechlorosoma* genus, reported to respire chlorate (which would not be expected to be present in the site groundwater) and to oxidize ferrous iron (Lack et al., 2002).

The DNA extracted from Samples A and B was subjected to nested PCR using universal bacterial primers followed by primers specific for the *Dehalococcoides* and *Dehalobacter* genera. While *Dehalobacter* sequences were detected in DNA from both Samples A and B, only Sample B returned a positive PCR for *Dehalococcoides*. This supports the findings from the clone library analysis that *Dehalococcoides* is more abundant inside the reactive barrier compared with 1 m upgradient from the barrier.

Although this phenomenon has been observed at the laboratory scale for other compounds (Scherer et al., 2000), this appeared to be the first documented instance for EDC. Exploiting this phenomenon (say, by augmenting the reactive iron barrier with a dehalorespiring enrichment culture) might broaden the capability of reactive iron barriers to reductively dechlorinate hitherto recalcitrant dissolved phase compounds.

## 12.11 Conclusions

At the time of the installation of the first reactive iron barrier in Australia in February 1999, it was certain whether it would be able to successfully degrade elevated concentrations of volatile CHCs in a low-pH, high-DOC aquifer. Nor was it envisaged that the hydrogeochemical conditions would be dramatically altered due to the subsequent installation and operation of a large hydraulic containment system. Yet, through these periods, the pilot-scale reactive iron barrier installed by Orica at its Botany site was able to achieve significant contaminant MR, and the mode of degradation evolved from abiotic to biological reductive dechlorination. The trial demonstrated the long-term efficacy and adaptability of reactive iron barriers, and—at least in the context of the Botany site—illustrated the engineering challenges faced when scaling up the technology for full-scale implementation.

## References

Duran, J.M., J. Vogan, and J.R. Stening. 2000. Reactive barrier performance in a complex contaminant and geochemical environment. In: G.B. Wickramanayake, A.R. Gavaskar, and A.S.C. Chen (eds.), *The Second International Conference on*

*Remediation of Chlorinated and Recalcitrant Compounds*, Vol. C2–6, Battelle Press, Columbus, OH, pp. 401–408.

EnviroMetal Technologies (ETI). 1998. *Bench Scale Treatability Report of the EnviroMetal Process Using Groundwater from the Orica Botany Facility*. Prepared for SHE Pacific Pty Ltd. EnviroMetal Technologies, Inc., Waterloo, ON (unpublished).

EnviroMetal Technologies (ETI). 2001. *Report on Analyses of Cores Obtained from the Pilot Scale Reactive Iron Barrier, Orica Botany Facility*. Prepared for SHE Pacific Pty Ltd. EnviroMetal Technologies, Inc., Waterloo, ON (unpublished).

Lack, J.G., S.K. Chaudhuri, R. Chakraborty, L.A. Achenbach, and J.D. Coates. 2002. Anaerobic biooxidation of Fe(II) by *Dechlorosoma suillum*. *Microb. Ecol.* 43:424–431.

Scherer, M., S. Richter, R. Valentine, and P. Alvarez. 2000. Chemistry and microbiology of permeable reactive barriers for in situ groundwater clean up. *Crit. Rev. Environ. Sci. Technol.* 30(3):363–411.

Stening, J.R., M. Manefield, M. Lee, A. Low, O. Zemb, A. Przepiora, and J. Vogan. 2008. Reductive dechlorination of 1,2-dichloroethane in a reactive iron barrier. In: *Proceedings of the Sixth International Conference on Remediation of Chlorinated and Recalcitrant Compounds*. May 19–22, 2008, Battelle Press, Monterey, CA.

Stening, J.R., J. Vogan, and J.M. Duran. 2000. Assessment of reactive iron barrier performance at a complex Australian site. In: C.D. Johnston, (ed.), *Contaminated Site Remediation Conference, from Source Zones to Ecosystems*. Centre for Groundwater Studies, CSIRO Land and Water, Wembley, WA.

United States Environmental Protection Agency (USEPA). 1998. *Permeable Reactive Barrier Technologies for Contaminant Remediation*. EPA/600/R-98/125. September 1998. Washington DC.

Woodward-Clyde Pty Ltd. 1996. ICI Botany Groundwater—Stage 2 Survey. *Technical Report Prepared for ICI Australia Engineering Pty Ltd*. Project No. 3392, Vols. 1–6. August 1996. Woodward-Clyde Pty Ltd, St Leonards, NSW.

# 13

## Permeable Reactive Barriers in Europe

Volker Birke and Harald Burmeier

### CONTENTS

## 13.1 Introduction

Permeable reactive barriers (PRBs) are an important component of established technologies for the remediation of contaminated groundwater. Since the first field-scale PRB was built in Sunnyvale, California in 1994–1995, more than 200 PRB systems have been installed worldwide (Gavaskar et al., 2000, 2002; Birke et al., 2003; Parbs and Birke, 2005; Birke and Parbs, 2006; ITRC, 2011; Bone, 2012; Burmeier et al., 2006; Birke and Burmeier, 2012a,b; see Chapter 2).

As discussed in Chapter 2, a PRB treats pollutants downstream from the source zone of the contamination ("hot spot"), which may be spread over a wide area or not accurately located (Figure 13.1). However, it is important to know the total contaminant mass in the source area and its approximate geometry to assess the operational features and lifetime of a PRB relatively accurately.

A PRB divides a polluted area into two sections, that is, it cuts off a contaminated groundwater plume from areas downgradient the barrier that have to be protected: a contaminated upstream segment and a remediated or noncontaminated downstream segment (Figure 13.1). The upper section contains the source of the contamination and the discharging contaminated

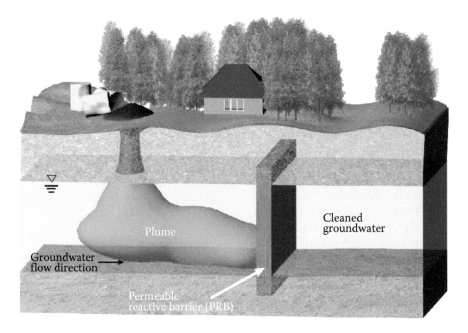

**FIGURE 13.1**
Fundamental principle of a PRB.

groundwater plume; hence, a continuously contaminated area upstream of the barrier is tolerated. The section downstream the barrier is affected by the cleanup effect of the barrier, and hence shows decreasing and eventually vanishing contaminant concentrations over time. PRBs are therefore chiefly regarded as protective measures for safeguarding communities and environments that would otherwise be exposed to the contaminant source, or in other words, jeopardized by contaminated groundwater discharging from the source in a certain distance downstream. In cases where the source becomes entirely depleted over the operational term of the PRB, the barrier may achieve an actual decontamination over time. Thus, PRBs can be regarded as both a safeguarding technique and as an actual decontamination technique, depending on the contamination scenario and its outcome during the barrier's operational life.

In summary, PRBs are not designed for swift remedial action of the source zone. They are designed for managing a source zone by eliminating the plume over a long period of time, accepting that the original source of the contamination and the upper part of its discharging plume are not tackled. The operational lifetime of a PRB may thus range from years to decades.

Two definitions for PRBs were issued in the United States in 2000 and in the United Kingdom in 2002, respectively. A German definition, being similar to that of the United Kingdom, was published in 2006 (Burmeier et al., 2006).

*The U.S. EPA defines PRBs as follows:* A PRB is an emplacement of reactive materials in the subsurface designed to intercept a contaminant plume, provide a flow path through the reactive media, and transform the contaminant(s) into environmentally acceptable forms to attain remediation concentration goals downgradient of the barrier (Gavaskar et al., 2000).

*The U.K. Environment Agency defines PRBs as follows:* A PRB is an engineered treatment zone of reactive material(s) that is placed in the subsurface to remediate contaminated fluids as they flow through it. A PRB has a negligible overall effect on bulk fluid flow rates in the subsurface strata, which is typically achieved by construction of a permeable reactive zone, or by construction of a permeable reactive "cell" bounded by low permeability barriers that direct the contaminant toward the zone of reactive media.

The emplacement of columns of reactive media, such as modified clays using soil-mixing techniques is not considered a PRB unless it is adequately demonstrated that the permeability of the mixed columns is not significantly less than that of the soil prior to mixing. Low permeability clay barriers that provide a degree of attenuation (typically sorption of contaminants to the clay minerals) are excluded from the definition of PRBs.

PRBs can be designed to treat any contaminated fluid, including contaminated soil gases, but are most commonly used to remediate contaminated groundwater within aquifers. The reactive media used in PRBs typically enhances the chemical or biological transformation of the pollutant(s), or retards its migration by sorption or immobilization of the pollutant onto the reactive media (Carey et al., 2002).

### 13.1.1 Engineered PRBs

The U.K. and German PRB guidance documents stress that a PRB should be engineered. The U.S. PRB guidance provides a more extended definition, which includes injection zones. Although the PRB concept was first developed in North America in the early 1990s, European projects have also played an important part in the development of the technology. The first full-scale zero-valent iron (ZVI) PRB in Belfast, U.K., had already been planned and erected in 1994. The first German full-scale ZVI PRB at Tübingen was erected in 1998, whereas the full-scale Austrian PRB at Brunn am Gebirge (a suburb of the Austrian capital Vienna) employed granular-activated carbon (GAC) in 1999. The total number of active European projects has reached around 50 by 2014, of which more than 25 are full scale. Some European suppliers offer sophisticated turnkey solutions for PRBs, while numerous research and development (R&D) trials have been performed since 2000, ranging in scale from pilot to full-scale applications.

During the evolution of the technology, two general types of construction emerged:

1. The continuous reactive barrier (CRB) entirely consists of a permeable zone of a reactive material that is installed in the path of a contaminant plume, for instance in a trench, and captures the entire plume. Manipulation of the groundwater flow or control over it or over the reactive material(s) is not possible using this type of construction (Figure 13.2).

2. The funnel-and-gate system (F&G) is characterized by impermeable walls that intercept the contaminated plume and direct it toward a permeable section loaded with the reactive material (gate) (see Chapter 3).

In the early 1990s, it was believed that both design types would work effectively for several decades in the subsurface, even without maintenance, once installed. For this reason, it was thought that PRBs did not require easy access to the reactive material or the groundwater, as malfunctions were thought to be unlikely.

During subsequent development of the PRB technology, the F&G design has been significantly modified at numerous sites, especially across Europe, to address special issues, such as handling a heterogeneous groundwater

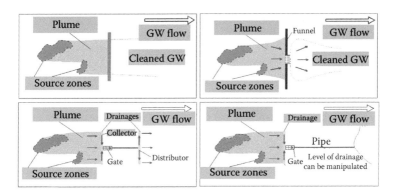

**FIGURE 13.2**
Comparison of the classical CRB (top, left) and F&G designs (top, right) with the D&G (bottom, left) and EC-PRB designs (bottom, right). A D&G as well as an EC-PRB may utilize drains instead of or in addition to cutoff walls to direct the groundwater flow toward vessels containing the reactive materials (or a pea gravel zone where microbiological degradation takes place). In an EC-PRB, the flow can be altered by actively changing the hydraulic head or even by pumping. The two latter types may best suit sites with low groundwater flows, low aquifer permeabilities and/or complex subsurface conditions, contamination scenarios, and so on.

flow or intercepting several plumes originating from different sources—in other words, to exert maximum control over the flow and parts of the PRB structure. Furthermore, entirely new design features have been developed, such as drain-and-gate (D&G) or trench-and-gate (T&G) systems, where the groundwater is directed toward a reactor chamber or in situ vessels (ISVs) equipped with inserted reactors and filter pipes, or the use of gravel drainages instead of cutoff walls (Figure 13.2).

In Europe, D&G or "Efficiently Controllable" PRBs (EC-PRBs), where the groundwater flow is directed and controlled by drainage, pipes, or even by active pumping are now common. This has led to a differentiation in the European design of PRBs from their North American counterparts, where CRBs and reactive injection zones are widely used.

European PRBs are pretty often equipped with ISVs or in ground reactors or removable cartridges, which are highly accessible via shafts, and which are directly connected to drains, pipes, and so on directing the contaminated groundwater to the vessels and the reactive materials. In addition, modified F&G technologies ("non-classical F&G") have been implemented, where some access to the reaction chambers is possible, that is, in the event of minor malfunctions; these may also facilitate inspection of the reactive zone to a certain degree. Also, modified F&Gs, where the hydraulics can be controlled by a discharge/outlet pipe or a similar measure, were erected in Europe (Parbs and Birke, 2005; Birke and Parbs, 2006; Burmeier et al., 2006). These PRBs can be configured to suit site-specific features and monitoring and maintenance can be controlled more effectively. Some technology providers propose a maintenance strategy based on annual operations that can range from a simple clearing of clogged sections to a full replacement of the reactive medium (particularly recommended for barriers that use the adsorption principle). For example, the French company Soletanche-Bachy uses a patented prefabricated three-chamber system that is inserted into a shaft or vault/cell. Its two lateral chambers are filled with gravel to passively guide/manipulate the natural groundwater flow and also permit access to the reactive material, which is usually placed in the middle chamber (Figure 13.3). A PRB in Amersfoort, the Netherlands, is an EC-PRB, which passively collects the groundwater through a cutoff wall, and discharges it by means of a pipe

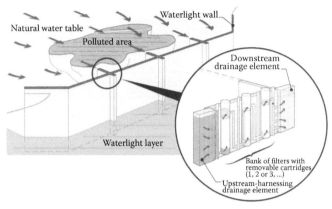

**FIGURE 13.3**
Panel-drain principle. (Courtesy by Soletanche-Bachy, Solenvironment, Paris, France.)

(that penetrates the cutoff wall) to a ZVI reactor located downstream (Parbs and Birke, 2005).

In PRBs, ZVI is used to degrade chlorinated volatile organic carbons (cVOCs) such as chlorinated ethenes (perchloroethene, PCE, trichloroethene, TCE, cis-dichloroethene, cis-DCE, and vinylchloride, VC) to chlorine-free degradation products (i.e., in the case of PCE or TCE via DCE and VC to halogen-free ethane, ethene, and/or ethine as the major degradation products). Besides chlorinated hydrocarbons (CHCs), certain heavy metals, such as chromium as chromate ($Cr^{VI}$), or arsenic (arsenite or arsenate) and certain radioactive elements such as uranium can also be treated successfully with ZVI that reduces these elements to a lower oxidation number/degree forming barely soluble compounds, thus precipitating inside the ZVI bed. Full-scale PRBs are now being used at former manufactured gas and coking plants in Europe (chiefly in the United Kingdom and Germany) to effectively treat polycyclic aromatic hydrocarbons (PAHs), hetero-(NSO)-PAHs, benzene, toluene, ethyl benzene and xylenes (BTEXs), phenols, and related compounds: either GAC or microbiological treatment or both of these approaches in a combined-treatment train have successfully been applied to adsorb and/or degrade these pollutants so far. Activated carbon is a very promising reagent for the adsorptive removal of PAHs, NSO-PAHs, and other contaminants such as highly persistent CHC, because PAHs as well as chlorinated aromatics (such as chlorobenzenes, chlorophenols, and polychlorinated biphenyls, PCBs) cannot be degraded by ZVI due to their relatively low reduction potentials.

## 13.2 Overview and Performance of Selected PRB Sites in Europe

### 13.2.1 PRB Projects in Germany

In 2012, the German long-term R&D program (cluster) for PRBs named "RUBIN" (German: "<u>R</u>eingungswände <u>u</u>nd-<u>b</u>arrieren <u>im</u> <u>N</u>etzwerkverbund") was concluded after 12 years of comprehensive work (Birke and Burmeier, 2012a,b, RUBIN, 2014). RUBIN has been funded by the Federal Government (the Federal Ministry of Education and Research, BMBF) with around 8.5 million € during that period. The authors coordinated, managed, and evaluated all work that had been implemented by 19 RUBIN member projects between 2000 and 2012. In addition, they had conducted an R&D project on the origin of differing reactivities of technical ZVI brands and production batches regarding the degradation of cVOCs, mainly chlorinated ethenes, in groundwater. The member projects overall delivered a significant and highly valuable new insight into the PRB technology. R&D work covered field scale, semitechnical, and laboratory investigations. Taking into account the results of the RUBIN projects as well as those of other external German PRB projects as well as approx. 100 PRB projects internationally, decisive scientific and technical knowledge regarding the current state of the art, potentials and limits, as well as drawbacks of the technology could be gathered, analyzed, evaluated, and disseminated. Especially, it could be achieved to identify which of the several existing variants of the technology (pertaining to design types, reactive materials, etc.) can be safely and most probably successfully applied in practical remediation in Germany and can therefore be recommended for further implementations in the future. Furthermore, the ongoing and potential future trends regarding further development of the technology were identified as well as predicted, respectively. This knowledge is pivotal for practitioners/the remediation market in Germany, because it had not been managed until around 2010 to establish the technology as a conventional remediation approach. This had been partially due to the fact that during RUBIN a few open issues had emerged at some German PRB sites, which could not entirely be resolved around 2000, however, at a higher degree by the work and results of the RUBIN program in the succeeding years. All results of RUBIN have been compiled in the German PRB handbook and guidance that was edited by the authors and some leaders of several other RUBIN member projects (Burmeier et al., 2006). The German PRB handbook consists of two volumes (in one book): volume #1 covers a comprehensive guidance for planning, erecting, operation and maintenance (O&M), and decommissioning of PRBs in Germany ($\approx$ 100 pages), taking into account German laws and regulations, such as the Federal Soil Protection Act and Ordinance ("Bundesbodenschutzgesetz", BBodSchG, und "Bundesbodenschutzgesetz", BBodSchV, respectively). Volume #2 covers a comprehensive state-of-the-art report comprising scientific and technical as well as economic, regulatory, and legal aspects ($\approx$ 400 pages plus reports of all RUBIN projects that have been implemented until 2006, overview of PRB sites and performance data worldwide, and an extensive list of references, Burmeier et al., 2006). Several further subjects/issues that had been encountered/identified during the first phase of RUBIN were comprehensively investigated in the second term of RUBIN ("phase #2") between 2006 and 2012:

1. Gas production/plugging and the varying reactivity of technical-grade ZVI brands in PRBs and their impact on the destruction efficacy of chlorinated volatile organic compounds (cVOCs) as well as on the overall performance/longevity.

2. Long-term performance and longevity of GAC PRBs regarding retardation (sorption) of polyaromatic hydrocarbons (PAHs) and polar NSO-PAHs.

3. Application and performance of EC-PRBs employing a Bio-PRB to treat a tar oil contamination, a "gas bubble"-PRB to treat acid mine

4. drainage or ammonia, and a D&G-PRB employing palladium and
hydrogen to treat cVOCs in situ.
4. Long-term monitoring of field-scale projects such as the ZVI-PRB at
Rheine, the GAC-PRBs at Karlsruhe, and Brunn am Gebirge (Vienna,
Austria), all running successfully for more than 10 years, and the
Bio-PRB at Offenbach (set up in 2007), which provided versatile and
highly valuable monitoring data over several years.

Thus, it could be concluded that PRBs represent a successful in situ remediation technology. Moreover, a comparison to long-term data obtained at
other PRB sites in Europe as well as in North America provided similar overall results. All missions, goals, and results of RUBIN phase #2 have been
compiled in a supplemental volume to the German PRB handbook and guidance (in German, published in 2012, Birke and Burmeier, 2012a,b).

### 13.2.2 A Review of German PRB Sites Comprising the First RUBIN Projects

#### 13.2.2.1 Bernau

Set up in 2001 on the premises of a former dry-cleaning facility of the former
Soviet army, funded by RUBIN, type: EC-PRB, a partly actively working system (lifting up groundwater by pumping, pilot-scale; one-reactor cell adjacent to the ground surface, accessible from top, equipped with 18 cylindrical
reactor vessels made of reinforced concrete), employs ZVI (chiefly Gotthart-
Maier) for treating high cVOC concentrations in two aquifers (75–350 mg/L
TCE). It is possible to run the reactors in parallel or series to control flow
length and residence times inside the reactive system. The PRB has con-
stantly achieved high degradation rates of more than 99% TCE removal, but
there is a low cis-DCE reduction. Hence, subsequent purification on activated
carbon can be applied to adsorb cis-DCE; there has been temporary clogging
of the iron by mineral precipitation and gas production ($N_2$ and $H_2$). A full
control over and accessibility to the system enable a relatively easy and swift
identification of problems and managing efficient solutions, such as clogging
of the ZVI bed by mineral precipitates and gas plugging (Birke et al., 2003,
2004; Weber et al., 2013) (Figure 13.4).

#### 13.2.2.2 Bitterfeld

Set up in 1999, the so-called "SAFIRA" test site, EC-PRB equipped with ISVs
(placed in five shafts, 3 m in diameter, 32 m deep), using active pumping;
different reactive materials and breakdown processes were tested between
1999 and 2004. Treatment of cVOCs and other CHCs, particularly chlorinated
aromatics (complex contaminant mixture inside a local aquifer) achieved
partly successful degradation of the main contaminants (e.g., GAC combined

**FIGURE 13.4**
Installation of single-reactor vessels/modules (top) and overview of the gate construction of the Bernau PRB (bottom, September 2001). At that time, modules had partly been charged only with ZVI and had not been equipped yet with cover plates (single plates can be seen at the left margin (top) of the bottom photo).

with natural attenuation (NA), enhanced NA, ultrasound, palladium, and ZVI–GAC combinations). It has to be emphasized that the system was neither primarily designed for a complete remediation of the local aquifer nor for the cleanup of the mega site Bitterfeld, but for testing novel approaches and reactive materials regarding complex mixtures of pollutants (Weiß et al., 2002, Parbs and Birke, 2005, SAFIRA, 2014).

### 13.2.2.3 Denkendorf

Set up in 2001, D&G, shaft reactor/ISV (full-scale, 90 m gravel/filter pipe drainage), and GAC for cVOC adsorption, clean up goal met, that is, several 10 up to hundreds of mg/L upstream the PRB decreased to just around 10 µg/L cVOCs in the discharging groundwater (Birke et al., 2003, 2004; Parbs and Birke, 2005).

At an industrial area/park at Denkendorf, a small town in the vicinity of Stuttgart, six different sources of cVOC contamination were detected in the 1990s. The main contaminants comprised TCE, PCE, cis-DCE, and 1,1,1-trichloroethane (1,1,1-TCA). The groundwater was also polluted by VC, which had probably been generated by NA processes (microbiological degradation) of PCE/TCE. The overall concentrations of the pollutants exceeded 200 mg/L in total within the hot spots/source zones; pure-phase cVOCs were present as well, although the average total concentration of cVOCs in the groundwater was below 30 mg/L. The groundwater exhibited very high carbonate hardness due to the abundant red and shelly limestone in the area. Sulfate levels were determined at 200 mg/L. The low hydraulic gradient of 2% prompted to design a full-scale D&G PRB that catches (collects) and drains the contaminated groundwater passively by means of a 90-m-long gravel drain equipped with additional filter pipes. The drain directs the groundwater toward a reactor loaded with GAC. The depth of the PRB is about 6 m. The Denkendorf D&G reactor was constructed as a shaft-shaped structure (ISV) employing standard civil-engineering techniques. The system meets its remediation goal of 10 µg/L cVOCs.

A bypass from the passively drained groundwater flow into the reactor was installed to test innovative reactive materials such as palladium on zeolite (palladium loading: 0.5% (w/w)) directly inside the shaft under field conditions. Investigations were implemented by a member project of the RUBIN R&D program (Burmeier et al., 2006). The catalyst had a "molecular design" (due to its "zeolite back bone" wherein the palladium was finely dispersed in three-dimensional (3D), molecular canals). This design was supposed to prevent sulfides from poisoning the palladium, being a frequent, serious problem when palladium is used in contaminated groundwater comprising relatively high sulfate concentrations (Birke et al., 2003, 2007; Parbs and Birke, 2005). Different other designs and types of catalysts had also been successfully tested using the bypass at the D&G PRB between 2007 and 2012 (Birke and Burmeier, 2012a,b).

### 13.2.2.4 Edenkoben

Set up in 1998 (pilot scale), extended to full scale in 2000. Type: F&G (six gates), restricted accessibility to the gates, ZVI for cVOC degradation. For detailed design features, see Rochmes and Woll (1998), Rochmes (2000), and Birke et al. (2003).

A groundwater contamination by cVOCs at Edenkoben was attributed to the former use of CHC solvents for production processes on-site. Several hot spots were found on the property, partly situated below some buildings. A heterogeneous cVOC plume that was more than 400 m wide, originating by at least three individual, partly overlapping plumes, was identified. These individual plumes contained different contaminants at varying concentrations. For example, the southern plume chiefly consisted of TCE and cis-DCE to a total of 8000 µg/L cVOCs. The middle plume comprised 1,1,1-TCA, TCE, and cis-DCE (up to 20,000 µg/L cVOCs), whereas the northern plume was contaminated predominantly by PCE ($\approx$ 2000 µg/L cVOCs). The average composition of cVOCs was 20% TCE, 50% cis-DCE, and 30% 1,1,1-TCA. The geology was characterized by infills and a highly heterogeneous sedimentology (silty overlying strata, 1–6 m thick). Two relevant aquifers had to be taken into account, the lower aquifer being nonpolluted. The polluted upper aquifer was split into two permeable layers, separated by silt and silt–sand strata varying in size (silty gravel–sand mixture referred to as "aquifer 1a," and medium-grained sand with a small portion of fine grain referred to as "aquifer 1b"). There are permeable layers with a varying thickness of 4–7 m. However, the overall permeability coefficient ($k_f$) was poor: $5 \times 10^{-5}$ m/s. Attempts to apply soil vapor extraction as well as preliminary experiments for hydraulic treatment of the hot spots failed to meet reasonable remediation targets. These results demonstrated that remediation could not be completely accomplished using conventional approaches alone. Furthermore, because of the relatively large plume, it was considered that protective measures downgradient of the hot spots were essential to protect the neighboring properties. A feasibility study including column experiments employing groundwater from the site along with groundwater modeling revealed that a PRB using ZVI and shaped as an F&G could be successfully employed. In 1998, a pilot-scale F&G was established for field testing in the middle of the plume and trialled 6 six months. Promising contaminant destruction rates of 99% were observed during that term. As a consequence, the site owner decided to have a full-scale PRB system designed and installed, despite the fact that it would be the very first (privately financed) full-scale PRB in Germany associated with a certain, increased risk compared to conventional techniques.

The full-scale F&G was installed in late 2000 and was put into service in February 2001. The Edenkoben gate type was designed for a diverted, vertical flow inside, whereby the groundwater is passively lifted by a vertical drain (gravel columns, diverting the groundwater flow by 90° relative to its natural horizontal flow) and directed through an ZVI bed that is installed just below ground level (Rochmes, 2000; Birke et al., 2003). The vertical drains thus provided a complete connection to the deeper, polluted groundwater areas. Six gates (each 10 m long and 1.25 m wide), constructed as a sheet pile caisson (open toward its bottom) and packed with granular ZVI (825 tons in total), reached down to approximately 8 m below ground level. A continuous sheet pile wall, 400 m long and more than 14 m deep (i.e., at around 14 m depth, it was pushed into the aquifer base) shaped the funnel that also ran through the gates, thus separating each gate into two chambers. However, inside the gates, the sheet pile wall is buried 1 m below the lowest groundwater level anticipated (5 m below ground level), hence serving as an overflow weir between the chambers. In other words, the flow path through the ZVI is intentionally doubled, due to the special gate design/construction. The sophisticated design of the PRB system in Edenkoben is based on two innovative principles:

1. Compared with a conventional gate construction, the width of a gate, that is, the actual (horizontal) thickness/extension of an iron layer required for a sufficient dehalogenation was significantly reduced by diverting the groundwater flow twice vertically through a given portion of ZVI.

2. Installing the gates near ground level, made access to and the potential recovery of ZVI significantly easier.

Since the erection of the full-scale PRB, no further information, especially no data regarding performance/cleanup efficacy have been published. It is believed that the PRB does not work as expected, probably because of a loss in permeability due to the accumulation of hydrogen gas within the ZVI bed at several gates, thus plugging the pores of the bed and perhaps rendering it impermeable.

### 13.2.2.5 Karlsruhe

A full-scale F&G equipped with GAC was erected in Karlsruhe in 2000/2001 to treat PAHs and BTEX (Schad et al., 2000; Birke et al., 2003, 2010; Burmeier et al., 2006). The site is located in the Rheine valley. During 79 years of operations, the former gas works plant had produced town gas, coke, tar, benzene, and ammonium sulfate from more than 4.3 million tons of coal in total. Soil and groundwater had been contaminated by PAHs and BTEX during that time, mostly because of several tar oil spills. A large PAH plume (about 200 m wide and 400 m long) stretching toward the city center of Karlsruhe originates from the property (covering around 100,000 m²). There was a partly accompanying vinyl VC plume, although its source was not definitively identified as yet (this plume originates upgradient of the site). Maximum contamination levels detected were 500–600 µg/L of PAHs (acenaphthene is the main component), 20 µg/L of benzene, and 2 µg/L of ammonium, whereas VC could be detected at up to 100 µg/L. The local aquifer is approximately 12 m thick consisting of sandy, densely bedded gravel, which is underlain by a clay layer at a depth of 16 m below the ground level. The groundwater flow rate was determined at about 12 L/s under natural conditions. A full-scale F&G barrier charged with about 150 tons of GAC in total, for which regeneration cycles of 5–15 years were expected (depending on the concentration of the contaminants), was planned and eventually set up in January 2001. The Karlsruhe PRB is about 240 m long and 17 m deep, arranged in an almost straight line, along which eight, nearly equidistant gates are positioned. The funnel consists of sheet piles that were pressed, not driven, into the ground using the "silent-piler-technique" to prevent damage to nearby buildings and gas supply pipelines. The gates consist of specifically perforated, cylindrical steel tubes that were set into the ground by means of large-diameter borings. Setting up the gates commenced by driving cylindrical, large-diameter (2.5 m) borehole casings (circular caisson installation) into the ground and excavating them to a final depth of 15–17 m below ground level (0.5 m below the aquifer base). Prefabricated, cylindrical gate segments were connected to each other and the whole construction was lowered into each shaft/borehole (≈ 18 m in length and 1.8 m in diameter). Monitoring wells were installed at the inflow and the outflow of the gates. Pea gravel was used as a filter medium to homogenize the flow of water through the gates, and loaded in front of and behind each gate. The central section of each gate was loaded with GAC. The total cost amounted to more than 4 million €; over an operational life of 50 years, the site owner expects up to 2 million € more to be spent, mainly on changing the GAC and on monitoring. The Federal State of Baden-Württemberg covered 3.5 million € of the total costs.

Over the first 3 years of operation, the Karlsruhe PRB cleanup results did not meet remediation goals. The main issues were groundwater bypassing the barrier at its northern edge and insufficient retardation of contaminants at some gates. In early 2004, an overflow of some gates was identified as a major issue, and therefore, all gates were equipped with extensions on their tops to raise the GAC layer above groundwater level, thus avoiding further overflow. That repair could be performed relatively readily because of the high accessibility of the design of the PRB. Shortly after that issue had been fixed, monitoring data revealed a consistently good cleanup performance for the first time. The remediation goal for benzene (1 µg/L) was met. The remediation goal for PAHs excluding naphthalene (0.2 µg/L) was still being exceeded at some gates but overall, since summer 2004, measurements demonstrate a good performance by the entire PRB installation. An overall degradation rate of 99% was achieved in August 2004.

Since April 2003, PAH concentrations at the northern edge have been decreasing and reached the remediation target value at the beginning of 2005. A modeling study carried out in 2004 proved that the circumvention of the northern edge of the funnel was caused by drainage measures in the course of a sewer construction 1 km north of the PRB over 2 years. Since April 2004, the cleanup efficacy achieved by the entire system has been close to 100%. A monitoring campaign recently (2007–2010) conducted by the RUBIN R&D program (Birke et al., 2010; Birke and Burmeier, 2012a,b) revealed an ongoing high performance regarding the adsorption of PAHs and BTEX as well as showing, for the first time, that NSO-PAHs were also removed from the groundwater successfully (Figure 13.5).

### 13.2.2.6 Offenbach

A pilot scale PRB was designed, erected, and investigated within the RUBIN R&D program in 2007 for the removal of tar oil pollutants from contaminated groundwater at an abandoned tar factory site in the city of Offenbach, Germany (Schad et al., 2005; Tiehm et al., 2008; Birke et al., 2010; Weingran et al., 2011). A three-step process was used, wherein the contaminated groundwater was treated inside the PRB comprising (i) sedimentation of ferric iron, (ii) aerobic biodegradation of the aromatic hydrocarbons (HCs) and heterocyclic compounds, and (iii) a subsequent zone packed with GAC for removing the remaining pollutants. Owing to the high pollutant concentration in the groundwater encountered at this site, hydrogen peroxide was selected as an oxygen carrier due to its higher water solubility compared to oxygen. Also, nitrate was added as an alternative electron acceptor.

Up to 180 mg/L hydrogen peroxide ($H_2O_2$) were added and did not have any toxic effect on the remedial bacteria. The feasibility of the concept was

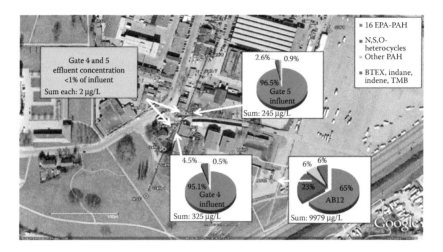

**FIGURE 13.5**
Aerial view and overview of the cleanup performance of the Karlsruhe F&G according to the latest monitoring campaigns conducted by the RUBIN R&D program between 2007 and 2009. Upstream concentration values are given on the right (influent), and downstream results are given in the box on the left (effluent). A very good retardation efficacy/cleanup performance regarding all contaminants of concern is verified. (Courtesy by Leuphana University of Lueneburg, Professor Dr. W. Ruck, Dr. W. Palm.)

confirmed in an on-site pilot-scale study performed with a sedimentation tank (removal of ferric iron), and two bioreactors in which it was found that more than 99.5% of the pollutants were degraded. This biodegradation activity corresponded well with a significant increase in the growth of the relevant bacteria. The repeated addition of moderate concentrations of $H_2O_2$ produced a more favorable result compared with the addition of high concentrations at a single dosing port and one point in time only. The modular design of the pilot-scale Bio-PRB equipped with three separated bioreactors is in accordance with the guideline concept for EC-PRBs that requires access to the reactors in the event of malfunction. An effective distribution of the water over the entire bioreactor length and depth was accomplished by an open water area. Within this, the groundwater flow is guided to a connecting pipe. The lamella separator is needed to achieve the sedimentation of precipitated iron. The construction of the pilot Bio-PRB at Offenbach initially caused a heavy disturbance to the local hydrogeochemical conditions. About 1 year after its erection, this disturbance had more or less disappeared. A zone with reduced hydraulic conductivity in the outflow of the system caused a significant reduction in the groundwater flow through the reactive zone: this varied from 30% to 65% of the expected value. The problem was successfully addressed by amending some design features. The devices for injecting $H_2O_2$ and nutrients did not work well during the first phase of the testing period (unstable dosage rates and several pump failures were encountered). The original pumps were then successfully replaced by another type of pump. Owing to these problems, the effort required for proper O&M of the system was much higher than expected. Originally, the test period was planned only for 1 year. To obtain a sufficient operation period under optimum conditions, the test phase was extended until September 2009 (around 2 years in total). It was subsequently decided that the system would be extended by longer funnels and (probably) a second gate for a full hydraulic containment of the site (Tiehm et al., 2008; Weingran et al., 2009, 2011; Birke et al., 2010; Birke and Burmeier, 2012a).

### 13.2.2.7 Rheine

In 1998, a pilot-scale CRB containing two types of ZVI in two separated segments (total length: 22.5 m) to treat cVOCs (mainly PCE, in 1998: around 10,000 μg/L, in 2012: around 2000 μg/L yet), was installed approximately 400 m downstream from a former dry-cleaning facility located in the small town of Rheine, about 30 km westward of Osnabrück (Ebert et al., 2001, 2007; Birke et al., 2003, 2004; Parbs and Birke, 2005; Burmeier et al., 2006; Birke and Burmeier, 2012a,b). Long-term monitoring of the performance of the PRB was an important part of the mission of the RUBIN network (RUBIN, 2014): the PRB was extensively used for regular in-depth monitoring campaigns over 12 years (2000 until 2012) including coring both ZVI types and groundwater modeling to determine performance and predict the long-term effect (Ebert et al., 2001). The Rheine pilot CRB is 22.5 m long, 0.6–0.9 m thick, and about 6 m deep. A single row of overlapping boreholes (diameter 0.9 m) was constructed by utilizing caisson installation; these were then filled with ZVI.

Two types (brands) of ZVI were employed: on the right side (viewed from upstream), the boreholes were loaded with 69 tons of granular iron ("Gotthart-Maier") mixed with pea gravel at a 1:2 volume ratio (34.5 tons each of iron and gravel) over a length of 10 m. Eighty five tons of "iron sponge" (ReSponge®, brand of Mull und Partner GmbH, Hanover, Germany) were applied on the left side (12.5 m long). A concrete-filled borehole separates the two segments. The name "iron sponge" is due to the material's appearance: the small, dark gray or black pellets (average diameter about 1 cm, Birke et al., 2004; Parbs and Birke, 2005) resemble a pumice-like material. They consist of reduced iron oxide and were supplied by the steelworks "ISPAT" (now Mittal Steel), Hamburg, Germany.

Monitoring results prove an effective performance of the ReSponge section over the whole PRB life span since 1998, with an overall degradation efficacy >99.5% while no decrease in reactivity has been indicated. In contrast to the efficient iron sponge section, the section containing a mixture of Gotthart-Maier ZVI and pea gravel showed significantly decreased reactivity only 6 months after installation. Its initial degradation efficacy for PCE was >98% decreasing to approx. 80% 1 year after installation. Since then, the performance has reached a relatively stable level, varying between 70% and 90%. Remediation efficacy is lower than expected from the known material properties and column experiments conducted prior to the setup of the pilot-scale PRB. The possible reasons include a flow bypass, fast passivation, or construction problems, namely layers of gravel with only small amounts of iron filling.

## 13.3 PRB Sites in Austria and Switzerland

### 13.3.1 Brunn am Gebirge

Set up in 1999, type: EC-PRB ("AR&B"-system, "Adsorptive Reactor and Barrier," full-scale, four ISVs in accessible shafts), loaded with GAC; contaminants and cleanup performance: PAHs, BTEX, cVOCs, benzene, concentrations of all contaminants below the detection limit, and remediation goals achieved since 1999 (Birke et al., 2004; Niederbacher and Nahold, 2005; PEREBAR, 2014).

A full-scale PRB system was installed at an abandoned site of a former tar and linoleum production and processing plant in Brunn am Gebirge (nearby Vienna), Austria in 1999 (Figures 13.6 and 13.7). The system consists of four adsorptive reactors packed with GAC and a hydraulic barrier ("Adsorptive reactor and barrier"). PAHs, phenols, BTEX, cVOCs (mainly TCE and cis-DCE), and HCs are the contaminants of concern. Extensive investigations of the site showed contamination both of the vadose and the saturated zone with concentrations up to 8.6 mg/L for PAH, 0.34 mg/L for phenols, 29 µg/L for benzene, 50 µg/L for toluene, 6.6 mg/L for HC, 0.8 µg/L for TCE, and 27 µg/L for cis-DCE. The total area involved covers more than 376,600 ft². The geological profile is characterized by 0–7 ft of anthropogenic deposits and alluvial sediments (sandy silty gravel) that are 10–20 ft thick below the ground surface. These sediments are underlain by shales of the mid-Pannonian age. There are intercalations of coarser layers, in which artesic water can be encountered. The groundwater table is 7–13 ft below ground surface (bgs). The base of the aquifer is 10–20 ft bgs. Tests indicated permeabilities ranging from $9.8 \times 10^{-3}$ to $3.3 \times 10^{-5}$ ft/s. The natural groundwater flow is west to east with a bend to the southeast, following an erosional depression. A migration of the plume toward the property's boundary has been verified. A site-adapted solution was developed including accommodating a pond fed with clean groundwater (at 5 ft below the actual groundwater level) so that it cannot be polluted. Four adsorptive reactors packed with 23 tons of GAC in total were combined with a 2–5-ft-thick hydraulic barrier. This west-to-east-directed, 720-ft-long barrier, made by jet grouting, cuts into the shoulder of tertiary shales at its Eastern edge. An L-shaped barrier was constructed, which efficiently keeps contaminated groundwater apart from the artificial

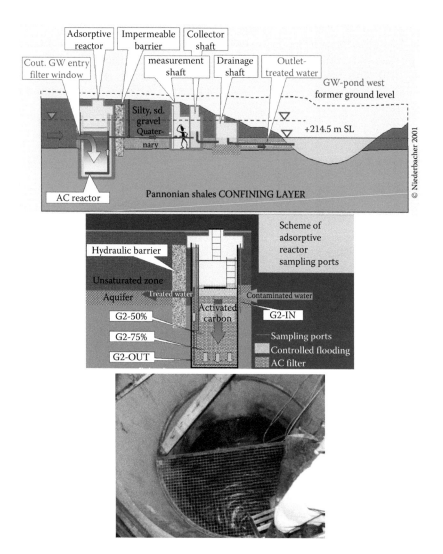

**FIGURE 13.6**
Elevation of the successfully working Brunn AR&B PRB (EC-PRB), Austria (top: elevation scheme of the entire system, middle: GAC shaft reactor ("adsorptive reactor," ISV), bottom: photo taken inside one shaft closely atop the reactor (February 2005)). (Courtesy of Dr. Peter Niederbacher, Klosterneuburg, Austria.)

pond. The adsorptive reactor units are positioned close to the barrier. Each reactor was placed in a drilled shaft, 9 ft in diameter and 26–30 ft deep. The reactor bodies were made of cylindrical glass fiber-fortified synthetic material equipped with filter screens. Each reactor was loaded with approx. 350–420 ft³ of GAC. The contaminated water enters the reactor through the filter screens, passes through the column reactor, and is collected at its bottom.

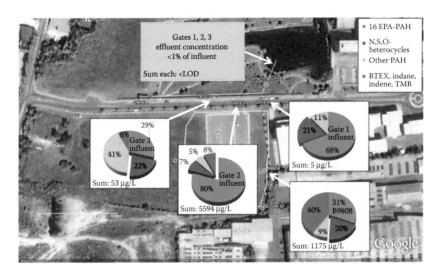

**FIGURE 13.7**
Aerial view and overview of the cleanup performance of the GAC-PRB at Brunn am Gebirge, Austria, according to recently recorded monitoring data (gained between 2007 and 2009, average values). Therefore, this PRB has been showing very good cleanup results since its erection in 1999. (Courtesy of Leuphana University of Lueneburg, Professor Dr. W. Ruck, Dr. W. Palm.)

The decontaminated groundwater is released from each reactor by means of a pipe that penetrates the barrier and is directed to another shaft downstream the barrier, where it is collected from all four reactors and mingled. Inside this shaft, monitoring is performed to validate the cleanup efficiency of the system. Design costs amounted to 100,000 $, and installation and construction came at 650,000 $. An additional investigation of the long-term behavior of the system is performed as part of the PEREBAR project of the European Community (PEREBAR, 2014). Protecting the GAC reactors from intruding oxygen helps avoid aerobic microbiological activity. A careful selection of materials for those parts that could come into contact with the groundwater and the reactor material is essential. Throughout around 15 years of operation, the regular monitoring has been verifying that all contaminants have been below detection limits (Figure 13.7).

### 13.3.2 Willisau

Set up in spring 2004, type: hanging CRB (full-scale, nonoverlapping large-diameter boreholes in two rows), ZVI is used for reductive chemical fixation of $Cr^{VI}$, and the first results gathered in 2005 proved an efficient removal of $Cr^{VI}$. Instead of a conventional continuous barrier, this PRB consists of large diameter cyclindrical boreholes (piles) installed in rows: a single row for lower $Cr^{VI}$ concentrations and an offset double row for higher $Cr^{VI}$ concentrations. The large diameter cyclindrical boreholes are filled with reactive cast iron shavings mixed with gravel to prevent precipitation of secondary phases in the pore space. The treatment of the contaminants takes place both within the large diameter cyclindrical boreholes loaded with ZVI and in the dissolved Fe(II)-plume generated downstream of the barrier. Monitoring over 3 years provided evidence of the mobilization, transport, and behavior of the contaminants in the aquifer. Tracer experiments revealed a rather complex hydrological regime at different scales, complicating the PRB's performance. Results from the large 3D hydrogeochemical dataset show that the double row of cylinders successfully treated the $Cr^{VI}$ contamination. Remediation by the single row was not effective enough due to insufficient lateral overlap of the cylinders and dissolving Fe(II)-plumes. The low amount of precipitated secondary phases observed in the pore space of the reactive material reduced the risk of clogging the system and suggested a favorable longevity of the barrier. Limiting factors to long-term operation are the availability and accessibility of Fe(II) within the cylinders and the concentration within the generated Fe(II)-plume (Flury et al., 2009).

### 13.3.3 Thun

A PRB for $Cr^{VI}$ reduction by gray-cast iron was installed in May 2008. It is composed of a double array of vertical piles containing iron shavings and gravel. The aquifer in Thun is almost saturated with dissolved oxygen and the groundwater flow velocities are 10–15 m/day. Two years after the PRB's installation, $Cr^{VI}$ concentrations were found to exceed the Swiss threshold value downstream of the barrier. Cr isotope measurements indicated that part of the $Cr^{VI}$ plume is bypassing the barrier. Using a Rayleigh fractionation model, a minimum overall $Cr^{VI}$ reduction efficiency of about 15% was estimated. A series of two-dimensional (2D) model simulations, including the fractionation of Cr isotopes, confirmed that the malfunction of the PRB was due to $Cr^{VI}$ contaminated groundwater partly bypassing the PRB. This might be probably due to insufficient permeability of the PRB piles. It was concluded that with such a special PRB design/construction, a complete and long-lasting $Cr^{VI}$ reduction was extremely difficult to achieve for $Cr^{VI}$ contaminations of oxygen- and calcium carbonate-saturated aquifers characterized by high groundwater velocities in addition (Wanner et al., 2012).

## 13.4  PRB Sites in Denmark

### 13.4.1  Vapokon

Set up in 1999, type: F&G (additional drainage system upstream to reduce groundwater flow rate through PRB), funnel length: 122 m, gate length: 15.2 m, depth: 0.6 m, and thickness: 9.1 m; a full-scale; ZVI (type/brand:

Connelly, Chicago, USA) is used for cVOCs (mainly TCA, TCE, and PCE) degradation. Monitoring results over 4 years of operation indicated an effective removal of cVOCs at 92.4%–97.5%. However, the continuous loss of porosity due to mineral precipitation was identified as a significant problem, given the high concentrations of calcium, which may significantly decrease the hydraulic performance of the PRB and may consequently limit its longevity to about 10 years only (Lo et al., 2003, Lai et al., 2006). A detailed monitoring campaign over 7.5 years after the barrier's erection showed very efficient removal (>99%) for the most important cVOCs (PCE, TCE, and 1,1,1-TCA). However, significant formation of cis-DCE within the PRB resulted in an overall insufficient removal of cis-DCE (≈ 80%). High concentrations of both TCE and cis-DCE upstream the PRB along with the significant formation of cis-DCE inside the PRB which gave rise to significant concentrations of cis-DCE downstream the PRB. This finding was not acceptable from a regulatory perspective and further remedial action was needed to remove the cis-DCE plume discharging into a small creek located about 100 m downstream the PRB. Another PRB located on the downstream side of the existing PRB (only treating the very narrow cis-DCE plume) was a possible solution. On the basis of the concentrations observed on the downstream side of the PRB, a relatively thick barrier was needed to meet target criteria of 10 µg/L. Enhanced NA by augmenting with cis-DCE degrading bacteria (*Dehalococcoides* species) in the downstream aquifer was another option (Muchitch et al., 2011).

### 13.4.2 Copenhagen Freight Yard

Set up in 1998; type: continuous trench (CRB), length: 15.2 m, depth: 6 m, thickness: 0.9 m, and full-scale, ZVI for treating cVOCs (up to 4 mg/L, mainly cis-DCE). Effective treatment of the upgradient concentration, however, a part of the plume that migrated around the barrier, hydraulic conductivity decreased significantly during an operational term of 18 months, probably due to gas evolution (Kiilerich et al., 2000; Vidic, 2001; Henderson and Demond, 2011).

## 13.5 First PRB Site in Italy

### 13.5.1 Avigliana, Near the City of Torino (Piemonte Region)

Set up in 2004; type: CRB, 120 m long, 13 m deep, and 0.6 m thick; full-scale; ZVI for cVOCs (up to 300 µg/L, mainly TCE, cis-DCE). Effective treatment of upgradient contamination within the first 3 years, but tests conducted within the PRB found a decrease of hydraulic conductivity by two orders of magnitude, due to the biopolymer (i.e., guar gum) applied during excavation of the barrier. This stimulated the microbial activity of sulfate reducers and methanogens intensively ($CH_4$ concentrations up to 5000 µg/L), so that biofouling and/or accumulation of gaseous methane became an issue (Zolla et al., 2006, 2009).

Excavation was performed in November 2004 and was carried out by a crawler crane equipped with a hydraulic grab and supported by guar gum slurry until the backfill with ZVI. The construction of the 120-m-long and 13-m-deep PRB was performed in 17 panels whose average length was 7 m. The decision to proceed by panels was a safety measure to avoid trench instability.

ZVI was supplied by "Gotthart Maier Metallpulver GmbH" (Rheinfelden, Germany) in the quantity of 1700 tons. The material, free from oils and other impurities was characterized by an iron content higher than 90% in weight and a carbon content lower than 4% (Di Molfetta and Sethi, 2005; Sethi et al., 2007).

Monitoring of the PRB, which started in November 2005, aimed to ensure that the plume was adequately captured and treated, and that downgradient concentrations of the target contaminants (and any by-product) were below the cleanup targets.

Monitoring activity includes

- Quarterly measurement of water levels, to indicate seasonal changes in groundwater flow.
- Chemical monitoring with the determination of groundwater field parameters (Eh, pH, dissolved oxygen, temperature, and conductivity), inorganic chemicals, and chlorinated organic compounds. Sample collection is conducted quarterly to indicate any seasonal changes in contaminant distribution or geochemistry.
- "Low flow purging" and "low flow sampling" methods are adopted to minimize chemical and hydrological disturbances in and around the well, to yield representative water samples.

Monitoring results gathered between 2005 and 2007 showed that output concentrations were chiefly below the limit of 30 µg/L of the total carcinogenic compounds; indeed, carcinogenic chlorinated aliphatic hydrocarbons (CAHs) were below detection levels in almost every water sample taken from downgradient wells. Reaction by-products (VC, 1,1-DCE, and 1,2-DCE) were nearly absent both inside and downgradient of the PRB, verifying that the barrier was able to perform a complete dehalogenation process (Sethi et al., 2007).

On the other hand, groundwater sampling found heavy sulfate depletion and the highest-reported methane concentrations (>5000 µg/L) of a ZVI PRB site. These were due to intense microbial activity by sulfate reducers and methanogens, whose proliferation was most likely stimulated by the use of a biopolymer (i.e., guar gum) applied as a shoring fluid during the excavation of the barrier. Slug tests within the barrier found an apparent hydraulic conductivity two orders of magnitude lower than the predicted value. This can be ascribed to biofouling and/or accumulation of methane inside the iron filings. This experience suggests that when biopolymer shoring is used, long-term column tests should be performed beforehand with initial bacterial inoculation and organic substrate dosing, to predict the effects of bacterial overgrowth and gas generation. During construction, particular care should be taken to minimize the amount of biopolymer used so that complete breakdown can be achieved (Zolla et al., 2009).

## 13.6 PRB Site in Czech Republic

### 13.6.1 Pardubice, East Bohemia

Set up in 2003; a pilot-scale, T&G system, in situ bioreactor, oxyhumolite (oxidized young lignite), various organic pollutants up to 30 mg/L BTEX, chlorobenzenes, naphthalene, nitro-derivatives, phenols, TCE, and total petroleum hydrocarbon (TPH) (Parbs and Birke, 2005; Vesela et al., 2006).

The concept of a biofiltration-permeable barrier was tested in the laboratory and in pilot scale. Oxyhumolite was used as an absorption material and biofilm carrier. During laboratory biofiltration experiments, it was established that naturally occurring microflora derived from contaminated water of the model pilot site had become adapted to local conditions and that it was possible to increase their activity by adding N and P nutrients. Laboratory column tests confirmed that a retention time of 15 h was sufficient for a 97% reduction of all contaminants in the groundwater (with the exception of poorly degradable substances such as nitrobenzene or N,N-diethylaniline), provided that other conditions (mainly oxygen and nutrient concentrations) were optimized. Two bacterial species were isolated from this contaminated groundwater. They were identified to be nitrate reducers, and some strains being able to fix nitrogen (Vesela et al., 2006).

Prior to installation, preliminary investigations of the unsaturated zone (boring up to 2.0 m, sampling) and saturated zone (drilling of five wells, pumping tests including sampling) were performed. The site consists of sand and sandy gravel–fluvial deposits of the Elbe River up to 2.5 m thick (Vesela et al., 2006). The aquifer is underlain by impermeable clays. On the basis of this investigation, the pilot biobarrier was designed as a T&G (D&G).

The system consisted of a single-drainage trench, installed perpendicular to the groundwater flow, which carried contaminated groundwater to an underground bioreactor/gate (Parbs and Birke, 2005). It was approx. 13 m long and 1.0–1.5 m wide, and it was keyed into the underlying impermeable clay (2.6–3.0 m below the ground). The bottom of the trench was sealed with a plastic liner; above, it was filled with permeable gravel and native soil. Treated groundwater flowed by gravity to the recharge gallery formed by two trenches filled with gravel approx. 10 m long and 1.0 m wide, to the depth of impermeable clay. This helped minimize pressure losses of groundwater flowing by gravity through the system and thus maximize the capture zone of the drainage trench. A reactive segment was constructed as an in situ bioreactor. The reactor was designed as a box measuring $2.0 \times 1.2 \times 4.8$ m. The first chamber (Chamber I) was equipped with an aeration segment at its base. This chamber had an internal size of $0.9 \times 1$ m; the water column fluctuated between 3.2 and 3.6 m (depending on hydrological conditions). The effective volume of Chamber I varied from 2.98 to 3.28 m$^3$ (average 3.13 m$^3$). Treated water flowed by gravity from the first chamber to the second and third chambers (Chambers II and III). These chambers were connected in parallel and both were equipped with a biofilter unit of 0.5 m$^3$. The filter of the second chamber was filled with "Keramzit" (ceramic granulate of LIAPOR, Lias Vyntirov, Czech Republic); the filter of the third chamber contained oxyhumolite (derived from the Vaclav mine near Duchcov, Czech Republic), with limestone (Vapenka Certovy schody, Czech Republic) as a pH buffer, and with gravel. Bullet valves regulated the water inflow into Chamber I and also its discharge to Chambers II and III. Piezometers were used to monitor the groundwater level in the drainage trench and the individual chambers, as well as in both the arms of the recharge gallery. They were also used for water sampling in all the chambers.

Pilot testing started in January 2004 and continued for 1 year. Tracer tests were performed to measure groundwater flux through the bioreactor segment under the current hydrological conditions and to determine retention times in the individual chambers of the in situ bioreactor; chemical and microbiological monitoring of decontamination effectiveness was also carried out.

Organic contaminants were removed with very high efficiency by the PRB biofiltration system. This varied from 20.5% to 97.5% in Chamber I. The lowest efficiency, 20.5%, was achieved for naphthalene; the highest efficiency, above 90%, was observed for BTEX (97.5%), TPH (96.2%), and nitro-derivatives (90.8%). A high decrease was also detected for other organic contaminants; chlorinated benzenes (86.6%), TCE (78.6%), and phenols (73.3%). In the case of Chamber II, a decrease of 9%–93% was observed. The lowest efficiency was achieved for the removal of TPH (8.7%), naphthalene (30.9%), and phenols (43.5%); other parameters showed a decrease above 50%; TCE of 56.7%, chlorinated benzenes of 71.7%, nitro-derivatives of 76.8%, and BTEX of 92.9%. Chamber III showed good efficiency in the range of 35%–98%. The lowest efficiency was observed for naphthalene (35.4%) and phenols (48.5%); other organic parameters showed decreases higher than 50%; chlorinated benzenes 61.4%, TPH 56.8%, TCE 52.5%, BTEX 98.2%, and nitro-derivatives 94.7%.

Along with chemical analyses of inorganic and organic contaminants, the total number of aerobic culturable psychrophilic bacteria in groundwater and the mineral nutrient content (N, P) were also measured. The results showed that the autochthonous microflora concentration increased by two orders. The concentration of nutrients (N, P) in the bioreactor unit was sufficient throughout the pilot testing. On the basis of the results of laboratory work, it was decided to use autochthonous microflora for the groundwater decontamination process. The biofilter units of the pilot PRB were not directly inoculated; autochthonous microflora present in groundwater was enhanced by aeration and added nutrients.

## 13.7  Europe's Oldest PRB (Belfast, Monkstown, Northern Ireland)

In 1994, Europe's first full-scale, first ZVI PRB and first PRB to use an in-ground reaction chamber (in-ground reactor) was designed and set up in the United Kingdom. Although the design has been widely adopted and developed, it should be recognized that the initial concept was designed to meet specific constraints of the original site. This was an operational industrial site in Belfast used for the manufacture of electronic components. Historic spillages of chlorinated solvents had led to an intense-though-localized contaminant source. Details of the site setting and initial performance of the reactor are given elsewhere (Jefferis, 2002, 2005; Parbs and Birke, 2005; Birke et al., 2007). The principal contaminant at the site was TCE and the highest identified concentration was 390 mg/L. Other chlorinated solvents were present but at much lower concentrations.

In Belfast, the site geology and location placed a number of restraints on the reactor design (Jefferis, 2005):

- The contaminant source extended to within a few meters of the site boundary. Beyond the boundary, there was a public road, and it was not practicable to extend the reactive zone into the road. The reactive treatment zone therefore needed to be very compact.

- The solvent source was underlain by a thin layer of clay that prevented its migration to a greater depth. If this layer was penetrated by a reactive gate, the free solvents would sink and pollute a lower aquifer stratum. In such an event, however, they would be ultimately retained by a thick clay layer at about 10 m depth that underlies the site and dips toward the proposed funnel.

- The groundwater perched on the thin clay layer was shallow and showed seasonal variations in depth. It would be difficult to achieve any significant depth of horizontal flow in a reactive treatment zone without deepening the gate and thus penetrating the underlying thin clay layer.

- A perched water table also existed in the fill covering the surface of the site. In wet seasons, if allowed to enter the reactive treatment zone, this water could dominate flow through it and unacceptably reduce the residence time. This perched water therefore had to be prevented from entering the reactive zone.

- Proximity to buildings and cost prevented the use of sheetpiles to form the reaction chamber and the funnel of the F&G system. At the time, all previous reactors had been formed within sheetpile boxes.

- If a slurry trench cutoff was used to form the funnel, then it was imperative that the iron filings should not be inundated and blocked by slurry. The iron would have to be contained or the slurry wall would have to be constructed first.

- Excavation next to a slurry wall, to install a reactive treatment zone, could cause local collapse of the cement–bentonite and/or a poor seal between the wall and the iron filings. It would be undesirable to have the possibility of a preferential flow path at this interface.

- The cleanup was being undertaken voluntarily and was not driven by regulatory requirements. It was therefore particularly important that those working on or adjacent to the project should not be exposed to contamination. Early in the design study, it was decided that there should be no hand excavation of contaminated soil or work near it, for example to form or fill the reactor. Personal protective equipment could have enabled hand excavation but the risks were deemed inappropriate for a voluntary remediation.

After consideration and rejection of many reactive treatment zone designs, the in situ reactor configuration was developed as best fitting the site constraints. In place of the previously used horizontal flow reactive treatment zones, the flow was arranged as vertical, in a 12 m tall by 1.2-m-diameter steel reactor shell, which was filled with iron filings as shown in the figure. This design enabled the reactor to be placed between the contamination and the site boundary. This could not have been achieved with a horizontal flow regime as the design calculations had shown that the flow path length needed to be at least 5 m plus entry and exit zones to collect and disperse the flow.

The reactor was placed in an enlargement in a cement–bentonite cutoff wall that was used to funnel the flow to the reactor. This wall was toed into the deep aquiclude layer and the enlargement was taken to a depth of slightly over 12 m to accommodate the reactor shell. The cutoff and enlargement penetrated through the clay layer on which the chlorinated solvents were retained. However, as the cutoff material was designed to have a permeability of $<10^{-9}$ m/s, minimal downward migration of the solvents would occur. The vertical flow direction within the reactor ensured that the full depth of the iron filings was saturated whatever the seasonal variation in groundwater level. The piping of the flow into the reactor and the change of direction from horizontal to vertical flow tended to homogenize the flow both in terms of concentration and across the cross-sectional area of the reactor. Flow heterogeneity across a reactor can seriously comprise its performance (Jefferis, 2002, 2005).

Because of the relatively low permeability and heterogeneity of the adjacent soil, it was decided that the flow to the reactor should be collected via an upstream, high-permeability, collector and that downstream of the reactor, there should be a similar distributor. The collector and distributor were formed from gravel-filled piles taken down to the top surface of the thin clay layer and capped with clay to prevent surface water ingress. A polymer-supported, gravel-backfilled, slurry trench was the preferred collector and distributor. In 1994, there were still concerns about the effect of a polymer remaining on the iron filings and as there was insufficient time to carry out the necessary research, augered piles were used instead of the slurry wall. The reactor was fitted with sampling points at 1.5 m intervals throughout the iron filings bed depth so that its performance could be monitored. Monitoring points were also installed in the collector and distributor piles.

Iron filings in contact with water in an oxygen-free environment produce hydrogen. This hydrogen was vented from the reactor via a vent tube fitted with a spark arrester and mounted in a tall lighting standard. Finally, the internal geometry of the reactor was arranged so that the pipework connections to the gravel-filled collector and distributor piles could be made from within the clean environment of the reactor shell without the need for any hand excavation or for anyone to enter the excavations.

The reactor has performed as designed and there has been substantial reduction in the source and in the downstream plume. A major uncertainty at the design stage was the flow through the reactor and field measurement proved difficult. Tests were undertaken with several tracer materials. These showed that there was spare flow capacity in the reactor because the first in-ground reactor had been designed with a reasonable margin of safety. However, this spare capacity was not wasted; rather, it was exploited to treat water pumped from the plume downstream of the reactor—a plume of contaminants that had developed prior to installation of the PRB. This proved very effective and significantly reduced the extent of the plume.

Those working on potential future PRB sites need to carefully consider the constraints of their sites. PRBs are not a "one design fits all technology." However, adoption of the in-ground reactor concept brings many engineering benefits and the design has since been widely used elsewhere. Although in-ground reactors have been considered since 1994, the benefits and problems of the ground reactor do not appear to have been fully recognized. These advantages include

- Providing a controlled reactor zone—a basic tenet of chemical engineering.

- Homogenization of the flow to provide a more uniform concentration—achieved by collecting the flow and piping it to the reactor—that is, separating the flow collector and the reactor. Achieving uniform flow across the full cross section of a reactor is extremely difficult at low bed velocities. Homogenization is also helped by changing the flow direction from horizontal to vertical.

- The vertical orientation allows the use of a greater ratio of flow path length to the cross-sectional area, thus reducing the potential for short circuiting—flow concentrating in high flow pathways due to slight heterogeneities in the bed. However, it must be accepted that short circuiting remains a serious issue in PRB design because of the very low flow rates (long residence times) required for many PRB reactive materials.

- Estimation of the input flow to PRBs is a major problem—especially for in-ground reactors as the reactor volume is likely to be constrained. The author's experience with Belfast and several other PRBs is that the current modeling techniques, although good at providing flow directions and groundwater contours, are soft when it comes to predicting flow rates. The PRB designer has still to accept a wide range of credible flow rates from the modeler—and design for this range.

- As the cost of a PRB is directly influenced by the flow rate, better prediction procedures are required. Also, site-pumping test protocols must be refined.

- In-ground reactors are good for sites where the flow is expected to be low to moderate. A current challenge is to design an in-ground reactor system for sites where the groundwater flow may be large—or large for part of a year.

- However, it should not be assumed that better modeling and testing will provide all the answers. The flow through a PRB will vary seasonally and over longer timescales because of changes to the groundwater regime resulting from developments around the site and in the watershed and climatic changes. The risk assessment for a PRB must consider these factors.

- A PRB may accumulate contaminants as well as destroy them. Decommissioning must be considered in the design at the outset (see Carey et al., 2002, for this and for much other advice).

- In Belfast, the reactor was installed in an enlargement in a cement–bentonite cutoff wall. This was necessary because the source of the contamination was very close to the site boundary. For later reactors with which the authors have been involved, the reactor has been placed inside the cutoff wall with only a pipe taken through the wall. This can be significantly cheaper and ensures that the contamination remains within the funnel.

- On some sites, it may be advantageous to pump the flow to the reactor. This can ensure a more uniform flow rate but it has to be demonstrated that there will be effective plume capture under all seasonal groundwater conditions.

The concept of an in-ground reactor adds flexibility to the design of reactive treatment zones, allowing more precise control of the reaction environment and easy chemical recharging or recovery and replacement of the active material should this be required. Also, several reactors may be linked in series to treat mixed contaminants.

The use of in-ground reactors allow the full armory of chemical engineering reactor technology to be applied to what is often regarded as a civil engineering/environmental science problem. This will bring many new ideas.

Significant remaining problems are: the design of in-ground reactors for high flow situations and the monitoring of PRB performance. If costs are to be kept to the minimum, monitoring intervals must be as long as possible—this requires confidence in PRB performance and the proactive design for long monitoring intervals. In steady state, PRB performance can be modeled and sampling within a reactor system may allow confidence that performance will remain satisfactory for months/years to come. However, there can be complicating factors such as desorption of contaminants as a result of competitive sorption between contaminant species leading to the release of sorbed contaminants as rather short spikes at concentrations higher than their original input concentrations. Procedures need to be developed to identify impending changes (Jefferis, 2005; Birke et al., 2007).

## 13.8 Conclusion

### 13.8.1 Long-Term Performance and Longevity of PRBs

Long-term performance studies and lessons learned from established systems over two decades (e.g., ITRC, 2005, 2011; Carey et al., 2002; Birke and Parbs, 2006; Birke et al., 2007) show that 80%–90% of all ZVI PRBs work successfully and many PRBs are performing well after more than a decade of operation. The life of a PRB is expected to range from 10 to 30 years. A PRB "failure" is usually attributed to poor site characterization and/or hydraulic design. The long-term performance data show that conventional F&G systems are more prone to performance limitations, caused by preferential flow paths, clogging of pore space due to mineral precipitation and/or gas accumulations (gas plugging by hydrogen, methane, etc.), and bypassing.

The generation of hydrogen and formation of a gas phase are significant processes that may occur in ZVI PRBs. The molar amount of available hydronium ions determines the extent of hydrogen generation. Since hydrogen carbonate and carbonic acid can be considered as the main source of hydronium ions in natural waters, the extent of gas formation is strongly associated with the dissolved inorganic carbon (DIC) concentration (Parbs et al., 2007; Ruhl et al., 2012). A gradually increasing accumulation of precipitates in ZVI PRBs together with the typical shift in pore size distribution toward smaller pores renders the barrier more prone to gas plugging. Gas-consuming processes play a major role in balancing gas production. Particularly, microorganisms are able to reduce the accumulation of gases by hydrogenotrophic sulfate reduction, denitrification, or acetogenesis. Thus, as no bioclogging events are known until now (e.g., Henderson and Demond, 2007), microbial colonization seems to favor the long-term performance of ZVI PRBs. Additionally, dechlorinating strains may even bear contaminant degradation in the case of cVOCs (Weber et al., 2013).

On this basis of past experience, CRBs are now the preferred solution in North America. One of the main arguments for preferring the CRB design is its lower sensitivity to design flaws. This means less risk when cleaning up groundwater at complex sites with heterogeneous flow of pollutants and contaminant distribution (which in turn simplifies the characterization phase). In Europe, however, modified F&G and EC-PRBs are the preferred solutions. The criteria supporting this technological option are that the PRB can be configured to suit site-specific features and that monitoring and maintenance can be controlled more effectively. Some vendors propose a maintenance strategy based on annual operations that can range from simple clearing of clogged sections to replacement of the reactive medium (particularly recommended for barriers based on the adsorption principle) or venting of gas accumulations. This approach can only be considered if the design of the barrier allows easy access to the treatment reactor. This optimized maintenance strategy is sometimes backed up with guarantees on the performance of the barrier, usually for periods of 10–30 years.

EC-PRBs, D&G PRBs, or modified (nonclassical) F&G PRBs have been erected at former manufactured gas plant (MGP) and related sites to treat PAHs and BTEX in more or less accessible reactive zones (in-ground/in situ reactors, cartridges, etc.), and they have successfully been operated at different sites across Europe (especially in the United Kingdom, France, Austria, and Germany) between approx. 5 and 15 years in 2014. They are packed with GAC and/or equipped with a biological treatment zone, where added nutrients and/or microbes enhance microbiological degradation (Bio-PRBs). Moreover, during the second working period of the German PRB R&D program "RUBIN" from 2006 to 2012, it was verified that even emerging novel contaminants, such as heterocyclic PAHs (NSO-PAHs), can effectively be retained and/or destroyed in such PRBs as well. Approximately 10–20 PRBs have been set up in Europe at MGP sites so far; in contrast, there are approximately 200 PRB sites worldwide, most of which are designed to treat cVOCs (PRBs employing ZVI). Therefore, treatment of PAHs, NSO-PAHs, and BTEX at MGP or related sites employing GAC and/or Bio-PRBs represent still a minority, despite their very promising long-term results.

# References

Birke, V., and Burmeier, H. 2012a. Anwendung von durchströmten Reinigungswänden zur Sanierung von Altlasten—Ergänzungsband zum Handbuch, 144 Seiten (Application of permeable reactive barriers to groundwater cleanup—Supplemental Volume, 144pp.), http://www.ptka.kit.edu/downloads/ptka-wte-w/Handbuch_Reinigungswaende-RUBIN_Ergaenzungsband_2012.pdf (accessed on June 20th, 2014).

Birke, V., and Burmeier, H. 2012b. Anwendung von durchströmten Reinigungswänden zur Sanierung von Altlasten—Ergänzungsband zum Handbuch (Permeable reactive barriers for groundwater remediation in Germany—Major results of investigations of the large scale R&D program "RUBIN" obtained during its second term (2006–2012)), *Altlastenspektrum*, 249–261 (in German, English abstract).

Birke, V., Burmeier, H., Jefferis, S., Gaboriau, H., Touzé, S., and Chartier, R. 2007. Permeable reactive barriers (PRBs) in Europe: Potentials and expectations. *Ital. J. Eng. Geol. Environ.*, Special Issue 2007 on TRANS-IT project, 31–38. http://www.ijege.uniroma1.it/rivista/special-2007/special-2007/ (accessed on June 26th, 2014).

Birke, V., Burmeier, H., Niederbacher, P., Hermanns Stengele, R., Koehler, S., Wegner, M., Maier, D. et al. 2004. PRBs in Germany, Austria, and Switzerland: Mainstreams, lessons learned, and new developments at 13 sites after six years. Paper 3A-14, In: Gavaskar, A.R., and Chen, A.S.C. (Eds.), *Remediation of Chlorinated and Recalcitrant Compounds—2004. Proceedings of the Fourth International Conference on Remediation of Chlorinated and Recalcitrant Compounds* (Monterey, CA; May 2004). ISBN 1-57477-145-0, published by Battelle Press, Columbus, OH.

Birke, V., Burmeier, H., and Rosenau, D. 2003. Design, construction and operation of tailored permeable reactive barriers. *Pract. Periodical Haz., Toxic, Radioact. Manag., ASCE* 7(4), 264–280.

Birke, V., and Parbs, A. 2006. *Lessons Learned Report, Outlook and Recommendations.* Deliverable reference number: D 6-3, D 6-5,—Project EuroDemo (Project no. GOCE 003985), University of Lueneburg. http://www.eurodemo.info/uploads/media/D6-3_D6-5_rev.pdf (accessed on June 20th, 2014).

Birke, V., Schuett, C., Burmeier, H., Weingran, C., Schmitt-Biegel, B., Müller, J., Feig, R., Tiehm, A., Müller, A., Mänz, J.S., Naumann, E.; Palm, W.-U., and Ruck, W.K.L. 2010. Permeable reactive barriers for groundwater remediation at contaminated former manufactured gasworks plants and related sites: Design features, performances achieved, and outlook. In: Fields, K.A., and Wickramanayake, G.B. (Chairs), *Remediation of Chlorinated and Recalcitrant Compounds—2010. Seventh International Conference on Remediation of Chlorinated and Recalcitrant Compounds*

(Monterey, CA; May 2010). ISBN 978-0-9819730-2-9, Battelle Memorial Institute, Columbus, OH, www.battelle.org/chlorcon.

Bone, B.D. 2012. *Review of UK Guidance on Permeable Reactive Barriers. Taipei International Conference on Remediation and Management of Soil and Ground Water Contaminated Sites*, Taipei, Taiwan, October 30–31, 2012. http://sgw.epa.gov.tw/public/upload_images/regsys/%E8%AB%96%E6%96%87%E9%9B%86_part_4.pdf (accessed on June 20th, 2014).

Burmeier, H., Birke, V., Ebert, M., Finkel, M., Rosenau, D., and Schad, H. 2006. Anwendung von durchströmten Reinigungswänden zur Sanierung von Altlasten, 471 Seiten, Anhänge (Application of permeable reactive barriers to groundwater cleanup, 471pp., 2 Appendices), http://www.rubin-online.de/deutsch/bibliothek/downloads/index.html (accessed on June 20th, 2014).

Carey, M.A., Fretwell, B.A., Mosley, N.G., and Smith, J.W.N. 2002. Guidance on the use of permeable reactive barriers for remediating contaminated ground water. Environment Agency, National Ground Water & Contaminated Land Centre Report NC/01/51, UK. https://www.gov.uk/government/uploads/system/uploads/attachment_data/file/290423/scho0902bitm-e-e.pdf (accessed May 05, 2014).

Di Molfetta, A., and Sethi, R. 2005. The first permeable reactive barrier in Italy. Consoil 2005—*Proceedings of the Ninth International FZK/TNO Conference on Soil Water Systems*, Bordeaux, October 3–7, 2005.

Ebert, M., Köber, R., Parbs, A., and Dahmke, A. 2007. Tracer experiments in zero-valent iron permeable reactive barriers. In: Lo, I.M.C., Surampalli, R., and Lai, K.C.K. (Eds.), *Zero-Valent Iron Reactive Materials for Hazardous Waste and Inorganics Removal*. American Society of Civil Engineers, Reston, VA, 282–308.

Ebert, M., Schäfer, D., and Köber, R. 2001. The use of column experiments to predict performance and long term stability of iron treatment walls.—2001 *International Containment and Remediation Technology Conference and Exhibition*, Orlando, FL. http://www.containment.fsu.edu/cd/content/ (accessed on June 22th, 2014)

Flury, B., Eggenberger, U., and Mäder, U. 2009. First results of operating and monitoring an innovative design of a permeable reactive barrier for the remediation of chromate contaminated groundwater. *Appl. Geochem.*, 24(4), 687–696.

Gavaskar, A., Gupta, N., Sass, B., Janosy, R., and Hicks, J. 2000. *Design Guidance for Application of Permeable Reactive Barriers for Groundwater Remediation*. Battelle, Columbus, OH, USA. http://www.itrcweb.org/Guidance/GetDocument?documentID=59 (accessed May 04, 2014).

Gavaskar, A., Sass, B., Gupta, N., Drescher, E., Yoon, W.-S., Sminchak, J., Hicks, J., and Condit, W. 2002. Final Report. *Evaluating the Longevity and Hydraulic Performance of Permeable Reactive Barriers at Department of Defense Sites*. Battelle, Columbus, OH, USA. http://www.serdp.org/Program-Areas/Environmental-Restoration/Contaminated-Groundwater/ER-199907 (accessed on June 20th, 2014).

Henderson, A., and Demond, A. 2011. Impact of solids formation and gas production on the permeability of ZVI PRBs. *J. Environ. Eng.*, 137(8), 689–696.

Henderson, A.D., and Demond, A.H. 2007. Long-term performance of zero-valent iron permeable reactive barriers: A critical review. *Env. Eng. Sci.*, 24, 401–423.

ITRC. 2005. *Permeable Reactive Barriers: Lessons Learned/New Direction.*—118p., The Interstate Technology & Regulatory Council, Washington, DC, USA, February 2005. http://www.itrcweb.org/Guidance/GetDocument?documentID=68 (accessed June 20th, 2014).

ITRC. 2011. *Permeable Reactive Barriers: Technology Update.* 179 p., The Interstate Technology & Regulatory Council, Washington, DC, USA, June 2011. http://www.itrcweb.org/Guidance/GetDocument?documentID=69 (accessed June 20th, 2014).

Jefferis, S.A. 2002. Engineering design of reactive treatment zones and potential monitoring problems. In: Simon, F.G., Meggyes, T., and McDonald, C. (Eds.), *Advanced Ground Water Remediation, Active and Passive Technologies.* Thomas Telford Publishing, London, ISBN 0727731211, 75–86.

Jefferis, S.A. 2005. Europe's first ZVI PRB and in-ground reactor—Thoughts from 10 years on. CONSOIL 2005, *Conference Proceedings,* Bordeaux, France, 2870–2873.

Kiilerich, O., Larsen, J.W., Nielsen, C., and Deigaard, L. 2000. Field results from the use of a permeable reactive wall. In: Wickramanayake, G.B., Gavaskar, A.R., and Chen, A.S.C. (Eds.), *The Second International Conference on Remediation of Chlorinated and Recalcitrant Compounds,* Monterey, CA; May 22–25, 2000. Battelle. 377–384.

Lai, K., Lo, I., Birkelund, V., and Kjeldsen, P. 2006. Field monitoring of a permeable reactive barrier for removal of chlorinated organics. *J. Environ. Eng.,* 132(2), 199–210.

Lo, I.M.C. 2003. *Field Monitoring of the Performance of a PRB at the Vapokon Site, Denmark. Summary of the Remediation Technologies Development Forum Permeable Reactive Barriers Action Team Meeting,* October 15–16, 2003, Niagara Falls, NY. http://www.rtdf.org/public/permbarr/minutes/101603/index.htm (accessed June 20th, 2014).

Muchitch, N., Van Nooten, T., Bastiaens, L., and Kjeldsen, P. 2011. Integrated evaluation of the performance of a more than seven year old permeable reactive barrier at a site contaminated with chlorinated aliphatic hydrocarbons (CAHs). *J. Contam. Hydrol.,* 126(3–4), 258–270.

Niederbacher, P., and Nahold, M. 2005. Installation and operation of an adsorptive reactor and barrier (AR&B) system in Brunn am Gebirge, Austria. In: Roehl, K.E., Meggyes, T., Simon, F.-G., and Stewart, D.I. (Eds.), *Long-Term Performance of Permeable Reactive Barriers.* Elsevier B.V., Genth, Belgium. ISBN 0-444-51536-4, 283–309.

Parbs, A., and Birke, V. 2005. *State-of-the-Art Report and Inventory on Already Demonstrated Innovative Remediation Technologies.* Deliverable D 6-2 of FP6-project EuroDemo (Project no. GOCE 003985), University of Lueneburg. http://www.eurodemo.info/fileadmin/inhalte/eurodemo/results/D6-2_Part1.pdf (accessed on June 20th, 2014).

Parbs, A., Ebert, M., and Dahmke, A., 2007. Long-term effects of dissolved carbonate species on the degradation of trichloroethylene by zero valent iron. *Environ. Sci. Technol.,* 41, 291–296.

PEREBAR. 2014. http://www.perebar.bam.de/(accessed on June 22th, 2014).

Rochmes, M. 2000. Erste Erfahrungen mit Reaktiven Wänden und Adsorberwänden in Deutschland (First experiences gained with PRBs and adsorptive permeable barriers in Germany). In: Franzius, V., Lühr, H.-P., and Bachmann, G. (Eds.), *Boden und Altlasten Symposium 2000,* Berlin (in German), 225–245.

Rochmes, M., and Woll, T. 1998. Sanierung eines LCKW-Schadens mit Funnel-and-Gate-Technik (Remediation of a cVOC contamination in groundwater by the F&G technology). *TerraTech* 5, 45–48 (in German).

RUBIN. 2014. http://www.rubin-online.de/ (accessed on June 20th, 2014).

Ruhl, A.S., Weber, A., and Jekel, M. 2012. Influence of dissolved inorganic carbon and calcium on gas formation and accumulation in iron permeable reactive barriers. *J. Contam. Hydrol.*, 142–143, 22–32.

SAFIRA, 2014. http://www.ufz.de/index.php?en=19788 (accessed June 20th, 2014).

Schad, H., Haist-Gulde, B., Klein, R., Maier, D., Maier, M., and Schulze, B. 2000. Funnel-and-gate at the former manufactured gas plant site in Karlsruhe: Sorption test results, hydraulic and technical design, construction. *Contaminated Soil 2000 (Proceedings of the Seventh International FZK/TNO Conference on Contaminated Soil September 18–22, 2000)*, Leipzig, Germany, 951–959.

Schad, H., Klein, R., Weiss, J., Tiehm, A., Müller, A., and Schmitt-Biegel, B. 2005. Biosorption barrier at a former tar factory in Offenbach: (1) An innovative concept for long-term in-situ treatment of highly contaminated groundwater. In: Uhlmann, O., Annokkée, G., and Aren, dt F. (Hrsg.) CONSOIL 2005, *Proceedings (CD) of the Ninth International FZK/TNO Conference on Soil–Water Systems*, Bordeaux, October 3–7, 2005, 1482–1486.

Sethi, R., Zolla, V., and Di Molfetta, A. 2007. Construction and monitoring of a permeable reactive barrier near the city of Torino, Italy. *Ital. J. Eng. Geol. Environ.*, Special Issue 1.

Tiehm, A., Müller, A., Alt, S., Jacob, H., Schad, H., and Weingran, C. 2008. *Development of a Groundwater Biobarrier for the Removal of PAH, BTEX, and Heterocyclic Hydrocarbons.* In: Water Science & Technology—WST, 1349–1355.

Vesela, L., Nemecek, J., Siglava, M., and Kubal, M. 2006. The biofiltration permeable reactive barrier: Practical experience from Synthesia. *Int. Biodeterioration Biodegradation*, 58, 224–230.

Vidic, R.D. 2001. Permeable reactive barriers: Case study review. *TE-010-01, Ground-Water Remediation Technologies Analysis Center (GWRTAC)*. Pittsburgh, PA.

Wanner, C., Zink, S., Eggenberger, U., and Mäder, U. 2012. Assessing the Cr(VI) reduction efficiency of a permeable reactive barrier using Cr isotope measurements and 2D reactive transport modeling. *J. Contam. Hydrol.*, 131, 54–63.

Weber, A., Ruhl, A., and Amos, R. 2013. Investigating dominant processes in ZVI permeable reactive barriers using reactive transport modeling. *J. Contam. Hydrol.*, 151, 68–82.

Weingran, C., Jacob, H., Feig, R., Tiehm, A., Müller, A., and Schad, H. 2009. *Full-Scale Construction and First Results of a Funnel and Gate Biobarrier at an Abandoned Tar Factory Site.* In Situ and *On-Site Bioremediation*. Battelle, Baltimore, USA.

Weingran, C., Schad, H., Tiehm, A., Müller, A., Müller, J., and Bartelsen, T. 2011. *Pilot Test and Field Construction of a Funnel and Gate Biobarrier on a Former Tar Factory Site.* SARCLE/CSME Gent, October 2011.

Weiß, H., Schirmer, M., Teutsch, G., and Merkel, P. 2002. Sanierungsforschung in regionalen kontaminierten Aquiferen (SAFIRA)–2. Projektüberblick und Pilotanlage (Research and development on remediation of contaminated local aquifers). *Grundwasser* 7(3), 135–139 (in German).

Zolla, V., Freyria, F.S., Sethi, R., and Di Molfetta, A. 2009. Hydrogeochemical and biological processes affecting the long-term performance of an iron-based permeable reactive barrier. *J. Env. Qual.*, 38(3), 897–908.

Zolla, V., Rolle, M., Sethi, R., and Di Molfetta, A. 2006. Performance evaluation of permeable reactive barrier using zero-valent iron at a chlorinated solvents' site. *XV International Symposium on Mine Planning Equipment Selection (MPES 2006)*, Torino, Italy, September 20–22, 2006.

# Index